TIME SERIES ANALYSIS:
Theory and Practice 4

TIME SERIES ANALYSIS:
Theory and Practice 4

Proceedings of the International Conference
held at Cincinnati, Ohio, August 1982

Edited by

O.D. ANDERSON
TSA&F, Nottingham, England

1983

NORTH-HOLLAND – AMSTERDAM · NEW YORK · OXFORD

ISBN: 0 444 86731 7

Published by:

ELSEVIER SCIENCE PUBLISHERS B.V.
P.O. Box 1991
1000 BZ Amsterdam
The Netherlands

Sole distributors for the U.S.A. and Canada:

ELSEVIER SCIENCE PUBLISHING COMPANY, INC.
52 Vanderbilt Avenue
New York, N.Y. 10017
U.S.A.

Library of Congress Cataloging in Publication Data
Main entry under title:

Time series analysis.

"The 8th International Time Series Meeting (ITSM) and 3rd American conference took place in Cincinnati, Ohio, 19-22 August 1982"--Introd.
1. Time-series analysis--Congresses. I. Anderson, O. D. (Oliver Duncan), 1940- . II. International Time Series Meeting (8th : 1982 : Cincinnati, Ohio)
QA280.T555 1983 519.5'5 83-11718
ISBN 0-444-86731-7 (Elsevier)

PRINTED IN THE NETHERLANDS

In grateful Memory of
Gwilym Jenkins,
1932-1982

CONTENTS

TIME SERIES ANALYSIS: Theory and Practice 4
O.D. Anderson (editor)

INTRODUCTION

A bit about the Meeting

The 8th International Time Series Meeting (ITSM) and 3rd American Conference took place in Cincinnati (Ohio) 19-22 August, 1982, in the Greater Cincinnati Convention Center and the immediately adjacent Headquarters Hotel, Stouffer's Cincinnati Towers; and was convened immediately after the conclusion of the 1982 Joint Statistical Meetings (American Statistical Association, Biometric Society and Institute of Mathematical Statistics) which were held at the same location. Thus, interested participants were able to conveniently and economically attend both events in a single trip without changing accommodation.

Our international conference featured both invited and contributed papers, the objects being to discuss recent developments in the theory and practice of Time Series Analysis and Forecasting (TSA&F), and to bring practitioners together from diverse parent disciplines, work environments and geographical locations.

Suitable time series topics for presentation included: Statistical Methodology; Applications to Economics and in Econometrics; Rational Expectations; Government, Business and Industrial Examples; Finance and Accountancy; the Hydrosciences, such as Limnology, Hydrology, Water Quality Regulation and Control, and the Modelling of Marine Environments; Persistence and Fractional Differencing; the Geosciences, especially such areas as Oil Exploration and Seismology; Civil Engineering and applied disciplines; Point Processes; Spatial and Space-Time Processes - their theory and application - especially in Geography and related areas, such as City Planning or Energy Demand Forecasting; Biology and Ecology; Environmental Studies - Air and River Pollution; Epidemiology; Medical Applications and Biomedical Engineering; Psychology; Irregularly Spaced Data (including Outliers and Missing Observations); Robust and Nonparametric Methods; Seasonal Modelling and Adjustment, Calendar Effects; Causality; Bayesian Approaches; Distributed Lags; Box-Jenkins Univariate ARIMA, Transfer-Function, Intervention and Multivariate Modelling; State Space; Nonlinear Modelling; Identification Problems; Estimation; Diagnostic Checking; Signal Extraction; Comparative Studies; Spectral Analysis, especially for the Physical Sciences; Business Cycle and Expectations Data; Data Revisions; Computer Software and Numerical Analysis; Forecasting (including new topics such as Traffic Forecasting and Safety, and Forecasting in Agriculture); and any other areas of the subject.

This introduction lists the participants, provides the technical programme, and prints abstracts of those papers not included in these Proceedings. It also acknowledges all individuals, who helped make the Meeting a success, and gives biographical sketches for those authors which filed appropriate details.

Conference Organisation (20 people)

Oliver D. Anderson (UK), General Chairman	TSA&F, Nottingham

<u>Technical Programme Committee</u>

Professor Houston H. Stokes (USA), Chairman	University of Illinois, Chicago
Dr Alan V. Cameron (USA), Session Organiser	State Space Systems, Irvine, California
Dr W. Robert Terry (USA), Session Organiser	University of Toledo, Ohio

Dr Shiv G. Kapoor (USA)	University of Illinois, Urbana-Champaign
Dr Sergio G. Koreisha (USA)	University of Oregon, Eugene
Dr Keh-Shin Lii (USA)	University of California, Riverside

Local Arrangements Task Force

Dr Michael S. Broida (USA), Chairman	Miami University, Oxford, Ohio
Mrs Julia Ali (USA)	Guest, Lexington, Kentucky
Mrs Barbara Broida (USA)	Miami University, Oxford, Ohio
Ms Anne E. Davenport (USA)	Miami University, Oxford, Ohio
Russell B. Kling (USA)	Guest, Oxford, Ohio
Dr Anne Koehler (USA)	Miami University, Oxford, Ohio
Gregory D. Long (USA)	Procter and Gamble, Memphis, Tennessee
Mrs Ellen F. Martin (USA)	Cincinnati Convention Bureau, Ohio
Mrs Barbara Mattei Smith (USA)	Square D Co, Florence, Kentucky
Dr Shahla Mehdizadeh (USA)	Miami University, Oxford, Ohio
Dr Dwight Smith (USA)	Miami University, Oxford, Ohio
Dr Chuck Wells (USA)	Miami University, Oxford, Ohio
Robert W. Wray (USA)	Square D Co, Florence, Kentucky.

Other Participants (a further 66 people from 12 countries)

Dr Bovas Abraham (Canada)	University of Waterloo, Ontario
Ali T. Akarca (USA)	University of Illinois, Chicago
Michael S. Alexander (USA)	Northern Trust Co, Chicago
Mohamed Al-Osh (Saudi Arabia)	University of Riyadh
Saleh M. Badran (USA)	University of Toledo, Ohio
Dr Adnan M. Barry (Saudi Arabia)	King Saud University, Damman
Professor Purnell H. Benson (USA)	Rutgers University, Newark, New Jersey
Lee R. Bishop (USA)	Hill Air Force Base, Utah
Dr Eddy W. Borghers (Belgium)	University of Antwerp
Dr John Brode (USA)	University of Massachusetts, Boston
Professor Fernand C.M. Broeckx (Belgium)	University of Antwerp
Dr Bernard L. Burtschy (France)	Ecole Nationale Supérieure des Télécommunications, Paris
Dr Phillip A. Cartwright (USA)	University of Illinois, Urbana-Champaign
Professor John S.Y. Chiu (USA)	University of Washington, Seattle
Ms Mee-Lum Chin (Malaysia)	Guest, Kuala Lumpur
Jan G. de Gooijer (Netherlands)	University of Amsterdam
Dr Kamal M. El-Sheshai (USA)	Georgia State University, Atlanta
A. Torbjorn Ericsson (Sweden)	Swedish Telecom HQ, Farsta
J. Ericsson (Sweden)	Guest, Farsta
Dr Robert A. Fildes (UK)	Manchester Business School
Dr David F. Findley (USA)	Bureau of Census, Washington DC
Dr Harry L.A.M. Geerts (USA)	University of Pittsburgh, Pennsylvania
Professor Z. Govindarajulu (USA)	University of Kentucky, Lexington
Dr Marc Hallin (Belgium)	Free University of Brussels
Dr Stephen E. Haynes (USA)	University of Oregon, Eugene
Lars Holmqvist (Sweden)	Linköping University
Gregory B. Hudak (USA)	Scientific Computing Associates, DeKalb, Illinois
Raymond R. Hyatt (USA)	Northeastern University, Boston, Massachusetts
Professor Kenneth J. Jones (USA)	Brandeis University, Waltham, Massachusetts

Dr Sergio G. Koreisha (USA)	University of Oregon, Eugene
Dr Robert M. Kuhn (USA)	Veterans Administration, Perry Point, Maryland
Dr Lon-Mu Liu (USA)	University of Illinois, Chicago
Dr Yih-Wu Liu (USA)	Youngstown State University, Ohio
Dr Greta Ljung (USA)	Boston University, Massachusetts
Dr Richard McCleary (USA)	Arizona State University, Tempe
Daniel P. McMillen (USA)	Argonne National Laboratory, Illinois
Dr Raman K. Mehra (USA)	Scientific Systems Inc, Cambridge, Massachusetts
Dr Jack Y. Narayan (USA)	State University of New York, Oswego
Ms Anne W. Nelson (USA)	University of Texas, Austin
Dr Gerald Nickelsburg (USA)	University of Southern California, Los Angeles
Dr Nicholas Noble (USA)	Miami University, Oxford, Ohio
Dr Julia A. Norton (USA)	California State University, Haywood
Michael J. O'Connor (USA)	University of Illinois, Urbana-Champaign
Professor J. Keith Ord (USA)	Pennsylvania State University
Dr Joe D. Petruccelli (USA)	Worcester Polytechnic Institute, Massachusetts
Professor Llad Phillips (USA)	University of California, Santa Barbara
Ms Nampeang Pingkarawat (USA)	University of Illinois, Chicago
Mr Pingkarawat (USA)	Guest, Chicago
Dr Susan Porter-Hudak (USA)	Guest, DeKalb, Illinois
David P. Reilly (USA)	Automatic Forecasting Systems Inc, Hatboro, Pennsylvania
Professor Roch Roy (Canada)	University of Montreal, Quebec
Sutaip L.C. Saw (USA)	University of Pennsylvania, Philadelphia
Dr A. Senthilselvan (Libya)	University of Garyounis, Benghazi
Mrs Haruko Shibata (Japan)	Guest, Tokyo
Dr Ritei Shibata (Japan)	Tokyo Institute of Technology
Chiaw-Hock Sim (Malaysia)	University of Malaya, Kuala Lumpur
Dr Lucille M. Terry (USA)	Bowling Green State University, Ohio
Dr W. Robert Terry (USA)	University of Toledo, Ohio
Dr Cheng-Jun Tian (China)	Shanxi University
Dr James M. Tipton (USA)	Baylor University, Waco, Texas
Dr William C. Torrez (USA)	University of California, Riverside
Ronald J. Usauskas (USA)	Northern Trust Co, Chicago
Professor H.D. (Rick) Vinod (USA)	Fordham University, New York City
Jean-Paul Wauters (France)	Union des Assurances de Paris
Professor John S. White (USA)	University of Minnesota, Minneapolis
You-Gin Zhong (China)	Shanghai Reactor Institute.

Technical Programme (40 contributions in a single stream)

<u>Thursday, 19 August</u>

<u>Session 1</u> 14.30-16.15

O.D. Anderson (UK) Welcome to 8th International Time Series Meeting (ITSM) and 3rd American Conference
B. Abraham (Canada) X-11 ARIMA Seasonal Adjustment and Forecasting Efficiency
J.K. Ord (USA) An Alternative Approach to the Specification of Multiple Time Series Models

J.S. White (USA) Discrete and Continuous Models for Serial Correlation

Session 2 16.45-18.30

P.A. Cartwright & P. Newbold (USA) A Time Series Approach to the Prediction of Oil Discoveries
L. Phillips & S. Ray (USA) Deterrence: A Rational Expectations Formulation
S. Koreisha (USA) Estimation and Forecasting of Equations with Expectations Variables using Multiple Input Transfer Functions
L.-M. Liu & G.B. Hudak (USA) An Integrated Time Series Analysis Computer Program: SCA System

Friday, 20 August

Session 3 8.30-10.15

E.W. Borghers (Belgium) The Influence of Japan on the Belgian Car Market: A Multivariate Transfer Function Approach
J.Y. Narayan (USA) Multiple Time Series Modelling of Macroeconomic Series
K.J. Jones (USA) The Use of Vector ARMA Models in Macroeconomic Forecasting
H.H. Stokes (USA) Output Fluctuations and Relative Price Adjustment: A Vector Model Approach

Session 4 10.45-12.30 *Arranged by Professor H.H. Stokes*

A.T. Akarca & T. Long (USA) The 1979 Oil Price Shock and Inflation in Five Industrial Countries: An Intervention Analysis
D.P. McMillen (USA) Vector-ARIMA Models and Oligopoly Behavior: OPEC and the World Oil Market
R.J. Usauskas & M.S. Alexander (USA) An Empirical Investigation into the Validity of Technical Charting Analysis
N. Pingkarawat (USA) Purchasing Power Parity Theory: A Vector Autoregressive Approach

Session 5 14.30-16.15

A.V. Cameron (USA) Recent Results in Forecasts and Models for Multiple Time Series using the State Space Forecasting Method
L.R. Bishop (USA) Optimizing an Air Traffic Control Tracker with State Space Forecasting
R.K. Mehra (USA) Similarities and Differences between the State Vector Representations for Single and Multiple Time Series
J. Brode (USA) On the Predictability of Long-Memory Time Series

Session 6 16.45-18.30

Z. Govindarajulu (USA) Nonparametric Tests for Randomness against Autocorrelated Normal Alternatives
H.D. Vinod & B.S. Hui (USA) A Canonical Correlations Approach to State Vector Analysis of Capital Appropriations and Expenditures
R. Shibata (Japan) A Theoretical View of the Use of AIC
W.C. Torrez (USA) Order Selection for Autoregression with Application to X-Ray Photoelectron Spectroscopy Data

Saturday, 21 August

Session 7 8.30-10.15

B.L. Burtschy (France) Predicting Ability of Business Surveys
J.S.Y. Chiu & T.P. Mauk (USA) Adaptive Forecasting versus Re-Estimation of Parameters: A Monte Carlo Comparison

Y.-W. Liu & R.H. Bee (USA) Forecasting Criminal Activity with ARIMA Models
J.M. Tipton & J.T. McClave (USA) Time Series Modeling: A Comparison of the Maximum χ^2 and Box-Jenkins Approaches

Session 8 10.45-12.30

R.R. Hyatt (USA) The Effect of Influential Observations on Prediction with a Time Series Model
J.A. Norton (USA) Fitting Joined Line Segments to Time Series Data: Urinary Estrogens as an Example
M. Hallin (Belgium) Nonstationary Second-Order Moving Average Processes II: Model Building and Invertibility
T.E. Kollintzas & H.L.A.M. Geerts (USA) Three Notes on the Formulation of Linear Rational Expectations Models

Session 9 14.30-16.15

P.H. Benson (USA) Progress in Forecasting Price Changes from Speculative Supply and Demand Functions
S.E. Haynes & J.A. Stone (USA) The Dynamic Links between Inflation and Unemployment: Some Empirical Evidence
C.-J. Tian & Y.-G. Zhong (China) On Evaluating an Estimate for the Seasonal Component of a Time Series
F.C.M. Broeckx (Belgium) Bayesian Estimation of Parameters in a Linear Regression Model, with Prior Information on their Bounds

Session 10 16.45-18.30 *Arranged by Dr W.R. Terry*

L.M. Terry & W.R. Terry (USA) A Multiple Time Series Analysis of the Relationship between Economic Activity and Women's Skirt Geometry
W.R. Terry & S.G. Kapoor (USA) A Vector Time Series Analysis of Cotton-Polyester Price Competition
S.G. Kapoor & W.R. Terry (USA) A Time Series Analysis of Tensile Strength in a Die Casting Process
M. O'Connor & S.G. Kapoor (USA) Time Series Analysis of Building Electrical Load.

Edited Abstracts (for presented papers not included in these Proceedings)

1. L.R. BISHOP (USA)
Optimizing an Air Traffic Control Tracker with State Space Forecasting

An Air Traffic Control tracker typically uses an A,B tracking algorithm (where A and B are smoothing constants for the X and Y components of target velocity) to predict the next position of the aircraft being tracked. A and B are permitted to increase or decrease as a function of track soundness, but the same A and B values are always used, irrespective of radar type, or distance from the radar. This practice is intuitively less than optimal because it does not consider the effect of distance from the radar on target reporting accuracy, and neglects for the purpose of general applicability that air routes are a fixed range of coordinates peculiar to a given radar.

This paper addresses the use of State Space Forecasting to automatically develop tracking predictions for given radars along a given air route. Tracking data (x, y and time) from representative tracks are fed into the State Space software and a State Vector is produced. The State Space determined model is then programmed for use within the area of the applicable air route. Addressed are the differences in predictive accuracy between the two methods and the size of the State Space realization needed to provide the best tracking algorithm for a given air route.

2. E.W. BORGHERS (Belgium)
The Influence of Japan on the Belgian Car Market: A Multivariate Transfer Function Approach

Prior to 1967, there were hardly any Japanese cars in Belgium. In 1980, however, almost 25% of new car registrations were of Japanese origin. Along with this spectacular growth of the Japanese market share came a drop in the French one, from 32% in 1970 to 24% in 1980.

Reacting to their decline in market share, and with government backing, the French companies urged (even pressured) the Belgians to restrict the imports of Japanese cars; assuming that their own share would increase as a result. It is this assumption that we test here. The analysis uses data up to 1980, and the predictions are compared with actuality during the post sample period 1981-1982.

3. J. BRODE (USA)
On the Predictability of Long-Memory Time Series

Granger & Joyeux proposed the use of fractional differencing to predict time series for which the autocorrelation coefficients decline slowly as the time interval goes to infinity. It will be shown that such series are notably badly behaved: 1) they have no defined variance; 2) they may not be predictable.

4. B.L. BURTSCHY (France)
Predicting Ability of Business Surveys

At a time of decreasing stability in the business cycle, it becomes more important to look for leading indicators such as the business climate index, which balances positive and negative survey responses to the actual business situation and the expected one in three months time, series T and T3 respectively. In this paper we investigate how well T3 predicts T(t+3).

Three approaches are described: Box-Jenkins transfer function modeling; relating T and T3 through an econometric model; combining T and T3 to create a new series for which a univariate Box-Jenkins model is obtained. This last approach is repeated with the end part of the combination, to test whether the relation between T and T3 is changing.

5. J.S.Y. CHIU & T.P. MAUK (USA)
Adaptive Forecasting versus Re-Estimation of Parameters: A Monte Carlo Comparison

We compare, through an analysis of variance, the predictive abilitites of two Box-Jenkins forecasting alternatives, adaptive forecasting and re-estimation of parameters, as applied to a collection of simulated time series with controlled structural characteristics. Both methods continually update forecasts as new data points become available.

The analysis of variance used to analyze the forecasting errors considered five factors: the level of differencing, the ARMA coefficient values, the two forecasting alternatives, the horizon length, and the time series length. The results tend to suggest the appropriateness of the more economic adaptive procedure, at least for short-term forecasting, rather than re-estimation of parameters.

6. R.R. HYATT (USA)
The Effect of Influential Observations on Prediction with a Time Series Model

An experiment was performed on a data set to assess the usefulness of identifying and treating influential data points when building a time series model. Influential data were identified using recently developed statistical methods as found in Regression Diagnostics, Wiley 1980.

The experiment consisted of first building a time series model for the data set holding aside the last 21 observations of the series. Parameters were estimated and influential observations identified. Two methods of conditioning the data were then employed to reduce or eliminate the effect of the influential points. The parameters were then re-estimated using the two new data sets created by the conditioning.

Finally, predictions made using each model were compared to the last 21 observations of the original data as well as to each other. A qualitative statistical and a non-parametric test were performed for each comparison to aid in assessing the merit of each method of conditioning as it relates to the quality of prediction.

7. T.E. KOLLINTZAS & H.L.A.M. GEERTS (USA)
Three Notes on the Formulation of Linear Rational Expectations Models

We present three extensions to the Hansen-Sargent linear rational expectations model by (a) allowing for a more general control-state interaction in the representative economic agent's objective functional, (b) representing the stochastic elements of the model as nonstationary ARMA rather than AR processes, and (c) allowing the stochastic elements of the model to be subject to interventions of the Box-Tiao type.

8. Y.-W. LIU & R.H. BEE (USA)
Forecasting Criminal Activity with ARIMA Models

The paper models and forecasts local criminal activities with ARIMA models. Two different quarterly time series pertaining to criminal activity are analyzed. Time series data from the first quarter of 1969 through the first quarter of 1980 serve in the estimation of the ARIMA parameters, and forecasts are then made for the period of the second quarter of 1980 through the first quarter of 1982, with comparisons made to actual observations. The original data sets include: violent crime index and property crime index. We conclude that the underlying processes which generate the realizations of these two quarterly crime series are not just pure white noise. Instead, they can be modeled by distinct seasonal ARIMA models, which reflect the underlying factors resulting in the occurrences of different criminal activities.

9. R.J. USAUSKAS & M.S. ALEXANDER (USA)
An Empirical Investigation into the Validity of Technical Charting Analysis

Technical charting analysis refers to the exercise of recording in graphic form the price and transactions history of trading in a certain stock or commodity and then deducing from that pictured history the probable future price trend.

Specifically, the validity of the bar chart (recording daily price highs, lows and closes, coupled with volume and open interest figures) is investigated through vector-ARIMA analysis.

10. J.S. WHITE (USA)
Discrete and Continuous Models for Serial Correlation

The AR(1) process $x_j = \alpha x_{j-1} + e_j$ may be thought of as a sampling at equally spaced time points of the diffusion process $dx(t) = \theta x(t)dt + dw(t)$. The relationships between the parametric estimators $\hat{\alpha} = \Sigma x_j x_{j-1}/\Sigma x_{j-1}^2$ and $\hat{\theta} = \int x(t)dx(t)/\int x(t)^2 dt$ are examined and various limiting distributions are obtained. It is noted that the limiting distributions of $\hat{\alpha}$ for $\alpha < 1$, $\alpha = 1$, and $\alpha > 1$ correspond to those for $\hat{\theta}$ when $\theta \to -\infty$, $\theta = 0$, and $\theta \to +\infty$, respectively.

These Proceedings

Regretably, rather less than half of the papers offered for this Conference have eventually made the present book. About a third were rejected before presentation, and the remainder failed to revise satisfactorily. We share the disappointment of the unpublished authors, but our aim to hold standards high must be maintained.

We hope the reader will not find too much amiss with the contributions as finally selected, revised and edited. On rereading the whole volume, only one paper in four gave the editor an impression that more work should have been done; and this proportion compares very favourably with material that appears in the leading journals. Whether the referees' recommendations should have been overridden for these seven papers is doubtful: when two or three experts give their thumbs up, an editor has to be a little arrogant (not just brave) to think he knows better.

The Referees

We are most grateful to the following 61 specialists (from 12 countries) who helped with the refereeing, and in the reviewing process, for this volume:

W. Acker (USA)
H. Akaike (Japan)
A.T. Akarca (USA)
M.M. Ali (USA)
A.P. Andersen (Australia)
O.D. Anderson (UK)
M.S. Bartlett (UK)
W.R. Bell (USA)
J.M. Bernardo (Spain)
J.S. Besag (UK)
L.R. Bishop (USA)
P. Bloomfield (USA)
W. Bruggeman (Belgium)
T.A. Buishand (Netherlands)
J.P. Burman (UK)
A.V. Cameron (USA)
C.W. Chan (UK)
C. Chatfield (UK)
S.B. Cohen (USA)
J.G. de Gooijer (Netherlands)
J.-M. Dufour (Canada)
K.M. El Sheshai (USA)
E.J. Godolphin (UK)
C.W.J. Granger (USA)
M.R. Grupe (USA)
E.J. Hannan (Australia)
L.O. Hansen (USA)
D.M. Hanssens (USA)
L.D. Haugh (USA)
I.T. Jolliffe (Canada)
J. Ledolter (USA)
L.M. Liu (USA)
I.B. MacNeill (Canada)
C.R. Mann (UK)
J.Y. Narayan (USA)
H.R. Neave (UK)
J.K. Ord (USA)
D.J. Pack (USA)
F. Palm (Netherlands)
S.M. Pandit (USA)
E. Parzen (USA)
D.A. Pierce (USA)
B.D. Ripley (UK)
T.J. Sargent (USA)
R. Shibata (USA)
R.C. Souza (Brazil)
H.J. Steudel (USA)
H.H. Stokes (USA)
T. Subba Rao (UK)
M. Taniguchi (Japan)
W.R. Terry (USA)
G. Thury (Austria)
J.M. Tipton (USA)
H. Tong (Hong Kong)
W. Torrez (USA)
T. Ulrych (USA)
H.D. Vinod (USA)
A.M. Walker (UK)
K.F. Wallis (UK)
W. Wasserfallen (Switzerland)
A. Zellner (USA).

The Authors

Dr BOVAS ABRAHAM is an Associate Professor of Statistics at the University of Waterloo, Ontario, Canada. He obtained his doctorate from the University of Wisconsin in 1975 and was an Assistant Professor of Statistics at Dalhousie University, Halifax, Canada from 1975-1977. He is a Fellow of the Royal Statistical Society, and a member of the American Statistical Association, the American Society for Quality Control, the Canadian Statistical Society, and the Time Series Analysis and Forecasting Society. He is also a Statistical Councilor for Region 4 of the American Society for Quality Control.

ALI T. AKARCA has a BS degree in Economics and Statistics from Middle East Technical University, Turkey, an MS degree in Economics from the University of Wisconsin-Madison and is currently a PhD candidate at Northwestern University. He has been teaching at the University of Illinois at Chicago since 1974. He has also taught at Northeastern Illinois University and worked as a research associate at the University of Chicago. He has authored several articles involving time series analysis, which have appeared in various publications including Resources and Energy, Journal of Energy and Development, and Research in Public Policy Analysis and Management.

OLIVER D. ANDERSON is a graduate of the UK Universities of Cambridge, London, Birmingham and Nottingham; with first class honours degrees in Mathematics and Economics, and master's degrees in Engineering, Science, and Statistics.

He has worked in Industry, for Government, and as a Consultant Statistician; and taught in Schools, Colleges and Universities. He has lectured in over 20 countries and published some 250 items. He is an active member of a dozen professional societies (in England and abroad) concerned with Education, Mathematics, Statistics, Operations Research, Management Science, Economics and Econometrics; and, in 1979, was honoured by election to the International Statistical Institute.

Professor PURNELL BENSON is on the faculty of the Graduate School of Management at Rutgers University, where he has been for the past 15 years. Previous to that he was a commercial market researcher in the business world. He received his PhD in sociology at the University of Chicago in 1952.

Dr FERNAND C.M. BROECKX is Professor of Mathematics, Faculty of Applied Economics, University of Antwerp, Belgium. He is Editor of North-Holland's Journal of Computational and Applied Mathematics and President of the Belgian O.R. Society.

ALAN CAMERON is President of State Space Systems Inc, a consulting and computer software firm specializing in forecasting and control applications. Dr Cameron has developed and marketed several successful computer programs for forecasting and model building. He has over ten years' experience in program development, systems analysis and management.

Dr Cameron is a lecturer on Forecasting Inventory Control and Production Planning at the University of California, Irvine; and previously served as Corporate Director of Systems and Procedures at Mattel Inc, and as Manager of Forecasting and Systems Development at Hunt Wesson Foods.

PHILLIP A. CARTWRIGHT holds BA and MA degrees from Texas Christian University, and an MS and PhD from the University of Illinois at Urbana-Champaign, all in Economics. He is currently Director of the Georgia Economic Forecasting Project, University of Georgia, Athens, USA.

Ms CHONDIRA CHATTERJEE has recently finished her MPhil degree in Statistics at the University of Waterloo, Ontario, Canada. She is continuing her graduate studies in computer science at the University of Calgary, Alberta.

Z. GOVINDARAJULU was born in India, 15 May 1933, and became a US citizen in 1971. He obtained a BA (1952) and an MA (1953), both in Mathematics, from Madras Christian College, India; and then took his PhD in Statistics (1961) from the University of Minnesota, USA. Dr Govindarajulu is currently Professor of Statistics at the University of Kentucky, Lexington, and has published 85 papers in statistics and a book on sequential analysis.

STEPHEN E. HAYNES received his PhD from the University of California at Santa Barbara in 1976, and has been with the University of Oregon, Eugene, since 1978. His interests include International Trade and Finance, as well as Time Series Analysis. He has published in the American Economic Review, the Review of Economics and Statistics, the Southern Economic Journal, and the Journal of Macroeconomics.

GREGORY HUDAK is a statistician and software developer with Scientific Computing Associates, DeKalb, Illinois. He received a BS in mathematics from Denison University and an MS in probability and statistics from Michigan State University. He is currently a doctoral candidate at the University of Wisconsin-Madison. As a program developer for the Wisconsin Multiple Time

Series package, he has also been involved in research projects to analyze the air quality of New Jersey, Portland and the Los Angeles region.

BALDWIN S. HUI is Senior Econometrician, Data Resources Inc, Lexington, Massachusetts. He received his PhD in Economic Statistics from the Wharton School, University of Pennsylvania. His principal areas of research are structural equations models, multiple time series analysis and multivariate statistics in marketing research. He has published in a number of journals including Journal of Marketing Research, Communications in Statistics, Proceedings of the American Marketing Association, Proceedings of the American Statistical Association, and chapters in books of collected technical papers. He is Editor of The DRI Statistician.

Dr SERGIO G. KOREISHA, a graduate of the University of California at Berkeley and Harvard University, is engaged in various research, consulting and educational activities. His primary areas of research include econometric and time series forecasting, and energy modeling. His papers have appeared in several professional journals. As a member of the Harvard Business School Energy Project he coauthored the best selling book Energy Future. His professional experience includes positions with the Weyerhaeuser Company as an industrial engineer; the Clorox Company as a planning analyst; and as a consultant for various firms. He is also engaged in executive education programs through the University of Oregon.

LON-MU LIU is an Assistant Professor in the Department of Quantitative Methods, University of Illinois at Chicago Circle. He received his Bachelor degree in Agronomy from the National Taiwan University in 1971; and then MS in Statistics (1975), MS in Computer Sciences (1977), and PhD in Statistics (1978), all from the University of Wisconsin at Madison. Before joining the University of Illinois, he served as a Senior Statistician in the Department of Biomathematics, University of California, Los Angeles.

DANIEL P. McMILLEN was a consultant at Argonne National Laboratory, where his responsibilities included modeling energy demand and OPEC production. He received a BS (1981) and MA (1982), both in economics, from the University of Illinois at Chicago Circle; and is presently a PhD candidate at Northwestern University.

JACK NARAYAN was born in Guyana, South America. He received his BSc (Honours in Math) from Mount Allison University (Canada) in 1966 and his PhD in Mathematics from Lehigh University in 1970. He is presently an Associate Professor of Mathematics at the State University of New York at Oswego, and an adjunct faculty member at the School of Management, Syracuse University, where he teaches a course in Forecasting. His research interests include Time Series Analysis, Numerical Analysis and Ordinary Differential Equations.

PAUL NEWBOLD has followed a glittering student and academic career. From top LSE student in 1966, he progressed via a PhD from Wisconsin (1970) to Professor of Economics at the University of Illinois, Urbana-Champaign, in 1979. (He has published two previous articles in our Proceedings Volumes; one in Forecasting, the other in Time Series Analysis: Theory and Practice 1.)

Dr JULIA A. NORTON is an Associate Professor of Statistics at the California State University, Hayward, where she has taught since 1974.

KEITH ORD has been Professor of Management Science at the Pennsylvania State University since 1981. Prior to that, he was Reader in Statistics at the University of Warwick, England. In addition to time series and forecasting, his research interests include spatial processes and distribution theory, together with work on applications in business, geography and ecology. Most recently he has been working on the revised edition of The Advanced Theory of Statistics, Volume 3, joining Sir Maurice Kendall and Professor Alan Stuart.

LLAD PHILLIPS is Professor of Economics and Chairman of the Economics Department at the University of California at Santa Barbara. He received his PhD from Harvard University. His research interests are in the economics of criminal justice, econometrics, and population economics. He is coeditor of, and a contributing author to, Economic Analysis of Pressing Social Problems. His recent work includes "Factor Demands in the Provision of Public Safety" in Economic Models of Criminal Behavior, "Some Aspects of the Social Pathological Behavior Effects of Unemployment Among Young People" in Legal Minimum Wages, and "The Criminal Justice System: Its Technology and Its Inefficiencies" in the Journal of Legal Studies.

NAMPEANG PINGKARAWAT has a BA from Chulalongkorn University, Thailand, and an MA from the University of Illinois at Chicago, both in Economics; and is currently a PhD candidate at the latter University. She also teaches at Chicago State University.

RITEI SHIBATA obtained his doctorate in 1981 from the Mathematics Department, Tokyo Institute of Technology, Japan, where he is now a Research Associate.

HOUSTON H. STOKES received his BA in Economics from Cornell University in 1962 and his MA and PhD from the University of Chicago in 1966 and 1969. He has written some 40 articles and a book and is the developer of the B34S Data Analysis Program. Dr Stokes is currently Professor of Economics and Director of Graduate Studies at the University of Illinois at Chicago. His fields of specialization are applied econometrics, time series, international trade and monetary theory and policy.

JOE A. STONE took his PhD from Michigan State University, East Lansing, in 1977, and is currently with the University of Oregon, Eugene. His interests include International Trade and Labor Economics, and he has published in the American Economic Review, the Journal of Political Economy, the Review of Economics and Statistics, the Southern Economic Journal, the Journal of Macroeconomics, and the Journal of Human Resources.

LUCILLE M. TERRY is an Assistant Professor in the Department of Home Economics at Bowling Green State University, Ohio (USA). She received her BA from Wartburg College, Waverly, Iowa; and her MS and PhD from the University of North Carolina at Greensboro. Her interests are in statistical modeling of consumer purchase behavior.

WILLIAM ROBERT TERRY is an Associate Professor in the Department of Industrial Engineering and a member of the graduate faculty in Systems Engineering at the University of Toledo, Ohio (USA). He received his BS and MS from Georgia Tech, Atlanta; and his PhD from North Carolina State University in Raleigh. His interests are in time series analysis and control systems.

CHENG-JUN TIAN was born in China on September 15, 1940. He graduated from Zhejiang University, China, in 1961 with a major in applied mathematics. From 1962 to 1964 he did postgraduate work in stochastic processes and probabilistic information theory, in the Department of Mathematics at Nankai University, China. From 1965 to 1979 he taught probability theory, mathematical statistics, multivariate analysis and information theory, in the Department of Mathematics at Shanxi University, China, where he held positions as assistant professor, then associate professor, of mathematics. From 1980 to 1982 he was a visiting associate professor in the Department of Mathematical Statistics, Columbia University, USA; and held a concurrent post in the Applied Mathematics Department, Brookhaven National Laboratory, USA. Dr Tian's research work has been in time series analysis, system identification and probabilistic information theory.

WILLIAM C. TORREZ has been Assistant Professor in the Department of Statistics at the University of California, Riverside, since 1979, working in the areas of applied probability and stochastic processes. He has also worked at the Jet Propulsion Laboratory in Pasadena, California, under the sponsorship of a NASA Summer Faculty Fellowship.

HRISHIKESH (RICK) D. VINOD (from Tenafly, New Jersey, USA) received his PhD in Economics from Harvard University and, earlier, an MA from the Delhi School of Economics, India. Currently, he is a tenured full professor of Economics at the Fordham University in New York, where he came from Bell Laboratories, Murray Hill, New Jersey. He has won several prizes and fellowships, including one from Harvard University. His publications (over forty) have appeared in most of the top journals in Econometrics, Statistics and Economics; and are on a wide variety of topics including integer programming, public utility economics, capital theory, maximum entropy models, heat equation models for inflation diffusion, advertising-sales response and regression methodology.

Dr Vinod is regarded as one of the experts on ridge regression, and is a co-author of a research monograph: Recent Advances in Regression Methods, Marcel Dekker, New York, 1981.

Dr YOU-GIN ZHONG graduated in Mathematics from Fudan University, Shanghai, China in 1965, where he then lectured until 1974. From 1970 he has been a Research Associate at the Shanghai Reactor Engineering Institute; and from 1980 to 1982 he was a visiting scholar at the Courant Institute of New York University.

Acknowledgements

The work of the Technical Programme Committee and Local Arrangements Task Force was much appreciated.

I would also like to thank all participants for coming, speakers for presenting their work, referees for assessing it, and authors for preparing final copy for publication. As usual, I think everyone will agree that Inez van der Heide, at North-Holland, has done an excellent job in preparing this volume for print.

Looking Forward

The following 1983 planned Meetings are already well in hand, and Proceedings for all three Conferences are expected:

General Interest ITSM, Nottingham (UK) 11-15 April, 1983
Special Topics ITSM, Toronto (Canada) 10-14 August, 1983
(Hydrological, Geophysical and Spatial Time Series)
General Interest ITSM, Toronto (Canada) 18-21 August, 1983.

OLIVER D. ANDERSON
TSA&F, 9 Ingham Grove, Lenton Gardens, Nottingham NG7 2LQ, England
January 1983

TIME SERIES ANALYSIS: Theory and Practice 4
O.D. Anderson (editor)

SEASONAL ADJUSTMENT WITH X-11-ARIMA AND FORECAST EFFICIENCY

Bovas Abraham and Chondira Chatterjee

Department of Statistics
University of Waterloo
Waterloo, Ontario
Canada N2L 3G1

Very often official time series are published in seasonally adjusted form. It is, then, important to see how forecasts obtained from seasonally adjusted data compare with those obtained from unadjusted data. In this paper we compare the forecasts obtained using the X-11-ARIMA programme of seasonal adjustment with those generated from stochastic difference equation (ARIMA) models. This comparison is performed empirically using several economic time series. It is shown that the two sets of forecasts behave similarly.

INTRODUCTION

Most economic and social time series exhibit seasonal variations which seemingly obscure understanding of what really goes on in the series. Thus there is a continued interest in government and industry to seasonally adjust (or remove the seasonal fluctuations from) data. It is believed (for example see Nerlove et al (1979)) that seasonal adjustment helps in many situations: for example in studying past business cycles; in the appraisal of current economic conditions, by the estimation of the trend or cyclical components so that cyclical changes can be revealed earlier; in the detection of turning points; in economic forecasting; and in studying the relationship between two economic variables. However, it is also argued that forecasts, relationships among variables, and even econometric modelling itself can by adversely affected by seasonal adjustments (see Wallis (1974), Plosser (1979)).

In this paper we wish to compare the forecasts obtained from the seasonal time series models of Box and Jenkins (1976) with those obtained through the X-11-ARIMA (Dagum (1975)) programme of seasonal adjustment. This study is done empirically by means of several economic time series. Section 2 gives an outline of the stochastic models and the seasonal adjustment procedure used. In section 3 we give the empirical comparisons; and section 4 presents a summary and some concluding remarks.

STOCHASTIC MODELS AND SEASONAL ADJUSTMENT

Stochastic models

Suppose $\{z_t, t=0, \pm1, \pm2, \ldots\}$ is a seasonal time series in which the observations become available at equally spaced time periods. Box and Jenkins (1976) have introduced the linear stochastic model

$$\phi(B)\Phi(B^s)\{(1-B)^d(1-B^s)^D z_t^{(\lambda)} - \mu\} = \theta(B)\Theta(B^s)a_t \qquad (2.1)$$

to represent a seasonal time series. In (2.1) B is a backward shift operator such that $Bz_t = z_{t-1}$, s is the period of seasonality, $z_t^{(\lambda)}$ is some appropriate transformation of z_t, $\mu = E\{(1-B)^d(1-B^s)^D z_t^{(\lambda)}\}$, $\{a_t, t=0,\pm1,\pm2,\ldots\}$ is a sequence of independent identically distributed normal random variables with mean zero and variance σ^2, and $\phi(B) = 1-\phi_1B-\ldots-\phi_pB^p$, $\Phi(B^s) = 1-\Phi_1B^s-\ldots-\Phi_PB^{Ps}$, $\theta_1(B) = 1-\theta_1B-\ldots-\theta_qB^q$ and $\Theta(B^s) = 1-\Theta_1B^s-\ldots-\Theta_QB^{Qs}$ are polynomial operators such that their roots are outside the unit circle. We refer to the model in (2.1) as a seasonal ARIMA model of order $(p,d,q)\times(P,D,Q)_s$. These models have proved themselves to be very useful and become very popular in forecasting a variety of time series.

It can be shown (see Box and Jenkins (1976)) that the minimum mean square error (mmse) forecast $\hat{z}_n(\ell)$ of a future observation $z_{n+\ell}$ $(\ell=1,2,\ldots)$ from the origin n is given by

$$\hat{z}_n(\ell) = E(z_{n+\ell}|z_n,z_{n-1},\ldots) \quad . \tag{2.2}$$

These forecasts can be computed directly from the model in (2.1) given the data $z_n,z_{n-1},\ldots$ and the parameters.

Seasonal Adjustment

Even with the increased research interest and effort in the area of seasonal adjustment, there is no precise definition of seasonality. Granger (1979) attributes the "causes" of seasonality to calendar, timing decisions, weather, and expectation. As in most of the literature we will take seasonal adjustment to consist of the estimation and removal of a seasonal component S_t from an observable series z_t, expressable (possibly after some transformation) as

$$z_t = S_t + N_t \tag{2.3}$$

N_t being the non-seasonal component with whose interpretation or analysis S_t is presumably interfering. We will refer to (2.3) as an additive decomposition and under this decomposition we take $z_t^* = z_t - \hat{S}_t$ as the seasonally adjusted series which is obtained by removing (subtracting) the estimated seasonal component from the observed series z_t. Usually (2.3) is also written as

$$z_t = S_t + T_t + I_t \tag{2.4}$$

where T_t represents a trend component and I_t is an 'irregular' (or noise) component (note here that $N_t = T_t + I_t$). For many time series an alternate decomposition

$$z_t = S_t \cdot T_t \cdot I_t \tag{2.5}$$

is also employed and often this is referred to as a multiplicative decomposition. In this paper we will only consider the additive model given in (2.4).

Currently several seasonal adjustment procedures and corresponding computer programmes are available. In general the procedures may be classified as regression, smoothing and signal extraction methods. For a review of these the reader is referred to Kuiper (1979), Burman (1979), and Pierce (1980).

One of the most widely used seasonal adjustment procedure is the X-11 variant of the Census Method-II programme (Shiskin et al., (1967)) developed at the U.S. Census Bureau in Washington, D.C.

An approximation to the net effect of this procedure is the (2k+1) moving average or linear filter given by

$$z_t^* = \sum_{j=-k}^{k} \alpha_j z_{t+j} \ , \ (\alpha_j = \alpha_{-j}) \tag{2.6}$$

where z_t^* and z_t represent the adjusted and original series respectively. Thus the adjusted series is obtained by passing the data through a symmetric filter where the linear filter or the adjustment coefficients $\{\alpha_j\}$ are obtained not explicitly but by a multi-step procedure (for example see Wallis (1974)).

One of the drawbacks of this method is that the observations at the ends of the series are smoothed by asymmetric moving averages and not by symmetric moving averages because of the unavailability of the observations. This leads to large revisions for the adjustments to the recent observations. The X-11-ARIMA programme is an attempt to correct this drawback.

X-11-ARIMA procedure

Step 1. Obtain a seasonal ARIMA model for the series. The automatic option of this programme chooses one of 3 pre-specified models using some built in criterion.

Step 2. Generate forecasts and backcasts to extend the observed series at both ends by one year using the ARIMA model from step 1.

Step 3. Seasonally adjust the extended series using the X-11 method.

This method is claimed to perform "better" than the X-11 and is now becoming popular (see Dagum (1975)).

EMPIRICAL COMPARISONS

In this study we compare the forecasts obtained from the X-11-ARIMA method of seasonal adjustment with those obtained from stochastic difference equation models. For this comparison we use 10 economic time series and the methodology is outlined below.

Methodology

For each of the series the first n observations z_t, $t = 1,2,\ldots,n$, (z_n is taken as a December observation) are used in model construction and the next 12m observations (m denotes the number of years) are considered as a hold out period for comparing the forecasts with the actuals.

Unadjusted form

The following steps are used in computing some statistics for forecast comparison.

Step 1. Using the iterative strategy of model building (Box and Jenkins (1976)) a model is constructed for the first n observations. Then the ℓ step ahead forecasts $\hat{z}_n(\ell)$ of $z_{n+\ell}$ and forecast errors $e_n(\ell) = \hat{z}_{n+\ell} - z_n(\ell)$, $\ell = 1,2,\ldots,12$ are generated for the first year of the hold out period.

Step 2: The average annual forecast error $E_1 = \sum_{\ell=1}^{12} e_n(\ell)/12$ is computed.

Step 3. Successive one step ahead forecasts $\hat{z}_{n+j-1}(1)$ of z_{n+j} and forecast errors $e_{n+j-1}(1) = z_{n+j} - \hat{z}_{n+j-1}(1)$ $j = 1,2,\ldots,12$ are obtained.

Step 4. The model obtained in step 1 is now re-estimated using the first $n+12$ observations (including the first year of the hold out period) and the errors $E_2 = \sum_{\ell=1}^{12} e_{n+12}(\ell)/12$ and $e_{n+12+j-1}(1)$, $j = 1,2,\ldots,12$ are generated. This process is repeated for all the m years in the hold out period. The re-estimation after each year of the hold out period allows the model to cope at least partially with certain parameter changes over the years.

For the whole hold out period we have the average annual forecast errors E_i, $i = 1,2,\ldots,m$ and the one step ahead forecast errors $e_{ij} = e_{n+12(i-1)+j-1}(1)$, $i = 1,2,\ldots,m$; $j = 1,2,\ldots,12$.

Step 5. From the e_{ij}'s we compute the statistics

$$\text{(i)}\ \ ME = \sum_{i=1}^{m} \sum_{j=1}^{12} e_{ij}/12m, \qquad \text{(ii)}\ \ MAD = \sum_{i=1}^{m} \sum_{j=1}^{12} |e_{ij}|/12m$$

$$\text{(iii)}\ \ MSE = \sum_{i=1}^{m} \sum_{j=1}^{12} e_{ij}^2/12m \tag{3.1}$$

Seasonally adjusted form

Step 1a. Using the X-11-ARIMA programme the first n observations are seasonally adjusted resulting in a seasonally adjusted series z_t^*, $t = 1,2,\ldots,n$ and a seasonal component series S_t with forecasts of the seasonal component for the next twelve months.

Step 2a. As in step 1, a model is constructed for the adjusted series and using this model forecasts $\hat{z}_n^*(\ell)$, for the adjusted observations $z_{n+\ell}^*$, are generated. From the forecasts of the seasonal factors and $\hat{z}_n^*(\ell)$, the forecasts $f_n(\ell)$ for $z_{n+\ell}$, and the forecast errors $e_n^*(\ell) = z_{n+\ell} - f_n(\ell), \ell = 1,2,\ldots,12$, and $E_1^* = \sum_{\ell=1}^{12} e_n^*(\ell)/12$ are obtained.

Step 3a. Using X-11-ARIMA, seasonally adjust the observations z_t, $t=1,2,\ldots,n+12$. This yields a set of seasonally adjusted observations z_t^* which include z_{n+j-1}^*, $j = 2,3,\ldots,12$.

In this step an unfair unadvantage is given to the forecasts produced throught the seasonal adjustment procedure because we are also using all the original observations z_{n+j} to generate z_{n+j}^* $(j=1,2,\ldots,12)$ which in turn are necessary to generate $\hat{z}_{n+j-1}^*(1)$ $(j=2,3,\ldots,12)$. Another alternative is to successively seasonally adjust all the observations up to z_{n+j-1}, and then generate $\hat{z}_{n+j-1}^*(1)$, $j = 2,3,\ldots,12$. However this route is extremely cumbersome computationally and not adopted here.

Step 4a. Using $\hat{z}_{n+j-1}^*(1)$ and the forecasts of the seasonal factors, generate the

The models obtained for each of the series in the unadjusted as well as adjusted form are given in Table 1 together with the initial and final estimates of the parameters. It should be pointed out that we did not let X-11-ARIMA choose one of the built in models to generate forecasts and backcasts prior to seasonal adjustment; we supplied the ARIMA model obtained after the model building phase. The estimates were obtained by an approximate maximum likelihood procedure (see McLeod (1977)). For all the series the parameter estimates were somewhat different for each of the years in the hold out period. It can also be noticed that, in general, the models for the seasonally adjusted data have the same form as the non-seasonal part of the models for the unadjusted data. However, the parameter estimates for both seem to be slightly different. Also, in the case of series (ix), the seasonally adjusted data show a negative correlation at lag 12 calling for an additional seasonal part in the model. This is probably due to some over adjustment. In an earlier empirical study with X-11, Plosser (1979) noticed that there were overadjustments in three of the five series he had considered. In this respect X-11-ARIMA may have a slight edge over X-11 since we notice overadjustment only in one series out of the ten considered here.

TABLE 1

SUMMARY OF MODELS FOR UNADJUSTED AND ADJUSTED DATA

				Parameter Estimates					
				Initial			Final		
Series		Transformation	Model	$\hat{\theta}$	$\hat{\Theta}$	$\hat{\sigma}^2$	$\hat{\theta}$	$\hat{\Theta}$	$\hat{\sigma}^2$
1.	UA	ℓn	For unadjusted $(0,1,1)(0,1,1)_{12}$	.21	.56	5.63×10^{-3}	.15	.57	4.90×10^{-3}
	A	ℓn	For adjusted $(0,1,1)(0,0,0)_{12}$	.27		3.30×10^{-3}	.24		2.81×10^{-3}
2.	UA	none		.87	.80	6.65×10^{-2}	.84	.82	7.60×10^{-2}
	A	none		.84		4.50×10^{-2}	.80		4.97×10^{-2}
3.	UA	ℓn		.67	.80	1.57×10^{-2}	.66	.80	1.52×10^{-2}
	A	ℓn		.66		1.13×10^{-2}	.63		1.06×10^{-2}
4.	UA	ℓn		.25	.58	2.01×10^{-4}	.26	.56	2.02×10^{-4}
	A	ℓn		.32		1.33×10^{-4}	.37		1.42×10^{-4}
5.	UA	ℓn		.17	.70	1.47×10^{-4}	.19	.70	1.51×10^{-4}
	A	ℓn		.14		1.13×10^{-4}	.18		1.01×10^{-4}
6.	UA	ℓn		.13	.58	6.77×10^{-5}	.20	.62	7.33×10^{-5}
	A	ℓn		.21		5.82×10^{-5}	.22		6.10×10^{-5}
7.	UA	ℓn		.41	.54	4.25×10^{-3}	.44	.54	4.84×10^{-3}
	A	ℓn		.45		2.43×10^{-3}	.47		2.78×10^{-3}
8.	UA	ℓn	$(1,1,1)(0,1,1)_{12}$	.92 ($\hat{\phi}$=.42)	.58	1.28×10^{-2}	.90 ($\hat{\phi}$=.48)	.64	1.25×10^{-2}
	A	ℓn	$(1,1,1)(0,0,0)_{12}$	.82 ($\hat{\phi}$=.35)		$.73\times10^{-2}$	.85 ($\hat{\phi}$=.39)		$.77\times10^{-2}$
9.	UA	ℓn	$(0,2,1)(0,1,1)_{12}$	.81	.79	2.72×10^{-4}	.75	.76	3.13×10^{-4}
	A	ℓn	$(0,2,1)(0,0,1)_{12}$	.84	.48	1.41×10^{-4}	.37	.37	1.65×10^{-4}
10.	UA	ℓn	$(1,1,0)(0,1,1)_{12}$	$\hat{\phi}$=-.15	.82	2.65×10^{-2}	$\hat{\phi}$=-.15	.83	2.51×10^{-2}
	A	ℓn	$(1,1,0)(0,0,0)_{12}$	$\hat{\phi}$=0.16		1.80×10^{-2}	$\hat{\phi}$=-.17		1.7×10^{-2}

One also notices that the residual variances corresponding to the seasonally adjusted series are smaller than those corresponding to the unadjusted series. This does not, however, imply that the forecasts from the former will be better than those from the latter. The residual variances corresponding to the seasonally adjusted series do not take into account the effect of estimation error involved in the estimation of the seasonal component. Hence one cannot base any comparisons on these residual variances. However, the statistics defined in (3.1) and (3.2) are helpful.

one step ahead forecasts $f_{n+j-1}(1)$ and forecast errors $e^*_{n+j-1}(1)$, $j=1,2,\ldots,12$.

Step 5a. Re-estimate the model obtained in step 2a using the seasonally adjusted series in step 3a. Using the procedures in steps 2a-4a obtain $E^*_2 = \sum_{\ell=1}^{12} e^*_{n+12}(\ell)/12$ and $e^*_{n+12+j-1}(1)$, $j=1,\ldots,12$. It is to be noted here that the generation of $e^*_{n+12+j-1}(1)$ requires the seasonal adjustment of all observations z_t, $t=1,2,\ldots,n+24$.

This process is repeated for all the m years in the hold out period and we end up with the forecast errors E^*_i, and

$$e^*_{ij} = e^*_{n+12(i-1)+j-1}(1) \qquad (i=1,2,\ldots,m;\ k=1,2,\ldots,12).$$

Step 6a. From the e^*_{ij}'s we compute statistics similar to those of step 4; $ME^* = \sum_{i=1}^{m}\sum_{j=1}^{12} e^*_{ij}/12m$, $MAD^* = \sum_{i=1}^{m}\sum_{j=1}^{12} |e^*_{ij}|/12m$, and $MSE^* = \sum_{i=1}^{m}\sum_{j=1}^{12} e^{*2}_{ij}/12m$. These statistics will be compared with those computed for the unadjusted data.

Data and Models

The following monthly time series will be used to compared the forecasts. These data are obtained from the Canadian Statistics Review and the Survey of Current Business.

(i) Unemployment rate for Canada, January 1958 - December 1975 (m=5; last 5 years considered as the hold out period)

(ii) Percentage change in the Consumer price index for Canada, January 1958 - December 1974 (m=4)

(iii) Total building permits issued in Canada (millions of dollars), January 1958 - December 1975 (m=5)

(iv) Index of industrial production for Canada, January 1961 - December 1975 (1961=100) (m=3)

(v) Total wages and salaries (millions of dollars) for Canada, January 1958 - December 1974 (m=5)

(vi) Total labour income (millions of dollars) for Canada, January 1958 - December 1973 (m=4)

(vii) Total revenue rail freight loadings (thousand tons) in Canada, January 1958 - December 1973 (m=5)

(viii) New passenger car sales in Canada (units), January 1958 - December 1975 (m=5)

(ix) Inventories of United States farm equipment and machinery, (millions of dollars) January 1958 - December 1975 (m=5)

(x) New dwelling units started (units) in Canada, January 1958 - December 1975 (m=5)

Forecast Comparison

As indicated in the methodology section we will consider the average annual forecast errors (E_i, and E_i^*), the biases (ME and ME*), the mean absolute deviations (MAD and MAD*), and the mean squared errors (MSE and MSE*) for forecast comparisons.

FORECAST ERRORS E_i AND E_i^* FOR THE TEN SERIES

——— E_i (Unadjusted); ------ E_i^* (Adjusted)

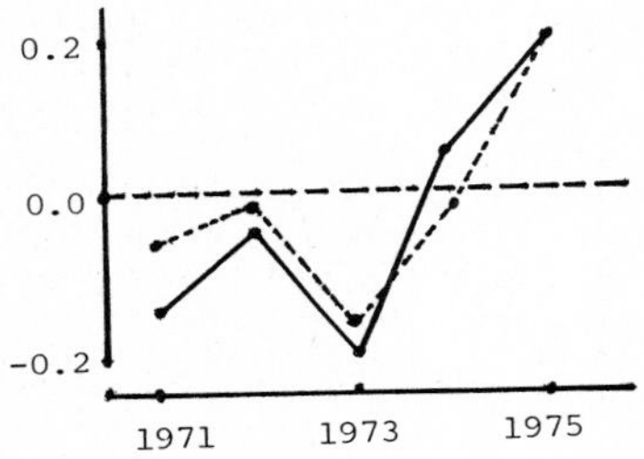

Figure 1. For Unemployment Rate.

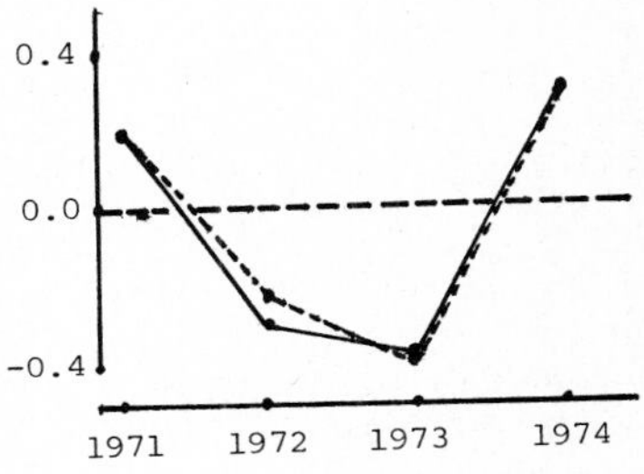

Figure 2. For Percentage Change in the CPI.

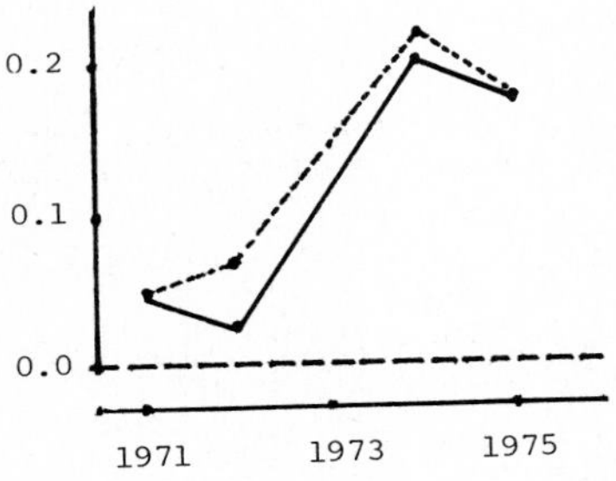

Figure 3. For Building Permits.

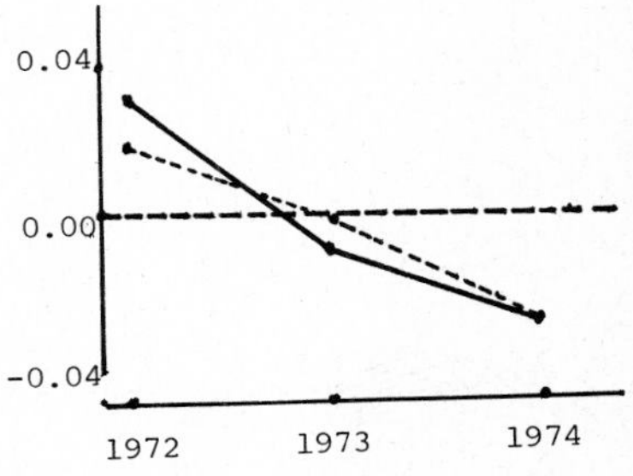

Figure 4. For Index of Industrial Production.

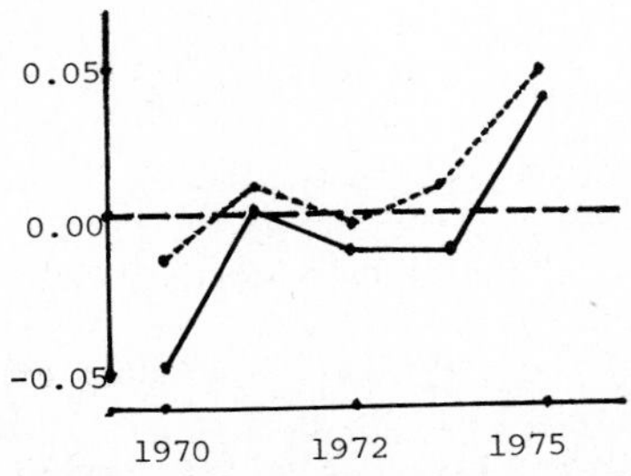

Figure 5. For Wages and Salaries.

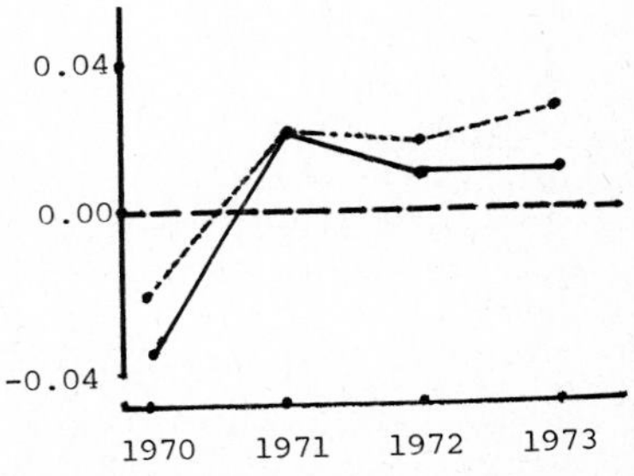

Figure 6. For Labour Income.

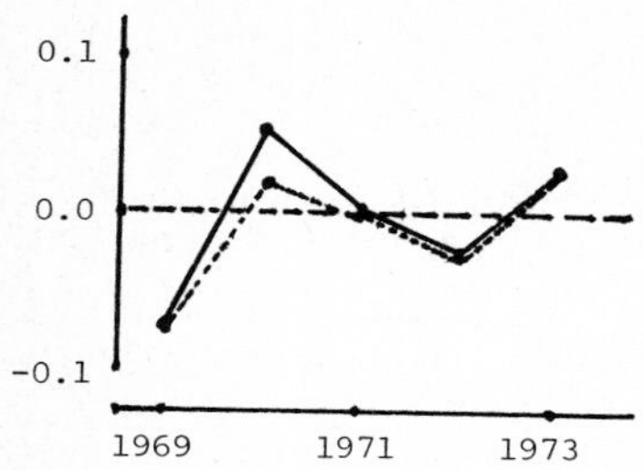

Figure 7. For Rail Freight Loadings.

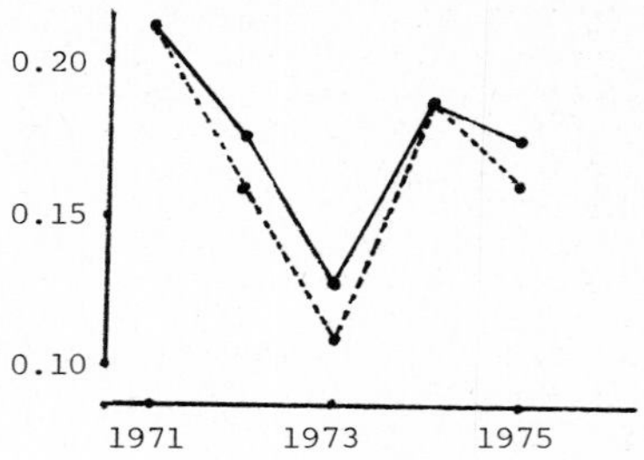

Figure 8. For New Passenger Car Sales.

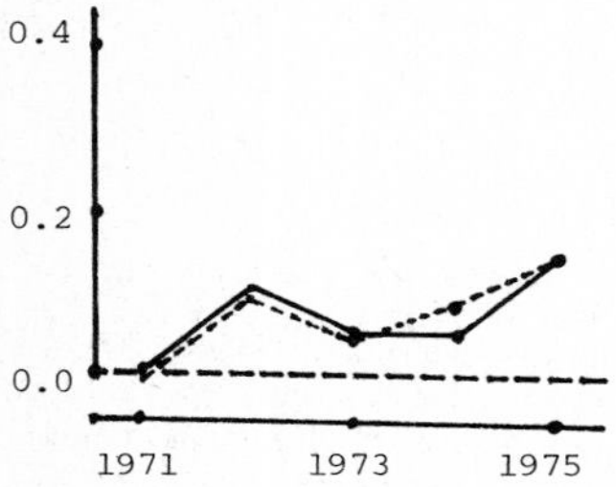

Figure 9. For Inventories of Farm Equipment and Machinery.

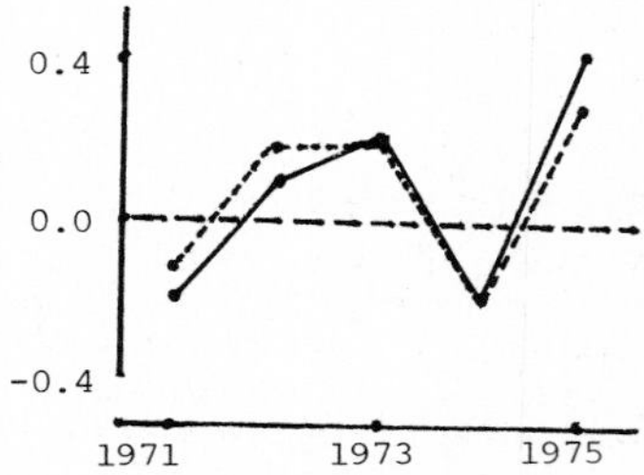

Figure 10. For New Dwelling Units Started.

TABLE 2

FORECAST COMPARISON

	1		2		3	
Series	ME	ME*	MAD	MAD*	MSE	MSE*
1	-.0058	.0021	.0427	.0373	.0026	.0020
2	-.0460	-.0246	.3032	.2969	.1360	.1311
3	.0382	.0216	.0948	.0943	.0136	.0133
4	-.0012	.0023	.0138	.0143	.0003	.0003
5	-.0001	.0026	.0097	.0094	.0002	.0001
6	.0003	.0030	.0073	.0084	.0001	.0001
7	.0000	.0005	.0612	.0569	.0078	.0068
8	.0535	.0483	.1248	.0837	.0201	.0192
9	-.0009	.0006	.0157	.0150	.0005	.0010
10	-.0001	.0010	.1077	.0977	.0217	.0183

For each series the forecast error E_i and E_i^* are plotted and shown in Figures 1 - 10. Both of these behave roughly the same way, overall, for each of the series. In series 1 there seems to be some slight over prediction in the initial years with a slight under prediction in the last year. In series 2 there is some under prediction in the first and last years with a small over prediction in the second and third years. There appears to be some under prediction in the initial years of series 3 and 9 and in all the years of series 8. In series 10 there is a slight over prediction in the first and fourth years followed by an under prediction in the final year.

The statistics ME, ME*, MAD, MAD*, MSE and MSE* are calculated using (3.1) and (3.2) and shown in Table 2. The bias seems to be small in each case; in series 4,5,6,7 and 10 the unadjusted form has an edge over the other form while the reverse can be argued in series 1 and 2. The mean absolute deviation is roughly the same for each form in all the series. However, it may be argued that the adjusted form is slightly better. Using mean squared error as a criterion one could argue that the unadjusted form does better in series 9 while the reverse could be said of series 5. In all other series MSE and MSE* are about the same. Thus, overall there do not seem to be very much difference between the two sets of forecasts. It appears, then, that the X-11 ARIMA forecasts are performing as well in practice as the forecasts produced by the theoretically optimal ARIMA models.

It is also interesting to note that an earlier empirical study by Plosser (1979) indicated that the forecasts from the ARIMA models were superior to those produced by the X-11 seasonal adjustment procedure. In the present study the seasonal adjustment procedure is not doing that badly. Hence it seems reasonable to conclude that X-11-ARIMA has a slight edge over X-11.

CONCLUDING REMARKS

We have outlined the main features of the X-11-ARIMA procedure and compared the forecasts obtained from it with those from the stochastic models. It was seen that the two sets of forecasts behave similarly. However, it should be pointed out that (i) the X-11-ARIMA procedure was given an unfair advantage in step 3a and (ii) instead of using the "built in" models of the automatic version, we forced the program to use the models which we obtained after the model builing stage. In any case, we have dealt with only a few series and so our conclusions must be tentative.

It is recognized that seasonal adjustment procedures are fully automatic and can be used to seasonally adjust a large number of series within a short time. Although this saves effort it can be a weakness because the same "model" is used to isolate seasonality from different types of series. However, X-11-ARIMA has more flexibility compared with other seasonal adjustment procedures since it uses the forecasts from ARIMA models to extend the series before the adjustment.

ACKNOWLEDGEMENTS

Bovas Abraham was partially supported by a Grant from the National Sciences and Engineering Research Council of Canada.

REFERENCES

BOX, G.E.P. and JENKINS, G.M. (1976). Time Series Analysis Forecasting and Control (Revised Edition). San Francisco: Holden Day.

BURMAN, J.P. (1979). "Seasonal Adjustment-A Survey", in Forecasting (TIMS Studies in the Management Sciences Vol. 12). Eds: S. Makridakis and S.S. Wheelwright. Amsterdam: North-Holland, 44-57.

CHATTERJEE, C. (1982). Deterministic Components and Seasonal Adjustments in Time Series Models. Unpublished M.Phil. Thesis, Department of Statistics, University of Waterloo, Waterloo, Ontario, Canada.

DAGUM, E.B. (1975). "Seasonal Factor Forecasts from ARIMA Models", paper presented at the 40th Session of the International Statistical Institute, Warsaw, Poland.

GRANGER, C.W.J. (1979). "Seasonality: Causation, Interpretation and Implications", in Seasonal Analysis of Economic Time Series. Ed: Arnold Zellner, Washington, D.C., U.S. Dept. of Commerce, Bureau of the Census, 33-46.

KENNY, P.B. and DURBIN, J. (1982). "Local trend estimation and seasonal adjustment of economic and social time series", JRSS A 145, 1-41.

KUIPER, J. (1979). "A Survey and Comparitive Analysis of Various Methods of Seasonal Adjustment", in Seasonal Analysis of Ecnomic Time Series. Ed: Arnold Zellner, Washington, D.C., U.S. Dept. of Commerce, Bureau of the Census, 59-76.

MCLEOD, A.I. (1977). "Improved Box-Jenkins estimators", Biometrika 64, 531-534.

NERLOVE, M., GRETHER, D.M., and CARVALHO, J.L. (1979). Analysis of Economic Time Series. New York: Academic Press.

PIERCE, D.A. (1980). "A Survey of Recent Developments in Seasonal Adjustment", The American Statistician 34, 125-134.

PLOSSER, C. (1979). "Short-Term Forecasting and Seasonal Adjustment", JASA 74, 15-24.

SHISKIN, J., YOUNG, A.H. and MUSGRAVE, J.C. (1967). "The X-11 Variant of the Census Method-II Seasonal Adjustment Program", Technical Paper No. 15, U.S. Bureau of the Census.

WALLIS, K.F. (1974). "Seasonal Adjustment and Relations Between Variables", JASA 69, 18-31.

WALLIS, K.F. (1982). "Seasonal Adjustment and Revision of Current Data: Linear Filters for the X-11 method", JRSS A 145, 74-85.

TIME SERIES ANALYSIS: Theory and Practice 4
O.D. Anderson (editor)

ON SEASONAL ADJUSTMENT EVALUATION FROM THE RESIDUALS*

C. J. Tian
Brookhaven National Laboratory
Upton, New York U.S.A.
Columbia University, New York U.S.A.
Shanxi University, China

Y.Q. Zhong
Courant Institute of Mathematical Sciences
New York University, New York U.S.A.
Shanghai Reactor Engineering Institute, China

In this paper, some properties of superposed periodic series are studied. The results indicate that we may evaluate an estimate for the seasonal component of a time series from the residuals on an interval with sufficient length. In particular, when the periods of the seasonal component are integers, we may inspect an estimate for the seasonal component from the residuals on an interval whose length is no less than the sum of those corresponding periods. On the other hand, if the inspection of an estimate for the seasonal component is carried out on an inadequate interval, then the evaluation is no longer assured.

I. INTRODUCTION

As is well known, most popular methods for seasonal adjustment are founded on some prior probabilistic assumptions. However, it is rather difficult to verify this kind of assumption from series of finite length. In fact, we usually do not have perfect knowledge of the structure or behavior of the series itself, i.e. we do not know the exact model of the time series, so the statistical testing of hypotheses often becomes a mere formality. When the prior probabilistic assumption cannot be verified precisely, this kind of testing is only the measurement of residuals and the result of statistical testing is no longer guaranteed.

For example, in the periodogram method (Anderson, 1971) we assume that the model is a linear combination of several harmonics and a white noise, and we also require that the series length of observations, m, is an integral multiple of the given period d. Since any estimate of the seasonal adjustment, generated by this method, must be a periodic series with period m, the extrapolated effect may be guaranteed by the residual if $m=kd$, where k is an integer. However, this conclusion no longer holds if $m \neq kd$, and the extrapolated effect may be worse even though the residuals are small.

Although seasonal adjustment, one of the most important topics in time series analysis, has been studied for a long time, there are many unsolved problems (Parzen, 1978; Pierce, 1978), including how

*Authored under contract DE-AC02-76CH00016 with the U.S. Department of Energy.

to estimate hidden periods from observed time series. In practice, there are two popular methods for estimating hidden periods: graphically and by spectral analysis (Bloomfield, 1976). Unfortunately, neither of these give correct results in many cases (Tian and Lu, 1982). Due to this kind of difficulty, there is a need to look for new approaches to seasonal adjustment (Akaike, 1980; Tian, 1982).

We have tried to study seasonal adjustment of time series from the viewpoint of numerical approximation: some probabilistic methods have been explained and methods extended. Especially, the successive average method was proposed in Tian (1982). Because no restrictions on time series structure are imposed in the convergence theorem of the successive average method, this just gives an approximation to the observed series, which is in the form of superposed periodic series, and it is not certain that we have obtained an estimate for the seasonal component of the time series. For example, given a set of observations from the series

$$x(t) = t, \qquad t=1,2,...,m; \qquad m=20$$

using the successive average method, we obtain the periodic superposed series

$$\hat{x}(t) = \bar{x} + f_1(t) + f_2(t) + \dots + f_5(t),$$

and

$$\bar{x} = \frac{1}{m}\sum_{t=1}^{m} x(t) = 10.5,$$

where $f_i(t)$ has the period $i+5$, $i=1,2,...,5$. Let us list the values of $f_i(t)$, $i=1,2,...,5$, and $\hat{x}(t)$ as follows:

t	$f_1(t)$	$f_2(t)$	$f_3(t)$	$f_4(t)$	$f_5(t)$	$\hat{x}(t)$
1	8.4750	-1.6667	14.1347	-2.4645	-27.9786	0.9999
2	-8.4750	-1.6667	-3.3251	2.4645	2.5023	2.0000
3	-32.3223	-1.6667	3.3251	-20.8566	44.0206	3.0001
4	-21.0223	1.6667	-14.1347	-10.2129	37.2033	4.0001
5	21.0223	1.6667	-41.4494	-9.1353	22.3959	5.0002
6	32.3223	1.6667	-16.0931	0.0000	-22.3959	6.0000
7		0.0000	16.0931	9.1353	-37.2033	7.0001
8			41.4494	10.2129	-44.0206	8.0000
9				20.8566	-2.5023	9.0000
10					27.9786	10.0000
11						11.0000
12						12.0000
13						13.0000
14						13.9999
15						15.0000
16						15.9998
17						16.9999
18						17.9999
19						19.0000
20						20.0001

The sum of squares is

$$Q = \sum_{t=1}^{20} [x(t) - \hat{x}(t)]^2 = 1.6 \times 10^{-7}.$$

Notice that although this periodic superposed series fits extremely well, the original time series does not have any cyclic trend at all.

In order to decide whether a time series has periodicity and to distinguish the real periods from the pseudo ones in the computational results, Tian and Lu (1982) suggested an empirical method. In the theoretical analysis, it is necessary and important to reconsider the problem of how to evaluate an estimate for the seasonal component of a time series from the residuals. For this purpose, we first discuss some related properties of the superposed periodic series; then the problem of seasonal adjustment evaluation is discussed in terms of them.

II. MAIN RESULTS

Suppose a function $x_k(t)$ is generated by superposition of k periodic functions $f_i(t)$, i=1,2,...,k, with respective periods τ_i , i.e.,

(1) $$x_k(t) = \sum_{i=1}^{k} f_i(t)$$

with $f_i(t) = f_i(t + \tau_i)$, i=1,...,k, t=1,2,... . Here, as a convention, the real functions $x_k(t)$ and $f_i(t)$ are defined on the set of positive integers.

We have the following results:

(A) For any $\varepsilon>0$, if $|x_k(t)|<\varepsilon$, $\forall t \leqslant T= \sum_{i=1}^{k} \tau_i$, then

(2) $$|x_k(t)|<B_k \varepsilon, \qquad \forall\ t$$

where B_k is a constant, dependent only on $\tau_1,\dots,\tau_k$.

(B) The constant B_k in Eq. (2) is given by

(3) $$B_k = 1 + \sum_{\ell=2}^{k} \prod_{j=\ell}^{k} \frac{M_j}{\tau_j}$$

where M_j is the lowest common multiple of $\tau_1, \tau_2,\dots,\tau_j$.

Remark 1. If the periods $\tau_1,\dots,\tau_k$ are mutually prime, the number of constraints T in (A) can be reduced to

(4) $$T' = \sum_{i=1}^{k} \tau_i - (k-1)$$

with a modified bound B_k',

(5) $$B_k' = 1 + \sum_{\ell=2}^{k} \prod_{j=\ell}^{k} \frac{2m_j}{t_j} .$$

However, if the number T of constraints is further decreased there exist counter examples to the proposition.

Example. For k=2, $\tau_1=3$ and $\tau_2=4$, if the number of constraints is reduced to $\tau_1 + \tau_2 - 2 = 5$, we observe the following equations

$$f_1(1) = f_1(2) = f_2(3) = 1,$$

and
$$f_1(3) = f_2(1) = f_2(2) = f_2(4) = -1.$$

It is easy to see that

$$x_2(t) = f_1(t) + f_2(t) = 0, \qquad t \leqslant 5,$$

however

$$x_2(6) = f_1(3) + f_2(2) = -2.$$

On the other hand, if the periods $\tau_1, \ldots, \tau_k$ have a common divisor d, the number of constraints T can be further reduced at least to

$$T'' = \sum_{i=1}^{k} \tau_i - (k-1)d. \tag{6}$$

Remark 2. For k=2, Eq. (3) becomes

$$B_2 = 1 + \frac{M_2}{\tau_2},$$

while there is a slight refinement for k=3,

$$B_3 = \frac{3}{4} \cdot \frac{M_2}{\tau_2} \cdot \frac{M_3}{\tau_3}. \tag{7}$$

These results indicate that we may evaluate a seasonal component estimate from the residuals on a certain interval with sufficient length. In particular, when the periods of the seasonal components are integers, we may inspect an estimate for seasonal component by means of residuals on an interval whose length is no less than the sum of those corresponding periods.

Generally speaking, if the residuals are quite small on that specified interval, then the seasonal component estimate may be considered acceptable. Especially, this inspection is useful in verifying the computational results of the successive average method.

On the other hand, the results here also show the fact that, if the inspection of the estimate is carried out on an inadequate interval, then no matter how small are the residuals on this interval, the evaluation of the estimate is not assured, even the time series periodicity is unverified.

This fact should be emphasized in the practice of seasonal adjustment.

III. LEMMAS

In order to prove the above properties, we need the following defintions and lemmas. (Griffiths and Hilton, 1980).

Suppose there are k linear spaces L_i, i=1,...,k, each with any dimension, and their direct sum space is L, say, then the vector v in L can be written

$$v = v_1 \oplus \ldots \oplus v_k,$$

where the vectors $v_i \in L_i$, i=1,...,k. Assume that there is a set of vectors $\{e_j^i,\ j=1,\ldots,\tau_i\}$ in space L_i, and for any positive integer t, the vector e_t^i is defined as

$$e_t^i = e_{t \bmod \tau_i}^i,$$

with the convention that $e_0^i = e_{\tau_i}^i$.

Lemma 1. For any positive integer t, the vector of the form

$$\varepsilon_t = e_t^1 \oplus \ldots \oplus e_t^k$$

in space L belongs to the linear subspace spanned by the set S of vectors

$$\{\varepsilon_j = e_j^1 \oplus \ldots \oplus e_j^k \mid j=1,\ldots,\sum_{i=1}^{k} \tau_i\}.$$

Moreover, the sum $A_k(t: \tau_1,\ldots,\tau_k\}$ of the absolute values of the coefficients in the linear expression of ε_t in terms of $\{\varepsilon_j\}$ satisfies the following recursive relations

$$A_k(t;\ \tau_1,\ldots,\tau_k) \leq \min\{1+2\sum_{i=1}^{I(t;k)} A_{k-1}(t-i\tau_k;\tau_1,\ldots,\tau_{k-1}), \quad 1+2\sum_{j=0}^{J(t;k)-1} A_{k-1}(t+k\tau_k;\tau_1,\ldots,\tau_{k-1})\} \tag{8}$$

where

$$I(t;k) = \left[\frac{(t-1) \bmod M_k(\tau_1,\ldots,\tau_k)}{\tau_k}\right], \tag{9}$$

$$J(t;k) = \frac{M_k(\tau_1,\ldots,\tau_k)}{\tau_k} - I(t;k), \tag{10}$$

and [x] is the largest integer less than or equal to the argument x, whereas $M_k(\tau_1,\ldots,\tau_k)$ is the lowest common multiple of integers $\tau_1,\ldots\tau_k$. It is trivial to point out that $A_1(t;\tau_1)=1$. Sometimes, $M_k(\tau_1,\ldots,\tau_k)$ is abbreviated to M_k, and $A_k(t;\tau_1,\ldots,\tau_k)$ to A_k. From this lemma it follows that

$$A_2 \leq 1 + 2\min\{\left[\frac{(t-1) \bmod M_2}{\tau_2}\right],\ \frac{M_2}{\tau_2} - \left[\frac{(t-1) \bmod M_2}{\tau_2}\right]\}$$

$$\leq 1 + 2\left[\frac{M_2}{2\tau_2}\right] \leq 1 + \frac{M_2}{\tau_2}.$$

And generally an upper bound B_k of A_k is given by

$$(11)\qquad A_k(t;\tau_1,\dots,\tau_k) \leq B_k(\tau_1,\dots,\tau_k) = 1 + \sum_{\ell=2}^{k} \prod_{j=\ell}^{k} \frac{M_j}{\tau_j} .$$

The following lemma will enable us to refine the estimate of the upper bound of A_k. (Vinogradov, 1975).

Lemma 2. If the greatest common divisor of positive integers r and s is f; then, for any integer d, the set of integers

$$(12)\qquad \{(d+k\,s) \bmod r \mid k=1,2,\dots,\tfrac{r}{f}\} = \{d \bmod f + i\,f \mid i=0,1,\dots,\tfrac{r}{f}-1\}.$$

Especially, when r and s are relatively prime, the following holds

$$\{(d+k\,s) \bmod r \mid k=1,2,\dots,r\} = \{0,1,\dots,r-1\}.$$

Corollary 1. Under the conditions of Lemma 2, for any positive integer $h \leq \frac{r}{f}$, the following inequality can be obtained

$$(13)\qquad \sum_{k=1}^{h} \min\{(d+k\,s) \bmod r,\; r-(d+k\,s) \bmod r\} \leq 2\,[\tfrac{r}{2}][\tfrac{h}{2}] - f[\tfrac{h}{2}]^2 + [\tfrac{r}{2}].$$

Corollary 2. Denoting $h' = \frac{r}{f} - h$, we have the inequality

$$(14)\qquad \min\{\sum_{k=1}^{h}, \sum_{k=1}^{h'}\} \leq \frac{3}{4}\frac{1}{f}[\tfrac{r}{2}]^2 + [\tfrac{r}{2}],$$

where the summations are the abbreviation of the sum in Corollary 1.

From the above corollaries, with $r = M_2(\tau_1,\tau_2)$ and $s=\tau_3$, we get

$$(15)\qquad A_3(t;\tau_1\,\tau_2,\tau_3) \leq 1 + \frac{M_3}{\tau_3} + \frac{3}{4}\frac{1}{f}\frac{M_2^2}{\tau_2} + \frac{2\,M_2}{\tau_2} \doteq \frac{3}{4}\frac{1}{f}\frac{M_2^2}{\tau_2} = \frac{3}{4}\frac{M_2}{\tau_2}\frac{M_3}{\tau_3} .$$

IV. PROOF OF LEMMAS AND THE MAIN RESULTS

(i) Proof of Lemma 1

Induction is used. Since the lemma is obviously true when k=1 with $A_1(t;\tau) \equiv 1$, we will investigate the situation of k=n assuming the lemma holds for k=n-1.

In each linear space L_i another set of vectors

$$\{\delta_j^i\,,\ j=1,\dots,\tau_i\} \qquad i = 1,\dots n-1$$

is formed by the formula $\delta_j^i = e_{j+\tau_n}^i - e_j^i$. It is directly seen that

$\delta_t^i = \delta_{t \bmod \tau_i}^i$ for any positive integer t.

By the assumption of induction, any vector $\hat{\delta}_t$ in space $L_1 \oplus \ldots \oplus L_{n-1}$ of the form

$$\hat{\delta}_t = \delta_t^1 \oplus \ldots \oplus \delta_t^{n-1},$$

where t is any positive integer, belongs to the linear subspace spanned by the set of vectors

$$\{\hat{\delta}_j = \delta_j^1 \oplus \ldots \oplus \delta_j^{n-1} \mid j = 1, \ldots, \sum_{i=1}^{n-1} \tau_i\},$$

and the absolute values of its linear coefficients in this expression sum to $A_{n-1}(t; \tau_1, \ldots, \tau_{n-1})$.

Denote by θ^n the zero vector in space L_n. It follows that

$$\delta_j = \varepsilon_{j+\tau_n} - \varepsilon_j = \hat{\delta}_j \oplus \theta^n, \qquad j = 1, \ldots, \sum_{i=1}^{n-1} \tau_i.$$

Notice that the vector δ_j can be generated by two vectors in the set $S = \{\varepsilon_j = e_j^1 \oplus \ldots \oplus e_j^n \mid j = 1, \ldots, \sum_{i=1}^{n} \tau_i\}$, while the vector $\delta_t = \hat{\delta}_t \oplus \theta^n$ in L belongs to the subspace spanned by the set $\{\delta_j \mid j = 1, \ldots, \sum_{i=1}^{n-1} \tau_i\}$; therefore, it also belongs to the subspace formed by the set S.

Since $\varepsilon_t = \varepsilon_{t \bmod M_n}$ for any positive integer t, we can assume $t \leq M_n$ without loss of generality.

Hence, we can write

$$\varepsilon_t = \varepsilon_{t \bmod \tau_n} + \sum_{i=1}^{I(t;n)} \delta_{t-i\tau_n}.$$

This proves that the vector ε_t lies in the subspace spanned by the set S. Moreover, the last equation gives rise immediately to the following inequality

$$A_n(t; \tau_1, \ldots, \tau_n) \leq + 2 \sum_{i=1}^{I(t;n)} A_{n-1}(t-i\tau_n : \tau_1; \ldots, \tau_{n-1}).$$

On the other hand, it is observed that

$$\sum_{i=1}^{I(t;n)} \sigma_{t-i\tau_n} + \sum_{j=0}^{J(t;n)-1} \sigma_{t+j\tau_n} = \varepsilon_{t+J(t;n)\times\tau_n} - \varepsilon_{t \bmod \tau_n} = 0$$

this is because

$$t + J(t;n)\times\tau_n - t \bmod \tau_n = M_n - I(t;n)\times\tau_n + t - t \bmod \tau_n = M_n$$

therefore the vector ε_t also can be be expressed as

$$\varepsilon_t = \varepsilon_{t \bmod \tau_n} - \sum_{j=0}^{J(t;n)-1} \sigma_{t+j\,\tau_n}. \qquad \text{Q.E.D.}$$

(ii) Proof of Lemma 2

It suffices to prove the following propositions with the assumptions of Lemma 2,

(P1). $(d+ks) \bmod r \bmod f = d \bmod f$;

(P2). $(d+ks) \bmod r < r$;

(P3). if there are integers k_1 and k_2, such that $(d+k_1s) \bmod r = (d+k_2s) \bmod r$, then it can be asserted that

$$k_1 - k_2 = 0 \bmod \frac{r}{f}.$$

To prove (P1), we use the fact that r has the divisor f, and so does s, hence

$$(d + ks) \bmod r \bmod f = (d + ks) \bmod f = d \bmod f.$$

The proposition (P2) is obvious. In order to prove (P3), we suppose that there exist integers m_1, m_2, and Δ satisfying the two equations

$$d + k_1s = m_1r + \Delta \text{ and } d + k_2s = m_2r + \Delta,$$

such that $(k_1 - k_2)s = (m_1 - m_2)r$.

If we write $s = af$, and $r = bf$, while the integer a and b are obviously prime then the last equation can be written

$$(k_1 - k_2)a = (m_1 - m_2)b.$$

Since a and b are co-prime, it follows that

$$k_1 - k_2 = 0 \bmod b.$$

However, $b = \frac{r}{f}$, which proves the proposition (P3). Q.E.D.

(iii) Proof of corollary 1.

By Lemma 2, for any positive integer $h \leqslant \frac{r}{f}$, the set of integers $\{(d + ks) \bmod r \mid k = 1,\ldots,h\}$ comprises only those with distinct values and with their differences being multiples of the integer f. Taking this fact into account as well as the fact that

$$\min\{j, r-j\} \leqslant [\tfrac{r}{2}],$$

we have the following inequality

$$\sum_{k=1}^{h} \min\{(d + ks) \bmod r, r-(d + ks) \bmod r\}$$

$$\leq 2 \sum_{j=0}^{[\frac{h}{2}]-1} ([\frac{r}{2}] - jf) + [\frac{r}{2}] - [\frac{h}{2}]f$$
$$= 2\,[\frac{r}{2}] \times [\frac{h}{2}] - f \times [\frac{h}{2}] \times ([\frac{h}{2}] - 1) + [\frac{r}{2}] - f[\frac{h}{2}]$$
$$= 2\,[\frac{r}{2}] \times [\frac{h}{2}] - f[\frac{h}{2}]^2 + [\frac{r}{2}]. \qquad \text{Q.E.D.}$$

(iv) Proof of Corollary 2

By observing the right-hand side of the above equation as a function of h for fixed r, its maximum can be reached at $h = [\frac{r}{2f}]$ with the range of $h \leq \frac{r}{f}$. Hence,

$$\min \{ \sum_{k=1}^{h}, \sum_{k=1}^{h'} \} \leq \sum_{k=1}^{[\frac{r}{2f}]} = \frac{3}{4} \times \frac{1}{f} \times [\frac{r}{2}]^2 + [\frac{r}{2}], \text{ Q.E.D.}$$

(v) Proof of the main results

The deduction from Lemma 1 to the main results in (II) runs as follows. Assume that $x_k(t) = \sum_{i=1}^{k} f_i(t)$, while functions $f_i(t)$ are real-valued with period τ_i, $i=1,\ldots,k$. Define k one-dimensional Euclidean spaces $E_1,\ldots,E_k$, and write $\vec{x}_i$ the unit vector in E_i, $i=1,\ldots,k$. Incidentally, the direct sum of these k spaces is the k-dimensional Euclidean space E.

In space E_i, the vectors e_t^i are defined as

$$(16) \qquad e_t^i = f_i(t)\, \vec{x}_i, \qquad i=1,\ldots,k;\ \forall t.$$

And in space E define

$$(17) \qquad \varepsilon_t = e_t^1 \oplus \ldots \oplus e_t^k = f_1(t)\, \vec{x}_1 \oplus \ldots \oplus f_k(t)\, \vec{x}_k.$$

Denote $T = \sum_{i=1}^{k} \tau_i$. By Lemma 1, it can be seen that, for any positive integer t, the vector ε_t has the following linear expression

$$(18) \qquad \varepsilon_t = \sum_{j=1}^{T} N_j(t)\, \varepsilon_j = \sum_{j=1}^{T} N_j(t)\, (f_1(j)\vec{x}_1 \oplus \ldots \oplus f_k(j)\vec{x}_k),$$

and the coefficients $N_j(t)$ in the above satisfy the following relationship

$$(19) \qquad A_k(t; \tau_1,\ldots,\tau_k) = \sum_{j=1}^{T} |\, N_j(t)|\,.$$

By comparing the two expressions for ε_t in Eq. (17) and (18), the coefficients attached to $\vec{x}_i$ should satisfy the equation

$$(20) \qquad f_i(t) = \sum_{j=1}^{T} N_j(t)\, f_i(j), \qquad i=1,\ldots,k.$$

Therefore, from Eqs. (20) and (19), we obtain:

$$
\begin{aligned}
|x_k(t)| &= \left| \sum_{i=1}^{k} f_i(t) \right| \\
&= \left| \sum_{i=1}^{k} \sum_{j=1}^{T} N_j(t)\, f_i(j) \right| \\
&= \left| \sum_{j=1}^{T} N_j(t)\, x_k(j) \right| \\
&\leq \sum_{j=1}^{T} |N_j(t)| \times |x_k(j)| \\
&= A_k(t;\tau_1,\ldots,\tau_k) \times |x_k(j)|. \qquad \text{Q.E.D.}
\end{aligned}
$$

Acknowledgement. We would like to express our appreciation to professors H. Robbins, T.L. Lai, M.H. Kalos and Drs. C.S. Kao and P.A. Whitlock for their help and encouragement. We are also indebted to the Applied Mathematics Department of Brookhaven National Laboratory, the Department of Mathematical Statistics of Columbia University and Courant Institute of Mathematical Sciences of New York University for providing the perfect environment for the endeavor.

REFERENCES

AKAIKE, H. (1974). A new look at the statistical model identification, IEEE Trans. on Automatic Control, AC-19, 116-123.

AKAIKE, H. (1980). Seasonal adjustment by a bayesian modeling, Journal of Time Series Analysis, 1, 1-13.

ANDERSON, T.W. (1976). The Statistical Analysis of Time Series. Wiley, New York.

BLOOMFIELD, P. (1976). Fourier Analysis of Time Series: An Introduction. Wiley, New York.

CLEVELAND, W.S. and TERPENNING, I.J. (1982). Graphical methods for seasonal adjustment, Journal of American Statistical Association, 77, 52-62.

GRIFFITHS, H.B. and HILTON, T.J. (1970). A Comprehensive Textbook of Classical Mathematics, A Contemporary Interpretation. Van Nostrand Reinhold, London.

PARZEN, E. (1978). Time series modelling, spectral analysis and forecasting, Directions in Time Series, Eds: Brillinger, D.R. and Tiao, G.C., Iowa State University, 80-111.

PIERCE, D.A. (1978). Some recent development in seasonal adjustment. Directions in Time Series, Eds: Brillinger, D.R. and Tiao, G.J. (1982). Iowa State University, 123-146.

TIAN, C.J. (1982). An approximation approach to seasonal adjustment of time series. In Applied Time Series Analysis (Proceedings of the International Conference Held at Houston, Texas, USA, August, 1981). Eds: Anderson, O.D. and Perryman, M.R., North-Holland, Amsterdam and New York, 445-464.

TIAN, C.J. and Lu, Z.J. (1982). Seasonal modeling with the successive average method, Proceedings of 1982 American Control Conference, 425-430.

VINOGRADOV, I.M. (1955). An Introduction to the Theory of Numbers. Pergamon, London and New York.

TIME SERIES ANALYSIS: Theory and Practice 4
O.D. Anderson (editor)

SENSITIVITY ANALYSIS IN SEASONAL DISTRIBUTED LAG MODELS

Wolfgang Polasek

University of Vienna
Institute of Statistics
A-1090 Vienna, Rooseveltplatz 6
Austria

This paper discusses local and global sensitivity analysis for seasonal distributed lag (SDL-) models in a hierarchical framework. Hierarchical regression models have been developed by Lindley and Smith (1972) and have been first applied to SDL-models by Trivedi and Lee (1980). It is shown how global sensitivity analysis in the line of Leamer (1978) can be applied to informative and non-informative 3-stage hierarchical SDL-models. Finally it is shown how local sensitivity analysis can be applied to the hyperparameters in a hierarchical SDL-model using the matrix by matrix derivatives of Balestra (1976) and Polasek (1980).

1. INTRODUCTION

Seasonal distributed lag (SDL-) models are extensions of ordinary distributed lag models which are given by

(1.1) $$y_t = \beta_0 + \beta_1 x_t + \beta_2 x_{t-1} + \ldots + \varepsilon_t, \qquad t = 1,\ldots,T.$$

Various seasonal distributed lag models can be derived by the assumption of different models for the pooling of information between seasonal and nonseasonal components.

In this paper the theory of hierarchical models is applied to pool time series and seasonal information in SDL-models. Hierarchical models have been developed by Lindley and Smith (1972) and Smith (1973) from a Bayesian point of view and are called "random coefficient models" by Swamy (1971) or type II-model by Box and Tiao (1973). For hierarchical models the data have to be arranged in the form

(1.2) $$y_t = \sum_{s=1}^{S} \sum_{i=1}^{N} \beta_{s,t} D_{s,t-i} x_{t-i} + \varepsilon_t , \qquad t = -N,\ldots,1\ldots,T$$

where $D_1,\ldots,D_S$ are (multiplicative) seasonal dummy variables of the zero-one-type.

Noninformative hierarchical SDL-models have first been analysed by Trivedi and Lee (1980). The three important models, the SDL * EB (exchangeability between seasons), the SDL * EW (exchangeability within seasons) and the seasonal smoothness model (SDL * SM) are summarized in Section 3. Global sensitivity analysis for these three models produces the same ellipsoids only differing in their prior location. Section 4 discusses a full informative 3-stage hierarchical model for the seasonal smoothness model. Global sensitivity analysis is applied as in Polasek (1982 a) and upper and lower bound ellipsoids are derived. These hierarchical ellipsoids are imbedded in the two stage ellipsoids of the noninformative models. The posterior means for the fist stage parameters lie closer to the maximum likelihood (ML-) estimates, while the posterior mean of the hyperparameter lies closer to the 3rd stage prior location. The Shiller-type smoothness model and the Gersovitz-McKinnon seasonal smoothness model are special cases of this model.

In Section 5 we derive prior-posterior sensitivities for the informative 3-stage hierarchical model using matrix by matrix derivative algebra. Sensitivity analysis of this kind was introduced by Leamer (1972) in connection with simple distributed lag models. The general formulas can be used for the seasonal smoothness model, while for the SDL * EB model special B-type derivatives (Polasek (1980)) are derived to take into account the special prior structures.

2. HIERARCHICAL MODELS

This section gives a short introduction into hierarchical models, based on the theory of Lindley and Smith (1972) and Smith (1973). The usual normal Bayesian linear model can be viewed as simplest full informative two stage model:

$$(2.1) \qquad y = X\beta + u, \qquad \beta \sim N(b^*, H_*).$$

The first stage is the linear model for the data, while the second stage represents the prior information. The posterior mean is given by

$$(2.2) \qquad b^{**} = (X'X + H_*^{-1})^{-1}(X'Xb + H_*^{-1}b^*).$$

This is a matrix weighted average between the ordinary least squares point $b = (X'X)^{-1}X'y$ and the prior location b^*. The general (diffuse) 3-stage hierarchical model developed by Lindley and Smith (1972) takes the form

$$(2.3) \qquad y \sim N(A_1\theta_1, C_1), \qquad \theta_1 \sim N(A_2\theta_2, C_2), \qquad \theta_2 \sim N(0,\infty).$$

It can be shown that the posterior mean for θ_1 is then given by

$$(2.4) \qquad \theta_1^{**} = (A_1'C_1^{-1}A_1 + C_2^{-1})^{-1}(A_1'C_1^{-1}A_1\hat{\theta}_1 + C_2^{-1}A_2\hat{\theta}_2).$$

This is a matrix weighted average between θ_1 and θ_2, i.e. generalised least squares (GLS-) estimates for the first stage and the second stage. This type of hierarchical structure is the basis for the local and global sensitivity analysis of the next three sections.

3. NONINFORMATIVE HIERARCHICAL SDL-MODELS

3.1 Models with exchangeable priors

Noninformative hierarchical 3-stage SDL-models were first studied by Trivedi and Lee (1980) using the theory developed by Lindley and Smith (1972) and Smith (1973). Simple hierarchical models follow from the assumption, that for the (SxK) coefficient matrix B, exchangeability between seasons EB (i.e. columns of B) or exchangeability within seasons EW (i.e. rows of B) is assumed. The first model is the SDL * EB-model

$$(3.1)\qquad \begin{aligned} \text{vec } Y &\sim N[(\text{diag}(X_s)\text{vec } B,\ D_\omega \otimes I_T)],\\ \text{vec } B &\sim N[1_S \otimes \xi,\ I_S \otimes \Sigma],\\ \xi &\sim N(0,\infty). \end{aligned}$$

With a diagonal covariance matrix $D_\omega = \text{diag}(\omega_1^2,\dots,\omega_S^2)$ the posterior mean can be divided into S equations

$$(3.2)\qquad b_s^{**} = (\omega_s^{-2}X_s'X_s + \Sigma^{-1})^{-1}(\omega_s^{-2}X_s'X_s b_s + \Sigma^{-1}\beta^*), \qquad s = 1,\dots,S,$$

where $b_s = (X_s'X_s)^{-1}X_s'y_s$ is the OLS-estimate. The prior location $\beta^* = \sum_{s=1}^{S} W_s b_s$ can be expressed by weighting the b_s-estimates with the following W_s-matrices:

$$(3.3)\qquad W_s = \Big(\sum_{s=1}^{S} (S_s'X_s\omega_s^{-2} + \Sigma^{-1})^{-1}X_s'X_s\omega_s^{-2}\Big)^{-1} \cdot (X_s'X_s\omega_s^{-2} + \Sigma^{-1})^{-1}X_s'X_s\omega_s^{-2}.$$

The second simple pooling model is the SDL * EW-model which is given by

$$(3.4)\qquad \begin{aligned} \text{vec } Y &\sim N[\text{diag}(X_s)\text{vec } B,\ D_\omega \otimes I_T],\\ \text{vec } B &\sim N[\gamma \otimes 1_N,\ \text{diag}(K_s) \otimes I_N],\\ \gamma &\sim N(0,\infty). \end{aligned}$$

This model differs only in the second stage from the SDL * EB model. Replacing the second stage by the transposed prior structure vec $B' \sim N[1_N \otimes \gamma,\ I_N \otimes D_K]$, then the posterior mean for this model can be derived as

$$(3.5)\qquad b_s^{**} = (\omega_s^{-2}X_s'X_s + K_s^{-1}I_N)^{-1}(\omega_s^{-2}X_s'X_s b_s + K_s^{-1}1_K\hat{\gamma}_s), \qquad s = 1,\dots,S,$$

where b_s is the OLS-estimate as before and the (Sx1) vector of hyperparameters is estimated by

$$(3.6)\qquad \hat{\gamma} = (1_N'(\omega_s^{-2}X_s'X_s + K_sI_N)^{-1}1_N)^{-1}\text{diag}(1_N'(\omega_s^{-2}X_s'X_s + K_sI_N)^{-1}b_s).$$

3.2 Smoothness models

Trivedi and Lee (1980) generalized Shiller's concept of smoothness priors to within and between (SBxSW) seasonal smoothness prior restrictions. In the following section we will analyse the informative 3-stage smoothness model.

$$\text{(3.7)} \quad \underset{(TxS)}{\text{vec } Y} \sim N[\text{diag}(X_s)\text{vec } B, \; D_\omega \otimes I_T], \quad (nxS)(SxS)(TxT)$$

$$\text{vec } B \sim N[J_q \otimes J_p)\text{vec } \Gamma, \; V \otimes W], \quad (Sxq)(nxp)(SxS)(nxn)$$

$$\text{vec } \Gamma \sim N(0,\infty),$$

where J_p: (nxp) and J_q: (Sxq) are Vandermonde matrices with $(i-1)^{j-1}$ in the i-th row and j-th column. They have the property, that they annihilate differencing matrices of the same order:

$$\text{(3.8)} \quad R_q J_q = 0_{S-q,q}.$$

Now R_q: (S-q)xS and R_p: ((n-p)x n) are the differencing matrices of order q and p as defined in Shiller (1973). If we multiply the second stage by $(R_q \otimes R_p)$, then the expression

$$\text{(3.9)} \quad (R_q \otimes R_p) \text{ vec } B = 0 + \text{vec } U$$

illustrates the prior knowledge, that the coefficients in B are "small".

The posterior mean for the SDL * SM-model is given by

$$\text{(3.10)} \quad \text{vec } B^{**} = \text{diag}(\omega_s^{-2}X_s'X_s + \sigma_s^{-2}W^{-1})^{-1}(\text{diag}(\omega_s^{-2}X_s'X_s)\text{vec } \hat{B} + (D_\sigma J_q \otimes W^{-1}J_p)\text{vec } \hat{\Gamma}).$$

This expression can be divided into equation by equation matrix weighted averages

$$\text{(3.11)} \quad b_s^{**} = (w_s^{-2}X_s'X_s + \sigma_s^{-2}W^{-1})^{-1}(\omega_s^{-2}X_s'X_s\hat{b}_s + \sigma_s^{-2}W^{-1}J_p \sum_{i=1}^{q} (J_p)_{si}\hat{\Gamma}_{\cdot i}), \quad s=1,\ldots,S,$$

where $(J_q)_{si}$ denotes the (s,i)-th element of matrix J_p. Furthermore, for matrices with the dimensions A: (mxn), B: (pxq) and C: (nqxmp) we use the relationship

$$\text{(3.12)} \quad (A \otimes B)\text{diag}(C_s)(A' \otimes B') = [B \sum_{s=1}^{S} a_{is}C_s a_{sj}B']_{ij}, \quad \begin{array}{l} i= 1,\ldots,m, \\ j= 1,\ldots,n, \end{array}$$

and $\hat{\Gamma} = (\hat{\Gamma}_{\cdot 1},\ldots,\hat{\Gamma}_{\cdot N})$ is the columnwise partitioned GLS-estimate of the hyper-parameter Γ, given by

$$\begin{aligned}
\text{(3.13)} \quad \text{vec } \hat{\Gamma} &= [(J_q' \otimes J_p')\,(\text{diag}(\omega_s^2(X_s'X_s)^{-1}) + D_\sigma \otimes W^{-1}(J_q \otimes J_p)]^{-1} \\
&\cdot [(J_{sq}' \otimes J_p')[(\text{diag}(\omega_s^2(X_s'X_s)^{-1}) + D_\sigma \otimes W]^{-1}\text{vec } \hat{B}] \\
&= [\sum_{s=1} q_{s.}'q_{s.} \otimes (J_p'(\omega_s^2(X_s'X_s)^{-1} + \sigma_s^2 W)^{-1}J_p)]^{-1} \\
&\cdot [\sum_{s=1}^{S} q_s' \cdot \otimes (J_p'(\omega_s^2(X_s'X_s)^{-1} + \sigma_s^2 W)^{-1}\hat{b}_s].
\end{aligned}$$

The second expression with sums of Kronecker products is more useful for computational purposes, in that it has less storage requirements. Note that simpler models are obtained for $J_q = 0$ (i.e. seasonal Shiller smoothness) and $J_p = 0$. The latter case results in the Gersovitz-McKinnon (1978)-model by setting q = s and using a singular covariance matrix.

3.3 Global sensitivity analysis in diffuse 3-stage models

The posterior means of the three models (3.2), (3.5), and (3.10) of the previous section can be divided into matrix weighted averages into every equation (because of the diagonal D_ω-structure). They allow, therefore simpler usual two stage sensitivity analysis and differ only through their prior moments:

model:	prior location b*:	prior variance H*:
SDL * EB	$\sum_{s=1}^{S} W_s b_s$	Σ
SDL * EW	$1_K \hat{\gamma}_s$	$K_s I_N$
SDL * SM	$J_p \sum_{i=1}^{q} (J_q)_{si} \hat{\Gamma}_{\cdot i}$	$\sigma_s^2 W$

Table 3.1 Prior moments of diffuse hierarchical SDL-models

Since the prior locations of the three models are determined by certain averages of GLS-estimates of the hyperparameters, only the specification of the covariance matrices in the second stage remains unknown. For this case Leamer (1978) has shown, that matrix weighted averages with a free covariance matrix in the second stage are bounded by ellipsoids. These ellipsoidal bounds can be calculated e.g. by the SEARCH-program (Leamer (1978)). For variance matrices known up to a constant, the set of estimates forms an information contract curve (ridge curve). This is only the case for the SDL * EW-model, as can be seen from Table 3.1.

4. THE INFORMATIVE HIERARCHICAL SEASONAL SMOOTHNESS MODEL (SDL * SM)

The SDL-models discussed in the previous section can be extended to informative 3-stage hierarchical models by replacing the noninformative 3rd stage by the prior distribution $\gamma \sim N(\gamma^*, H^*)$, where the *-index indicates known last stage. The informative SDL * EB and SDL * EW have been discussed in Polasek (1982 a) together with the hierarchical sensitivity analysis. In this section we analyse the informative 3-stage seasonal smoothness (SDL * SM)-model. Since the information pooling process is more complicated for a smoothness model we have to extend the analysis in the following way.

The seasonal smoothness model (3.7) becomes the form

$$(4.1) \qquad \begin{aligned} \text{vec } Y &\sim N[\text{diag}(X_s)\text{vec } B,\ D_\omega \otimes I_T] \\ \text{vec } B &\sim N[(J_q \otimes J_p)\text{vec } \Gamma,\ D_\sigma \otimes W] \\ \text{vec } \Gamma &\sim N[0,\ V_L^* \otimes V_R^*] \end{aligned}$$

by adding an informative 3rd stage. The prior location is 0 and the known covariance matrix $V^* = V_L^* \otimes V_R^*$ represents our prior knowledge of smoothness, i.e. that appropriate differences of B are "small".

The general theoretical analysis of an informative 3-stage model can be found in Smith (1973). Inserting into the theorem of Smith and noting that the prior location is zero, we obtain the expressions for the posterior means for B and Γ.

$$(4.2)\quad \text{vec } B^{**} = [\text{diag}(\omega_s^{-2}X_s'X_s) + ((D_\sigma \otimes W) + (J_q \otimes J_p)V^* \cdot (J_q \otimes J_p'))^{-1}]^{-1}.\text{diag}(\omega_s^{-2}X_s'X_s) \text{ vec } \hat{B},$$

$$(4.3)\quad \text{vec } \Gamma^{**} = [\sum_{s=1}^{S} q_{s.}'q_{s.} \otimes (J_p'(\omega_s^{-2}(X_s'X_s)^{-1} + \sigma_s^2 W)^{-1}J_p + V^{*-1}]^{-1} \cdot [\sum_{s=1}^{S} g_{s.}'q_{s.} \otimes (J_p'(\omega_s^{-2}(X_s'X_s)^{-1} + \sigma_s^2 W)^{-1}J_p)\text{vec } \hat{\Gamma}],$$

where $q_{s.}$ is the s-th row-vector of $J_q = (Sxq)$. It was shown in Polasek and Leamer (1980), that the variances in the matrix weighted averages of an informative 3-stage model are bounded, therefore a special global sensitivity analysis for unknown 2nd-stage covariance matrices can be derived, resulting in upper bound and lower bound hierarchical ellipsoids. The parameters of these ellipsoids are given in the next two theorems:

Theorem 4.1 3-stage SDL * SM-model: lower bound for the posterior mean vec B**. The posterior mean vec B** in (4.2) of model (4.1) lies in the ellipsoid

$$(4.4)\quad (\text{vec } B^{**}-f_*)'H_*(\text{vec } B^{**}-f_*) \leq c_*$$

with the parameters

$$(4.5)\quad H_* = \text{diag}(\omega_s^{-2}X_s'X_s)(J_q \otimes J_p)V^*(J_q' \otimes J_p')\text{diag}(\omega_s^{-2}X_s'X_s) + \text{diag}(\omega_s^{-2}X_s'X_s),$$

$$(4.6)\quad f_* = \text{vec } B/2 + [(J_q V_L^* J_q' \otimes J_p V_R^* J_p')^{-1} + \text{diag}(\omega_s^{-2}X_s'X_s)]^{-1}\text{diag}(\omega_s^{-2}X_s'X_s)\text{vec } \hat{B}/2,$$

$$(4.7)\quad c_* = \text{vec}'B(J_q V_L^* J_q' \otimes J_p V_R^* J_p' + \text{diag } \omega_s^2(X_s'X_s)^{-1})^{-1}\text{vec } B/4.$$

Proof: Insert into the general theorem 3.3 of Polasek (1982 a).

Theorem 4.2 3-stage-SDL * SM-model: Upper bound for the hyperparameter vec Γ**.

The posterior mean (4.3) of the hyperparameter Γ of model (4.1) lies in the ellipsoid

$$(4.8)\quad (\text{vec } \Gamma^{**} - \text{vec } \hat{\Gamma})'H_u(\text{vec } \Gamma^{**} - \text{vec } \hat{\Gamma}) \leq c_u,$$

$$(4.9)\quad H_u = [(J_q' \otimes J_p')\text{diag}(\omega_s^2(X_s'X_s)^{-1})(J_q \otimes J_p) + V^{*-1}]^{-1},$$

$$(4.10)\quad c_u = \text{vec}'\,\hat{\Gamma}(V^*H_u^{-1}V^*)^{-1}\text{vec } \hat{\Gamma}/4.$$

Proof: Insert into Theorem 3.4 of Polasek (1982 a).

Note that vec Γ** can be expressed as a matrix weighted average between vec $\hat{\Gamma}$ and the limiting case vec Γ*(II):

$$\text{vec}\ \Gamma^{**} = [\sum_{s=1}^{S} q_{s.}' q_{s.} \otimes (J_p'(\omega_s^{-2}(X_s'X_s)^{-1} + \sigma_s^2 W)^{-1} J_p) + V^{*-1}]^{-1} \quad (4.11)$$

$$\cdot\ [(V^{*-1} + H_J)\text{vec}\ \Gamma^*(II) + (J_q'D\ J_q \otimes J_p'WJ_p)\text{vec}\ \hat{\Gamma}],$$

$$\text{vec}\ \Gamma^*(II) = [V^{*-1} + H_J]^{-1}\ \text{vec}\ \hat{\Gamma}, \quad (4.12)$$

with

$$H_J = (J_q' \otimes_S J_p')\text{diag}(\omega_s^{-2}X_s'X_s)(J_q \otimes J_p) \quad (4.13)$$

$$= [J_p' \sum_{s=1}^{S} (J_q')_{is}\,\omega_s^{-2}X_s'X_s(J_q)_{si}\,J_p]_{i,j}, \qquad i= 1,\dots,S,\ j= 1,\dots,q.$$

4.1 Hierarchical global sensitivity analysis

The geometrical interpretation of the upper and lower bound hierarchical ellipsoids is given in Fig. 4.1

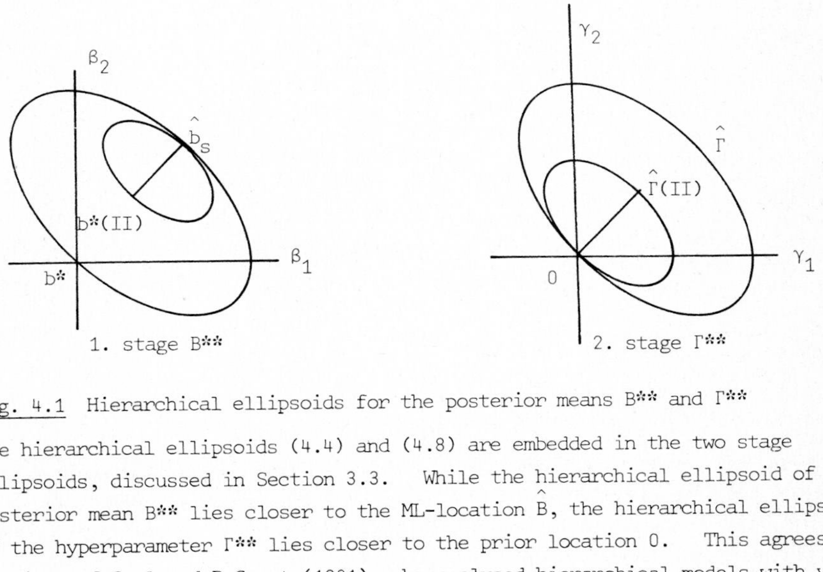

Fig. 4.1 Hierarchical ellipsoids for the posterior means B** and Γ**

The hierarchical ellipsoids (4.4) and (4.8) are embedded in the two stage ellipsoids, discussed in Section 3.3. While the hierarchical ellipsoid of the posterior mean B** lies closer to the ML-location $\hat{B}$, the hierarchical ellipsoid of the hyperparameter Γ** lies closer to the prior location 0. This agrees with findings of Goel and DeGroot (1981), who analysed hierarchical models with various information measures.

4.2 Summary: Posterior moments of the SDL * EB and SDL * EW-models

a) Posterior moments for the informative SDL * EB-model with 3rd-stage specification $\gamma \sim N(\xi^*, H^{*-1})$ are given by

$$\text{vec}\ B^{**} = [\text{diag}(\omega_s^{-2}X_s'X_s) + G]^{-1}[\text{diag}(\omega_s^{-2}X_s'X_s)\text{vec}\ B + G(1_S \otimes \xi^*)], \quad (4.14)$$

$$G = (1_S 1_S' \otimes H^* + I_S \otimes \Sigma)^{-1},$$

$$\xi^{**} = (H^* + H\_ + S\Sigma)^{-1}(H^*\xi^*/S + (H\_ + S\Sigma)\hat{\xi}), \quad (4.15)$$

$$H_- = \sum_{s=1}^{S} \omega_s^2 (X_s'X_s)^{-1}.$$

b) The posterior moments for the informative SDL * EW-models with 3rd stage specification $\gamma \sim N(0, V^{*-1})$ are given by

(4.16) $$\text{vec } B^{**} = [\text{diag}(\omega_s^{-2} X_s'X_s) + (V^* \otimes 1_N 1_N' + \text{diag}(K_s) \otimes I_N)^{-1}]^{-1} \text{diag}(\omega_s^{-2} X_s'X_s) \text{vec } \hat{B},$$

(4.17) $$\gamma^{**} = [H^* + \text{diag}(1_N'\omega_s^2 (X_s'X_s)^{-1} 1_N + K_s)]^{-1} \cdot \text{diag}(\omega_s^2 1_N'(X_s'X_s)^{-1} 1_N + K_s)\, \hat{\gamma}.$$

A detailed derivation is given in Polasek (1982 a).

5. LOCAL PRIOR-POSTERIOR ANALYSIS

A problem often associated with a full Bayesian analysis is the public usefulness of specific prior information. Since it is primarily designed to cover individual needs for inferences, are the results useful for a larger class of applications? Local priorposterior analysis is one way to solve this problem: How do the posterior moments change, when the prior moments are changed. This reflects partly (local) stability of the results and partly derivatives implying local (linear) mappings. "Hopefully the neighborhood of adequate approximation is large enough to encompass all readers' opinion of interest" (Leamer (1972)).

Local sensitivity analysis can also help to specify a multivariate prior distribution. Keyparameters can be identified in this way and lead to a more careful consideration of their values.

5.1 Local sensitivities for the hyperparameters

It is easier to start the local sensitivity analysis with the analysis of hyperparameters since the general structure of the posterior mean is given by (see Smith (1973))

(5.1) $$\theta_2^{**} = (\hat{D}_2^{-1} + C_3^{-1})^{-1} (\hat{D}_2^{-1} \hat{\theta}_2 + C_3^{-1} A_3 \theta_3^*)$$

where $\hat{D}_2^{-1} = \text{Var}(\hat{\theta}_2)$ is the variance of the GLS-estimate of θ_2 and is independent of C_3. Therefore the structure is equivalent to a ordinary two stage (Bayesian) sensitivity problem, analysed by Leamer (1972) or with matrix calculus by Polasek (1981). The local sensitivity of the posterior moments $(\theta_2^{**}, H_2^{**})$ of θ_2 in (5.1) with respect to the prior moments (θ_3^*, C_3^*) are given by

(5.2) $$\partial\theta_2^{**} // \partial\theta_3^* = \text{vec}(\hat{D}_2^{-1} + C_3^{-1})^{-\mu} C_3^{-1} A_3,$$

(5.3) $$\partial H_2^{**} // \partial C_3^{-1} = \text{vec}^2 (\hat{D}_2^{-1} + C_3^{-1})^{-1} C_3^{-1} A_3,$$

(5.4) $\partial\theta_2^{**}//\partial C_3^{-1} = \text{vec } H_2^{**-1} C_3^{-1} A_3 \text{ vec}' C_3^{-1}(\theta_2^{**} - A_2\theta_2)$,

(5.5) $\text{vec}^2 A = \text{vec } A \text{ vec}' A, \quad H_2^{**} = \hat{D}_2^{-1} + C_3^{-1}$.

Using the A- or B-type derivative, this can be divided into elementwise sensitivities (see Polasek (1981)).

5.2 Local sensitivities for the parameters

The sensitivities of the posterior moments of θ_1 require a separate derivation because the prior structure is more complicated. Following Smith (1973) the general posterior mean is given by $\theta_1^{**} = H_{**}^{-1} g$ where H_{**} and g are given by

(5.6a) $H^{**} = A_1' C_1^{-1} A_1 + (C_2 + A_2 C_3 A_2')^{-1}$,

(5.6b) $g = A_1' C_1^{-1} A_1 \hat{\theta}_1 + (C_2 + A_2 C_3 A_2')^{-1} A_2 A_3 \theta_3^{*}$.

Using $H_{**} = H^{**}$ interchangeably and the term $G = (C_2 + A_2 C_3 A_2')^{-1}$ the local sensitivities can be derived as

(5.7) $\partial\theta_1^{**}/\partial\theta_3 = (H_{**}^{-1} G\, A_2 A_3 \otimes I_n)\partial\theta_3/\partial\theta_3 = \text{vec}(H_{**}^{-1} G\, A_2 A_3)$,

(5.8)
$$\begin{aligned}\partial H_{**}^{-1}/\partial C_3 &= -(H_{**}^{-1} G \otimes I_n)(\partial H_{**}/\partial C_3)(H_{**}^{-1} \otimes I_n)\\ &= (H_{**}^{-1} \otimes I_n)(\partial A_2 C_3 A_2'/\partial C_3)(GH_{**}^{-1} \otimes I_n)\\ &= (H_{**}^{-1} \otimes I_n)\ \text{vec}^2 A_2 (GH_{**}^{-1} \otimes I_n)\\ &= \text{vec}^2 H_{**}^{-1} G A_2 .\end{aligned}$$

Note that the alternative derivatives with respect to C_3^{-1} differs only slightly by the factor C_3:

(5.9) $\partial H_{**}^{-1}/\partial C_3^{-1} = \text{vec}^2 H_{**}^{-1} G\, A_2 C_3$,

(5.10)
$$\begin{aligned}\partial\theta_1^{**}/\partial C_3 &= H_{**}^{-1}/\partial C_3 (g \otimes I_n) + (H_{**}^{-1} \otimes I_n)\ \partial g/\partial C_3\\ &= (H_{**}^{-1} G \otimes I_n)\ \text{vec}^2 A_2 (G\, H_{**}^{-1} g \otimes I_n) -\\ &\quad - (H_{**}^{-1} G \otimes I_n)\partial A_2 C_3 A_2/\partial C_3 (G A_2 A_3 \theta_3^{*} \otimes I_n)\\ &= \text{vec } H_{**}^{-1} G\, A_2 \text{ vec}' A_2 G(\theta_1^{**} - A_2 A_3 \theta_3^{*}).\end{aligned}$$

For $\theta_1 = \text{vec } B$, $\theta_2 = \text{vec } \Gamma$, $A_1 = \text{diag}(X_1, \ldots, X_S)$, $A_2 = (R_q \otimes R_p)$ and $A_3 = I_n$ these formulas can be directly used to calculate the local prior-posterior sensitivities for the informative SDL * SM-model. The forlulas can be simplified for the SDL * EB- and SDL * EW model since there are fewer prior parameters involved. For this purpose, we use the A- and B-type derivatives (see Polasek (1980)).

5.3 Sensitivities for the exchangeable SDL-models

The hierarchical SDL * EB-model requires the prior moments $1_S \otimes \xi^* = \theta_3^*$ $1_S \otimes \xi^* = \theta_3^*$ and $C_3^* = 1_S 1_S' \otimes H^*$ and, therefore, the B-type derivative is more appropriate:

(5.11) $\partial \text{vec } B^{**}/\partial\xi^* = (H_{**}^{-1}G \otimes I_N)(1_S \otimes \text{vec } I_n),$

(5.12) $\partial H_{**}^{-1}/\partial H_*^{-1} = (H_{**}^{-1}G \otimes I_N)(1_S 1_S' \otimes \text{vec}^2 H_*^{-1})(G\, H_{**}^{-1} \otimes I_N),$

(5.13) $\partial \text{vec } B^{**}/\partial H_*^{-1} = (H_{**}^{-1}G \otimes I_N)(1_S 1_S' \otimes \text{vec}^2 H_*^{-1})\ .$

$. (G(\text{vec } B^{**} - 1_S \otimes \xi^*) \otimes I_N).$

The hierarchical SDL * EW-models have the prior moments $\theta_3^* = \gamma \otimes 1_N$ and $C_3^* = V^* \otimes 1_N 1_N'$ and, therefore, the A-type derivative leads to simpler results

(5.14) $\partial \text{vec } B^{**}//\partial\gamma = (I_S \otimes H_{**}^{-1}G)(\text{vec } I_S \otimes 1_N),$

(5.15) $\partial H_{**}^{-1}//\partial V_*^{-1} = (I_S \otimes H_{**}^{-1}G)(\text{vec}^2 I_S \otimes 1_N 1_N')(I_S \otimes G\, H_{**}^{-1}),$

(5.16) $\partial \text{vec } B^{**}//\partial\gamma = (I_S \otimes H_{**}^{-1}G)(\text{vec}^2 I_S \otimes 1_N 1_N')\ .\ (G(\text{vec } B^{**} - \gamma \otimes 1_N) \otimes I_S)$

5.4 Elementwise sensitivities

To demonstrate the structure of the elementwise derivatives for the exchangeable SDL-models we take the case of the $\partial \text{vec } B^{**}/\partial\xi$ derivative. With the relationship $\partial B//\partial A = P_{q,m}(\partial B/\partial A)P_{n,q}$ we have

(5.17) $\partial \text{vec } B^{**}//\partial\xi^* = (I_n \otimes H_{**}^{-1}G)(\text{vec } I_n \otimes 1_S)$

$$= \begin{pmatrix} H_{**}^{-1}G(e_1^n \otimes 1_S) \\ \vdots \\ H_{**}^{-1}G(e_n^n \otimes 1_S) \end{pmatrix}$$

where e_i^n is the i-th unity vector of order n. If we partition G appropriately into (SxS) blocks G_{ij} we have

(5.18) $$\partial \text{vec } B^{**}/\partial\xi_k^* = H_{**}^{-1} \begin{pmatrix} G_{11} \dots G_{1N} \\ \vdots \quad\quad \vdots \\ G_{N1} \dots G_{NN} \end{pmatrix} \begin{pmatrix} 0 \\ \vdots \\ 1_S \\ \vdots \\ 0 \end{pmatrix} = H_{**}^{-1} \cdot \begin{pmatrix} G_{1k}1_S \\ \vdots \\ G_{Nk}1_S \end{pmatrix}, \qquad k = 1,\dots,N.$$

Because the hyperparameter take the structure $1_S \otimes \xi^*$, the matrix G is particioned into (SxS) blocks and the row-sums of the k-th block column have to be calculated. The usual $\partial b^{**}/\partial b_k^*$ derivative uses only column by column elements and, therefore no sums of derivatives are necessary.

REFERENCES

BALESTRA, P. (1976). La Dérivation Matricielle. Collection de L'IME No.12, Sirey Paris.

BOX, G.E.P. and TIAO, G. (1973). Bayesian Inference in Statistical Analysis. Addison Wesley, Reading, MA.

DHRYMES, P.J. (1971). Distributed Lags: Problems of Estimation and Formulation. Holden Day, San Francisco.

GOEL, P. and DEGROOT, M.H. (1981). Information about hyperparameters in hierarchical models. Annals of Statistics.

GERSOVITZ, M. and MACKINNON, J.G. (1978). Seasonality in regression. An application of smoothness priors. JASA 73, 264-273.

LEAMER, E.E. (1972). A class of informative priors and distributed lag analysis. Econometrica, Vol.40, No.6, 1059-1081.

LEAMER, E.E. (1978). Specification Searches. John Wiley, New York.

LINDLEY, D.V. and SMITH, A.F.M. (1972). Bayes estimate for the linear model. JRSSB, Vol.34, 1-41, with discussion.

MADDALA, G.S. (1971). The use of variance components models in pooling cross section and time series data. Econometrica 37, 341-358.

POLASEK, W. (1980). Local sensitivity analysis in the general linear model. Department of Economics, University of Southern California, Los Angeles, MRG NO. 8006.

POLASEK, W. (1981). Sensitivity analysis and matrix derivatives. Preprint 32, Institute of Statistics and Computer Science, University of Vienna.

POLASEK, W. (1982a). Hierarchical bounds in seasonal distributed lag models. In O.D. ANDERSON (ed.) "Time Series Analysis. Theory and Practice 1", North Holland Publ.Co., 497-514.

POLASEK, W. (1982b). Two kinds of pooling information in cross sectional regression models. In W. Grossmann et al. (1982) eds. "Probability and Statistical Inference", 297-306.

POLASEK, W. and LEAMER, E.E. (1980). Bounds for parameter and hyperparameter in linear hierarchical models. Institute of Statistics and Computer Science, University of Vienna.

SHILLER, R.J. (1973). A distributed lag estimator derived from smoothness priors. Econometrica 41, 775-789.

SMITH, A.F.M. (1973). A general bayesian linear model. JRRSB, Vol.35, 67-75.

SWAMY, P.A.V.B. (1971). Statistical Inference in a Random Coefficient Regression Model. Springer Verlag, New York.

TRIVEDI, P.K. and LEE, B.M.S. (1980). Seasonal variability in a distributed lag model. Australian National University.

ZELLNER, A. (1962). An efficient method of estimating seemingly unrelated regressions and tests for aggregation bias. JASA 57, 348-368.

TIME SERIES ANALYSIS: Theory and Practice 4
O.D. Anderson (editor)

BAYESIAN ESTIMATION OF PARAMETERS IN A LINEAR REGRESSION MODEL WITH NORMALLY DISTRIBUTED PRIOR INFORMATION

Fernand C.M. Broeckx

Universiteit Antwerpen (RUCA)
Middelheimlaan 1 2020 Antwerpen, Belgium

Consider a set of n parameters ($n > 2$) to be estimated in a linear or linearized regression model. Let there be prior information available about exactly two of these parameters in the form of double inequalities. Then Bayesian estimation can take care of this kind of information in order to estimate the parameters, especially when the Least Squares Method fails, due to multicollinearity. A micro economic application, based on a long series of quarterly data for an oilfirm is given, and the numerical results discussed. A comparison is made between using uniform and normal prior density functions.

1. INTRODUCTION

In the problem of estimating N parameters, $\underline{\lambda} = (\lambda_1, \lambda_2, \ldots, \lambda_N)$, assumed to be independent in a definition space, Bayesian statistics suppose that prior (to the sample) information about these parameters is given in the form

$$\phi(\underline{\lambda}) = \prod_{i=1}^{N} \phi_i(\lambda_i) \tag{1}$$

where the ϕ_i represent density functions (d.f.) in the continuous case.

Let further the parameters belong to a probability law $f_X(x|\underline{\lambda})$ of a random variable X, so that we can think of the sample information contained in the likelihood function :

$$L(\underline{X}|\underline{\lambda}) = \prod_{i=1}^{n} f_X(X_i|\underline{\lambda}) \tag{2}$$

based on a sample $\underline{X} = (X_1, X_2, \ldots, X_n)$. The so-called posterior information $p(\underline{\lambda}|\underline{X})$ is obtained by Bayes-rule :

$$p(\underline{\lambda}|\underline{X}) \propto L(\underline{X}|\underline{\lambda}) . \phi(\underline{\lambda}) . \tag{3}$$

The Bayesian viewpoint being to take in account both prior information and sample information. In Zellner (1971) different kinds of loss-functions are proposed. Taking a quadratic one, e.g. like :

$$V(\underline{\lambda}, \hat{\underline{\lambda}}) = k(\underline{\lambda} - \hat{\underline{\lambda}})(\underline{\lambda} - \hat{\underline{\lambda}})', \quad k > 0 \tag{4}$$

gives rise to the Bayesian Estimator of $\underline{\lambda}$, which is then the mean of the posterior d.f. :

$$B.E.(\lambda_i) = \int_\Lambda \lambda_i \cdot p(\underline{\lambda}|\underline{X})d\underline{\lambda} \Big/ \int_\Lambda p(\underline{\lambda}|\underline{X})d\underline{\lambda} \quad \text{for all } i = 1,2,\ldots N. \tag{5}$$

Let it be known "a priori" that the parameters, for external mathematical reasons, belong to an action space : $\underline{\lambda} \varepsilon D \subset \Lambda$. (6)
Then it is easy to show that (5) becomes :

$$B.E.(\lambda_i) = \int_D \lambda_i p(\underline{\lambda}|\underline{X})d\underline{\lambda} \Big/ \int_D p(\underline{\lambda}|\underline{X})d\underline{\lambda} \quad \text{for all } i = 1,2,\ldots,N. \tag{7}$$

When, on the other hand, "nothing" is known about some of the λ_i, (3) shows that their prior d.f. has still to be written down. An answer to this question was first given by Jeffreys (1967). Many others have derived afterwards so called "diffuse priors".

$$\left.\begin{array}{lll} \text{if } \lambda_i & (-\infty, +\infty) \text{ take } \phi(\lambda_i) \propto & \text{constant} \\ \text{if } \lambda_i \varepsilon & (0, +\infty) & \text{constant} / \lambda_i \\ \text{if } \lambda_i \varepsilon & (0, 1) & 1 / \left[\lambda_i (1 - \lambda_i)\right] . \end{array}\right\} \tag{8}$$

2. ASSUMPTIONS

In Broeckx et al. (1976), a solution is given for the general linear regression model, in the case that for external mathematical reasons exactly two parameters are known to be uniformly distributed in a given double inequality region. Based on (6) this model can be

$$\left.\begin{array}{l} Y_t = \underline{\beta}' \underline{x}_t + u_t \; ; \; t = 1,2,\ldots,T \\ u_t = \rho u_{t-1} + \varepsilon_t, \; t = 2,3,\ldots,T \end{array}\right\} \tag{9}$$

with : $\underline{\beta} = (\beta_0, \beta_1, \ldots, \beta_{N-1})'$, parameters to be estimated

$\underline{x}_t = \left(1, x_t(1), x_t(2), \ldots, x_t(N-1)\right)'$, the t-th observation on the N-1 exogenous variables

y_t =t-th observation on the endogenous variable

u_t = t-th error term

ρ = autoregressive factor

ε_t = error term to be normally distributed $(0,\sigma^2)$

$$m_1 \leqslant \beta_1 \leqslant M_1 \; , \; m_2 \leqslant \beta_2 \leqslant M_2 \tag{10}$$

It is known that Ordinary Least Squares cannot solve this problem, while Ridge estimation possibly can, see Moulaert F.(1979). The Bayesian method can, when following expression can be calculated :

$$\text{B.E.}(\beta_i) = \frac{\int_{-\infty}^{+\infty} d\rho \int_0^{+\infty} \frac{d\sigma}{\sigma} \int_{m_1}^{M_1} d\beta_1 \int_{m_2}^{M_2} d\beta_2 \int_{-\infty}^{+\infty} d\beta_3 \ldots \int_{-\infty}^{+\infty} d\beta_{N-1} L(X,y \mid \underline{\beta},\rho,\sigma)\cdot\beta_i}{\int_{-\infty}^{+\infty} d\rho \int_0^{+\infty} \frac{d\sigma}{\sigma} \int_{m_1}^{M_1} d\beta_1 \int_{m_2}^{M_2} d\beta_2 \int_{-\infty}^{+\infty} d\beta_3 \ldots \int_{-\infty}^{+\infty} d\beta_{N-1} L(X,y \mid \underline{\beta},\rho,\sigma)}$$

for all $i = 1,2,\ldots,N$ (11)

X = matrix of T observation vectors as in (9)

$\underline{y}$ = vector of T observations as in (9).

In the integrations, (6) is applied taking in account (10). Following A. Zellner, we take a priori $-\infty < \rho < +\infty$, for the first order autoregressive factor. We can prove afterwards that $0 \leqslant |\text{B.E.}(\rho)| \leqslant 1$. Further $0 < \sigma < +\infty$ which gives rise to prior d.f.'s as announced in (8).

We will suppose multicollinearity between exactly two of the regressors in the model. This in view of the application in §3. Let it be between β_1 and β_2. Divergences in integrations are avoided, exactly by the prior information, present in (10), which is of the type (6).

The specific likelihood function for (2) is based here on the normality of the ε_i. Variances of β_i, and covariances between β_i and β_j, are calculated as expressions similar to (11), for $\text{B.E.}(\beta_i^2)$ and $\text{B.E.}(\beta_i\beta_j)$, $i \neq j$.

3. AN ECONOMETRIC APPLICATION

Based on a long series of quarterly data of a Belgian oilfirm, we try now to estimate the parameters in a Cobb-Douglas production function :

$$P = b.L^{\alpha}C^{\beta}e^{\gamma t} \tag{12}$$

where L = labour ; C = capital ; t = time.

Assuming a multiplicative error term in (12), we can linearize the model, by taking logs as follows :

$$\ln P = \ln b + \alpha \ln L + \beta \ln C + \gamma t + u \tag{13}$$

In this time series the assumption of autocorrelated error terms is plausible, as is the possibility of collinearity between long series of data concerning labour and capital. Neo-classical economic theory puts forward :

$$0 < \alpha < 1 \quad \text{and} \quad 0 < \beta < 1 \tag{14}$$

These inequalities represent the external mathematical prior information, of type (10).

In appendix A, the quarterly data on an eight years period are presented. For production the oilfirm can express different types of crude in one common measure (barrels), labour is expressed in hours of all types of workers, and for capital a common measure

was found, after long discussions. The factor $e^{\nu t}$ in (11) stands for technical change in time. We converted these figures to index-figures, based on the first period, and took logarithms.
In those figures multicollinearity is high, and autocorrelation exists, but is rather low. The O.L.S.-method estimates :

$$\ln P = -\underset{(.2888)}{6.7068} + \underset{(.5273)}{2.1047}\ln L + \underset{(3.8130)}{1.3080}\ln C - \underset{(.0164)}{.0192}\,t \qquad (15)$$

Between brackets are given the standard deviations for the estimations. It is especially the multicollinearity in the data, which is responsible for those bad estimations.

With the method described in § 2, we get much better results, using uniform priors :

$$\ln P = -\underset{(.0318)}{.1145} + \underset{(.0774)}{.5672}\ln L + \underset{(.0676)}{.6332}\ln C + \underset{(.0082)}{.0381}\,t \qquad (16)$$

4. USING NORMAL PRIORS, FOR AN UNLIMITED NUMBER OF PARAMETERS IN THE MODEL

We now want to introduce priors for an arbitrary number of parameters, in the model (8). The uniform-prior-assumption for exactly two parameters in (9), is changed now, as following : for s ($s \leqslant N-1$) parameters we assume for external mathematical reasons that they are normally distributed with ($100 - p_j$) % confidence around the center C_j of an interval (m_j, M_j).
We introduce Z-scores, Z_{p_j} and corresponding standard deviations σ_j by :

$$P\left[-Z_{p_j} < \frac{\beta_j - C_j}{\sigma_j} < Z_{p_j}\right] = (100 - p_j)\% \ , \ (j=1,\dots,s) \qquad (17)$$

4.1. Changes in the numerical work are presented in Broeckx et al. (1980). The mathematical treatment is much easier : normality-assumption gives rise to $(-\infty, +\infty)$ integrations. The incorporation of terms of the exponential type in the likelihood function asserts now for convergence.

4.2. Applied on the given econometric problem it is clear that Z_{p_j}-values vary between values corresponding to great importance of normality, and values giving practically no weight to the specific priors. So these are in a certain sense, corresponding to the Least Squares Method.

Numerical results are shown in appendix B, using for both parameters α and β, the same confidence-level $Z_{p_j} = Z_p$, for $j = 1,2$.

Big Z_p-values show a too important attraction by the center of given a priori interval. That means that normality assumptions are then too strong, and sample information has practically no influence.

Small Z_p-values generate estimations with too big standard deviations-values. In a certain sense they correspond to a Least Squares Method-solution, without influence of normality. This can be explained as follows :

Let, for each β_j $(j \leqslant s)$, the normal prior be of the form :

$$\exp\left[-\frac{(\beta_j - c_j)^2}{2\sigma_{N,j}^2}\right]$$

In first approximation, the part of the likelihood function concerning that β_j, is of the form :

$$\exp\left[-\frac{(\beta_j - \hat{\beta}_j)^2}{2\sigma_{L,j}^2}\right]$$

It is easily seen that the resulting distribution, used in the posterior d.f., is normal with a variance Σ_j^2 determined by :

$$\frac{1}{\Sigma_j^2} = \frac{1}{\sigma_{N,j}^2} + \frac{1}{\sigma_{L,j}^2} \qquad (18)$$

4.3. Trying to diminish the influence of normality, we now suggest, from moment theoretic viewpoint, taking :

$$\sigma_{N,j} = \frac{M_j - m_j}{\sqrt{3}} \qquad (j \leqslant s) \qquad (19)$$

Trying at the same time to diminish the importance of the center of the interval (m_j, M_j) we propose, in order to reach an optimal Z_{p_j}-value for each β_j, changing the simple normality assumption as follows :

$\beta_j (j \leqslant s)$ is $N(\mu_j, \sigma_{L,j})$ with μ_j being itself $N(c_j, \sigma_{N,j})$.

It is clear that the β_j is itself normally distributed around c_j with

$$\sigma_j^2 = \sigma_{L,j}^2 + \sigma_{N,j}^2 \qquad (20)$$

4.4. Application to our given problem :

$$\sigma_{N,1} = \sigma_{N,2} = \frac{1}{\sqrt{3}} \sim .58.$$

After a run on our data of the normal-prior-method, we learn from appendix B that the standard deviations in Least-Squares-sense are : $\sigma_{L,1} = 1.6378$ and $\sigma_{L,2} = 1.0981$

The final standard deviation with which normal-prior-method has to work, gives, from (20) : $\sigma_1 = .29$ and $\sigma_2 = .40$
The solution is then :

$$\ln P = -2.1095 + \underset{(.2507)}{.7534} \ln L + \underset{(.1036)}{.6563} \ln C + \underset{(.0177)}{.3319}\, t \qquad (21)$$

4.5. Conclusion

The normality assumptions can guarantee convergence in integrations, and treat as many parameters as wanted in a generalized linear model. But our specific application shows that the quality of the estimations by this second algorithm is not in comparison with the advantages of its mathematical treatment, which is on the other hand very attractive.

APPENDIX A

Period	Production	Labour	Capital
1	100.000	100.000	100.000
2	95.903	105.844	100.228
3	105.551	106.697	100.486
4	99.677	113.232	103.125
5	107.762	118.180	116.875
6	114.394	120.211	117.158
7	118.797	123.672	118.252
8	114.742	124.088	120.572
9	110.112	122.868	127.798
10	122.436	123.107	127.672
11	132.798	127.207	128.087
12	141.712	134.597	129.606
13	145.708	140.941	130.106
14	186.470	153.223	131.705
15	197.175	163.138	133.577
16	320.581	169.106	154.676
17	306.099	168.638	152.416
18	352.165	188.339	153.774
19	331.548	188.419	156.521
20	376.920	189.457	156.386
21	396.478	190.516	176.119
22	385.441	191.670	176.428
23	452.692	197.723	176.414
24	492.244	205.910	176.748
25	458.770	212.753	173.359
26	459.239	218.984	173.401
27	459.708	220.401	175.331
28	460.489	219.882	181.954
29	506.981	223.672	175.183
30	517.003	235.300	175.321
31	588.789	242.809	176.489
32	544.597	256.782	177.128

APPENDIX B

Z_p	α	(S.D.)	β	(S.D.)	γ	(S.D.)
.4	.5022	(.0024)	.5023	(.0011)	.0417	(.0055)
3.09	.5036	(.0039)	.5039	(.0019)	.0404	(.0055)
2.58	.5052	(.0056)	.5057	(.0027)	.0404	(.0061)
2.33	.5064	(.0068)	.5070	(.0033)	.0424	(.0046)
1.96	.5091	(.0096)	.5098	(.0046)	.0419	(.0046)
1.64	.5129	(.0136)	.5139	(.0065)	.0417	(.0046)
1.28	.5209	(.0219)	.5225	(.0103)	.0413	(.0046)
.67	.5716	(.0727)	.5741	(.0337)	.0387	(.0054)
.33	.7347	(.2310)	.7195	(.1294)	.03096	(.0110)
.1	1.2868	(1.0305)	1.0425	(.7275)	.0084	(.0404)
.05	1.4898	(1.4526)	1.1160	(.9912)	.0011	(.0545)
.04	1.5227	(1.5267)	1.1263	(1.0346)	-.000023	(.0569)
.03	1.5500	(1.5896)	1.1346	(1.0708)	-.00097	(.0590)
.02	1.5705	(1.6378)	1.1407	(1.0981)	-.0017	(.0657)

REFERENCES

Broeckx F., Goovaerts M., Van den Broeck J., (1976), Numerical estimation of Bayesian parameters involving uniform priors, Revue Belge de Statistique, d'Informatique et de Recherche Opérationnelle, vol. 16, pages 1-13.

Broeckx F., (1975), Contributions to Bayesian estimation, Ph.D. thesis, University of Ghent, Belgium.

Broeckx F., Goovaerts M., Van den Broeck J., (1980), Use of normal priors in estimation of Cobb-Douglas production function models, working paper 80/5, State University Centre, Antwerp.

Moulaert F., (1979), A "conservative" Ridge Estimator, Economic Letters, vol. 18, pages 159-164.

Jeffreys H., (1967), Theory of Probability, Oxford at the Claverton.

Zellner A., (1971), An Introduction to Bayesian Inference in Econometrics, J. Wiley, N.Y.

TIME SERIES ANALYSIS: Theory and Practice 4
O.D. Anderson (editor)

NONSTATIONARY SECOND-ORDER MOVING AVERAGE PROCESSES
II : MODEL BUILDING AND INVERTIBILITY

Marc Hallin

Institut de Statistique
Université Libre de Bruxelles
1050 Bruxelles, Belgium

Recent results (Hallin, 1982b), on a matrix generalization of *positive definite continued fractions*, allow a more precise and explicit solution to the nonstationary MA(2) theoretical model-building problem, treated in Hallin (1982a). The invertible solution is also made explicit; the nonstationary extension of *borderline noninvertible processes* is characterized.

1. THE THEORETICAL MODEL-BUILDING PROBLEM.

Denote by $\gamma_{tj} = E(z_t z_{t-j})$, $j = 0,1,2$, $t \in \mathbb{Z} = \{0, \pm 1, \pm 2, \ldots\}$ the autocovariances of some MA(2) process $\{z_t; t \in \mathbb{Z}\}$, i.e. (cf. Hallin (1982b) *Theorem 7*) real numbers such that the infinite symmetric band matrix with rows

$$\ldots 0 \quad 0 \quad \gamma_{t+2,2} \ \gamma_{t+1,1} \ \gamma_{to} \ \gamma_{t1} \ \gamma_{t2} \quad 0 \quad 0 \quad \ldots \qquad t \in \mathbb{Z} \tag{1}$$

is positive definite.

The *theoretical model-building problem*, treated in Hallin (1982a), is that of obtaining, in terms of γ_{tj}'s, the set of all possible MA(2) *models* for z_t, hence the set of all linear difference operators of order two

$$A_t(L) = a_{to} + a_{t1}L + a_{t2}L^2 \qquad a_{to} > 0 \qquad a_{t2} \neq 0 \qquad t \in \mathbb{Z} \tag{2}$$

such that (L denotes the *lag operator* and $\{\varepsilon_t; t \in \mathbb{Z}\}$ a second-order white noise : $E(\varepsilon_t) = 0$, $E(\varepsilon_s \varepsilon_t) = \delta_{st}$)

$$z_t = A_t(L)\varepsilon_t \ (= a_{to}\varepsilon_t + a_{t1}\varepsilon_{t-1} + a_{t2}\varepsilon_{t-2}) \qquad t \in \mathbb{Z}$$

in a quadratic mean sense, with $\text{var}\,(z_t - A_t(L)\varepsilon_t) = 0$.

In our (1982a) paper, we showed that this model-building problem admits a three-parameter family of solutions, which are given in terms of recursions whose admissible initial values were not obtained explicitly - although the matrix generalization of the notion of *positive definite continued fractions*, Hallin (1982b), now makes this possible.

2. AN EXPLICIT SOLUTION TO THE MA(2) MODEL-BUILDING PROBLEM

Let

$$\Sigma_t = \begin{pmatrix} \gamma_{to} & \gamma_{t1} \\ \gamma_{t1} & \gamma_{t-1,0} \end{pmatrix} \qquad \Gamma_t = \begin{pmatrix} \gamma_{t2} & 0 \\ \gamma_{t-1,1} & \gamma_{t-1,2} \end{pmatrix} \qquad t \in \mathbb{Z} . \tag{3}$$

From Hallin (1982b), we know that the matrices

$$\Phi_t^+ = \Sigma_t - \lim_{k=\infty} \Gamma'_{t+2}(\Sigma_{t+2}-\Gamma'_{t+4}(\Sigma_{t+4}-\Gamma'_{t+6}(\dots\Gamma'_{t+2k}\,\Sigma^{-1}_{t+2k}\Gamma_{t+2k}\dots)^{-1}\Gamma_{t+6})^{-1}\Gamma_{t+4})^{-1}\Gamma_{t+2} \tag{4}$$

and

$$\Phi_t^- = \lim_{k=\infty} \Gamma_t(\Sigma_{t-2}-\Gamma_{t-2}(\Sigma_{t-4}-\Gamma_{t-4}(\dots\Gamma_{t-2k}\,\Sigma^{-1}_{t-2(k+1)}\Gamma'_{t-2k}\dots)^{-1}\Gamma'_{t-4})^{-1}\Gamma'_{t-2})^{-1}\Gamma'_t \tag{5}$$

exist, are positive definite, and such that $\Phi_t^+ - \Phi_t^-$ is itself positive semi-definite. The infinite expressions appearing in (4) and (5) are what we call *positive definite matrix continued fractions* (p.d. M-fractions). As we shall see, Φ_t^+ and Φ_t^- play an essential role in the definition of the possible models for $\{z_t ; t \in \mathbb{Z}\}$; moreover, Φ_t^- will be associated with the only invertible model (if any). We give here two equivalent formulations of the solution of the model-building problem. In *Theorem 1*, the coefficients a_{to}, a_{t1} and a_{t2} of the possible models are expressed by means of nonlinear first-order recursions; in *Theorem 2*, the same coefficients are expressed in terms of appropriate solutions of the *autocovariance difference equation*

$$\gamma_{t+2,2}\psi_t + \gamma_{t+1,1}\psi_{t-1} + \gamma_{to}\psi_{t-2} + \gamma_{t1}\psi_{t-3} + \gamma_{t2}\psi_{t-4} = 0 \qquad t \in \mathbb{Z} .$$

We denote by sign (x) the *sign* of any real x (as in (7) : sign $(a_{t-1,2})$ = sign $(\gamma_{t-1,2})$ means that $a_{t-1,2}$ should be $\sqrt{a^2_{t-1,2}}$ whenever $\gamma_{t-1,2} > 0$, and $-\sqrt{a^2_{t-1,2}}$ whenever $\gamma_{t-1,2} < 0$).

Theorem 1. The operator $A_t(L) = a_{to} + a_{t1}L + a_{t2}L^2$ $(t \in \mathbb{Z})$ provides an adequate MA(2) model for z_t iff

(i) the coefficients a_{tj} $(j = 1,2;\ t \in \mathbb{Z})$ satisfy the nonlinear first-order recursions

$$\begin{cases} a_{t+2,2} = \gamma_{t+2,2}/(\gamma_{to} - a^2_{t1} - a^2_{t2})^{1/2} & t \in \mathbb{Z} \\ a_{t+1,1} = (\gamma_{t+1,1} - a_{t+1,2}a_{t1})/(\gamma_{to} - a^2_{t1} - a^2_{t2})^{1/2} & t \in \mathbb{Z} \end{cases} \tag{6}$$

hence

$$\begin{cases} a_{t-1,1} = (\gamma_{t1} - a_{t1}\gamma_{t2}/a_{t+1,2})/a_{t2} & t \in \mathbb{Z} \\ a^2_{t-1,2} = \gamma_{t-1,0} - a^2_{t-1,1} - (\gamma_{t2}/a_{t+1,2})^2 , & \\ \quad \text{sign}\,(a_{t-1,2}) = \text{sign}\,(\gamma_{t-1,2}) & t \in \mathbb{Z} \end{cases} \tag{7}$$

(ii) with initial values (for some arbitrary $t_o \in \mathbb{Z}$) $a_{t_o 1}=x$, $a_{t_o 2}=y$, $a_{t_o+1,2}=z$ belonging to the nonempty set (p.d. : *positive definite* - p.s.d. : *positive semidefinite*)

$$\left\{x,y,z \in \mathbb{R}^3 \;\middle|\; \Phi = \begin{pmatrix} z^2 & xz \\ xz & x^2+y^2 \end{pmatrix} \text{ p.d, } \Phi^+_{t_o} - \Phi \text{ and } \Phi - \Phi^-_{t_o} \text{ p.s.d.}\right\} \tag{8}$$

(iii) and

$$a_{to} = (\gamma_{to} - a^2_{t1} - a^2_{t2})^{1/2} \qquad t \in \mathbb{Z} .$$

Theorem 2. Let ψ^1_t and ψ^2_t be two linearly independent solutions of the homogeneous difference equation of order four (the *autocovariance difference equation*)

$$\gamma_{t+2,2}\psi_t + \gamma_{t+1,1}\psi_{t-1} + \gamma_{to}\psi_{t-2} + \gamma_{t1}\psi_{t-3} + \gamma_{t2}\psi_{t-4} = 0 \qquad t \in \mathbb{Z} . \tag{9}$$

The operator $A_t(L) = a_{to} + a_{t1}L + a_{t2}L^2$ ($t \in \mathbb{Z}$) provides an adequate MA(2) model for z_t iff

(i)

$$a_{to} = \left(\gamma_{t+2,2} \begin{vmatrix} \psi^1_t & \psi^2_t \\ \psi^1_{t-1} & \psi^2_{t-1} \end{vmatrix} \begin{vmatrix} \psi^1_{t-1} & \psi^2_{t-1} \\ \psi^1_{t-2} & \psi^2_{t-2} \end{vmatrix}\right)^{1/2}$$

$$a^2_{t2} = \gamma_{t2} \begin{vmatrix} \psi^1_{t-3} & \psi^2_{t-3} \\ \psi^1_{t-4} & \psi^2_{t-4} \end{vmatrix} \Big/ \begin{vmatrix} \psi^1_{t-2} & \psi^2_{t-2} \\ \psi^1_{t-3} & \psi^2_{t-3} \end{vmatrix} , \quad \text{sign}\,(a_{t2}) = \text{sign}\,(\gamma_{t2})$$

$$a^2_{t1} = \gamma_{t+1,2} \begin{vmatrix} \psi^1_{t-1} & \psi^2_{t-1} \\ \psi^1_{t-3} & \psi^2_{t-3} \end{vmatrix}^2 \Big/ \begin{vmatrix} \psi^1_{t-1} & \psi^2_{t-1} \\ \psi^1_{t-2} & \psi^2_{t-2} \end{vmatrix} \begin{vmatrix} \psi^1_{t-2} & \psi^2_{t-2} \\ \psi^1_{t-3} & \psi^2_{t-3} \end{vmatrix}$$

$$\text{sign}(a_{t1}) = -\text{sign} \begin{vmatrix} \psi^1_{t-1} & \psi^2_{t-1} \\ \psi^1_{t-2} & \psi^2_{t-2} \end{vmatrix} \begin{vmatrix} \psi^1_{t-1} & \psi^2_{t-1} \\ \psi^1_{t-3} & \psi^2_{t-3} \end{vmatrix} \qquad t \in \mathbb{Z} \tag{10}$$

(ii) the solutions ψ^1_t and ψ^2_t take on initial values (for some arbitrary $t_o \in \mathbb{Z}$)

$$\psi^1_{t_o-1} = x \quad \psi^1_{t_o-2} = 1 \quad \psi^1_{t_o-3} = 0 \quad \psi^1_{t_o-4} = z$$

$$\psi^2_{t_o-1} = y \quad \psi^2_{t_o-2} = 0 \quad \psi^2_{t_o-3} = 1 \quad \psi^2_{t_o-4} = -(\gamma_{t+1,2}\,x + \gamma_{t1})/\gamma_{t2}$$

with (x,y,z) belonging to the nonempty set

$$\{x,y,z \in \mathbb{R}^3 \mid \quad \Phi = \begin{pmatrix} -(\gamma_{t_o+1,2})\frac{1}{y} & -(\gamma_{t_o+1,2})\frac{x}{y} \\ -(\gamma_{t_o+1,2})\frac{x}{y} & -(\gamma_{t_o+1,2})\frac{x}{y} - (\gamma_{t_o 2})z \end{pmatrix} \text{ p.d.,}$$

$$\Phi^{+}_{t_o+1} - \Phi \text{ and } \Phi - \Phi^{-}_{t_o+1} \text{ p.s.d.}\} \tag{11}$$

Proof of Theorem 1. Let $z_t = A_t(L)\varepsilon_t$ be a MA(2) model for z_t; putting

$$\underline{z}_t = \begin{pmatrix} z_t \\ z_{t-1} \end{pmatrix} \quad \underline{\varepsilon}_t = \begin{pmatrix} \varepsilon_t \\ \varepsilon_{t-1} \end{pmatrix} \quad A_{to} = \begin{pmatrix} a_{to} & a_{t1} \\ 0 & a_{t-1,0} \end{pmatrix} \quad A_{t1} = \begin{pmatrix} a_{t2} & 0 \\ a_{t-1,1} & a_{t-1,2} \end{pmatrix} \quad t \in \mathbb{Z}, \tag{12}$$

$\{\underline{z}_t;\ t \in 2\mathbb{Z}\}$ is, obviously, a bivariate MA(1) process, since

$$\underline{z}_t = A_{to}\,\underline{\varepsilon}_t + A_{t1}\,\underline{\varepsilon}_{t-2} \qquad t \in \mathbb{Z}, \tag{13}$$

with $E(\underline{z}_t\,\underline{z}_t') = \Sigma_t$ and $E(\underline{z}_t\,\underline{z}_{t-2}') = \Gamma_t$ (cf. (3)). According to *Theorem 8* in Hallin (1982b), any p.d. Φ_{t_o} such that $\Phi^{+}_{t_o} - \Phi_{t_o}$ and $\Phi_{t_o} - \Phi^{-}_{t_o}$ are p.s.d. generates a sequence of p.d. matrices Φ_t

$$\Phi_t = \Gamma_t\,(\Sigma_{t-2} - \Phi_{t-2})^{-1}\Gamma_t' \quad t \in \mathbb{Z} \iff \Phi_t = \Sigma_t - \Gamma_{t+2}'\,\Phi_{t+2}^{-1}\,\Gamma_{t+2} \quad t \in \mathbb{Z}. \tag{14}$$

To each of these sequences corresponds a class of bivariate MA(1) models (13) with

$$A_{t1}\,A_{t1}' = \Phi_t \quad \text{and} \quad A_{to} = \Gamma_{t+2}'\,A_{t+2,1}'^{-1},$$

among which one and only one yields a lower diagonal A_{t1} (as in (12)) such that

$$\text{sign}\,(A_{t1})_{11} = \text{sign}(\gamma_{t2}),\quad \text{sign}\,(A_{t1})_{22} = \text{sign}\,(\gamma_{t-1,2}) \qquad t \in \mathbb{Z},$$

thus providing a valid univariate MA(2) model for z_t. Conversely, any MA(2) model for z_t can be obtained this way.

Now, if A_{to} is lower diagonal as in (12), Φ_{t_o} is of the form

$$\Phi_{t_o} = \begin{pmatrix} a^2_{t_o+1,2} & a_{t_o 1}\,a_{t_o+1,2} \\ a_{t_o 1}\,a_{t_o+1,2} & a^2_{t_o 1} + a^2_{t_o 2} \end{pmatrix} = \begin{pmatrix} z^2 & x\,z \\ x\,z & x^2+y^2 \end{pmatrix}$$

(in the notations of (8)). (6) and (7) are equivalent to (14), and (i) and (ii) are thus established. (iii) is obvious. □

Proof of Theorem 2. Part (i) immediately follows from Hallin (1982a). It remains to assign adequate initial values to ψ_t^1 and ψ_t^2 ; being solutions of a difference equation of order four, they require four initial values each, at time $t_0-1, \ldots, t_0-4$, say. (10) however only depends on the vectorial space spanned by ψ_t^1 and ψ_t^2; thus we always can fix $\psi_{t_0-2}^1 = \psi_{t_0-3}^2 = 1$ and $\psi_{t_0-2}^2 = \psi_{t_0-3}^1 = 0$. Still from (Hallin, 1982a condition (8)), we know that

$$\gamma_{t+1,2}\begin{vmatrix}\psi_{t-1}^1 & \psi_{t-1}^2\\ \psi_{t-3}^1 & \psi_{t-3}^2\end{vmatrix} + \gamma_{t1}\begin{vmatrix}\psi_{t-2}^1 & \psi_{t-2}^2\\ \psi_{t-3}^1 & \psi_{t-3}^2\end{vmatrix} + \gamma_{t2}\begin{vmatrix}\psi_{t-2}^1 & \psi_{t-2}^2\\ \psi_{t-4}^1 & \psi_{t-4}^2\end{vmatrix}$$

has to vanish for $t \in \mathbb{Z}$, which holds iff it holds for one arbitrary $t_0 \in \mathbb{Z}$; let therefore $\psi_{t_0-4}^2 = -(\gamma_{t_0+1,2}\,\psi_{t_0-1}^1 + \gamma_{t_01})/\gamma_{t_02}$. Three initial values, $\psi_{t_0-1}^1$, $\psi_{t_0-1}^2$ and $\psi_{t_0-4}^2$ remain thus undetermined, each triple of them defining, through (10), an MA(2) model. Such a model will be a valid MA(2) model (real coefficients a_{tj} etc.) for z_t iff the triple $(\psi_{t_0-1}^1, \psi_{t_0-1}^2, \psi_{t_0-4}^2)$ is such that the corresponding Φ_t matrix (cf. (16)) is p.d. for any $t \in \mathbb{Z}$; and Φ_t is p.d. for any $t \in \mathbb{Z}$ iff (cf. Hallin, 1982b)

$$\Phi_{t_0+1} \text{ p.d.} \quad , \quad \overset{+}{\Phi}_{t_0+1} - \Phi_{t_0} \quad \text{and } \Phi_{t_0+1} - \overset{-}{\Phi}_{t_0+1} \text{ p.s.d.}$$

for some arbitrary $t_0 \in \mathbb{Z}$.

Computing Φ_{t_0+1} in terms of $(\psi_{t_0-1}^1, \psi_{t_0-1}^2, \psi_{t_0-4}^2)$ yields

$$\Phi_{t_0+1} = \begin{pmatrix} -\gamma_{t_0+1,2}/\psi_{t_0-1}^2 & -\gamma_{t_0+1,2}\,\psi_{t_0-1}^1/\psi_{t_0-1}^2 \\ -\gamma_{t_0+1,2}\,\psi_{t_0-1}^1/\psi_{t_0-1}^2 & -\dfrac{\gamma_{t_0+1,2}\,\psi_{t_0-1}^1}{\psi_{t_0-1}^2} - \gamma_{t_02}\,\psi_{t_0-4}^2 \end{pmatrix}$$

$$= -\begin{pmatrix} \dfrac{\gamma_{t_0+1,2}}{y} & \dfrac{\gamma_{t_0+1,2}}{y}x \\ \dfrac{\gamma_{t_0+1,2}}{y}x & \dfrac{\gamma_{t_0+1,2}}{y}x + \gamma_{t_02}\,z \end{pmatrix}$$

(11) is thus established, which completes the proof. □

3. INVERTIBILITY

It follows from *Theorems 1* and *2* that a given MA(2) stochastic process $\{z_t;\ t \in \mathbb{Z}\}$ can be adequately described by

- a three-parameter infinite family of MA(2)models whenever $\overset{+}{\Phi}_t - \overset{-}{\Phi}_t$ is of rank two,

- a one-parameter infinite family of MA(2) models whenever $\Phi_t^+ - \Phi_t^-$ is of rank one,
- an unique MA(2) model whenever $\Phi_t^+ = \Phi_t^-$

(notice that the rank of $\Phi_t^+ - \Phi_t^-$ does not depend on t); indeed, there exists a one-to-one correspondence between the sets of initial values (8) and (11) and the set of all possible models for $\{z_t;\ t \in \mathbb{Z}\}$.

If, however, a model has to be used for forecasting purposes, it needs to be an *invertible model*, i.e. a model such that ε_t can be expressed as a mean-square-convergent linear combination of $z_t, z_{t-1}, \ldots$. It is therefore important to be able to select, among the infinity of possible models for a given MA(2) process, the invertible model - if any.

This characterization is provided by the following theorem. Denote by $\varphi_{ij}^-(t)$ $(i,j = 1,2;\ t \in \mathbb{Z})$ the entries of Φ_t^- .

Theorem 3. (i) If the infinite product

$$\Phi_{t_o}^- \Gamma_{t_o}^{\prime -1} \Phi_{t_o-2}^- \Gamma_{t_o-2}^{\prime -1} \Phi_{t_o-4}^- \Gamma_{t_o-4}^{\prime -1} \cdots \tag{15}$$

vanishes (converges, componentwise, to $0_{2\times 2}$) for some $t_o \in \mathbb{Z}$, it vanishes for any $t \in \mathbb{Z}$, and there exists one and only one invertible MA(2) model for z_t :

$$z_t = a_{to}^- \varepsilon_t + a_{t1}^- \varepsilon_{t-1} + a_{t2}^- \varepsilon_{t-2} \qquad t \in \mathbb{Z} \tag{16}$$

with(1)

$$\begin{cases} a_{to}^- = \gamma_{to} - \varphi_{11}^-(t) - (\varphi_{12}^-(t+1))^2/\varphi_{11}^-(t+1) \\ (a_{t1}^-)^2 = (\varphi_{12}^-(t+1))^2/\varphi_{11}^-(t+1) \qquad \mathrm{sign}(a_{t1}^-) = \mathrm{sign}(\varphi_{12}^-(t+1)\gamma_{t+1,2}) \\ (a_{t2}^-)^2 = \varphi_{11}^-(t) \qquad \mathrm{sign}(a_{t2}^-) = \mathrm{sign}(\gamma_{t2}) \qquad t \in \mathbb{Z} \end{cases} \tag{17}$$

or, equivalently, a_{to}^-, a_{t1}^- and a_{t2}^- given by (10), where ψ_t^1 and ψ_t^2 take on initial values

$$\psi_{t_o-1}^1 = \frac{\varphi_{12}^-(t_o+1)}{\varphi_{11}^-(t_o+1)} \quad \psi_{t_o-2}^1 = 1 \quad \psi_{t_o-3}^1 = 0 \quad \psi_{t_o-4}^1 = \frac{1}{\gamma_{t_o2}}(\varphi_{12}^-(t_o+1)-\varphi_{22}^-(t_o+1))$$

$$\psi_{t_o-1}^2 = -\frac{\gamma_{t_o+1,2}}{\varphi_{11}^-(t_o+1)} \quad \psi_{t_o-2}^2 = 0 \quad \psi_{t_o-3}^2 = 1 \quad \psi_{t_o-4}^2 = -(\gamma_{t+1,2}\frac{\varphi_{12}^-(t_o+1)}{\varphi_{11}^-(t_o+1)} + \gamma_{t1})/\gamma_{t2}. \tag{18}$$

---

(1) Recall that $\Phi_{t+2}^- = \Gamma_{t+2}(\Sigma_t - \Phi_t^-)^{-1}\,\Gamma_{t+2}'$ and $\Phi_{t-2}^- = \Sigma_{t-2} - \Gamma_t'(\Phi_t^-)^{-1}\,\Gamma_t$ (cf. Hallin, 1982b).

(ii) If the infinite product (15) is not equal to 0_{2x2}, there exists no invertible model for z_t; z_t is then a nonstationary *borderline noninvertible process* (cf. Anderson (1977)).

The proof of this theorem requires some further properties of *positive definite matrix continued fractions* : let us first recall some of the results of Hallin (1982a).

Put $t_o = 0$; Φ_o^- is defined as

$$\Phi_o^- = \lim_{s=\infty} \Gamma_o(\Sigma_{-2}-\Gamma_{-2}(\Sigma_{-4}- \cdots \Gamma_{-2s}\, \Sigma_{-2(s+1)}^{-1}\Gamma'_{-2s}\cdots)^{-1}\Gamma'_{-2})^{-1}\Gamma'_o$$

$$= \lim_{s=\infty} \Phi_o^{-(s)} \quad ; \qquad (19)$$

$\Phi_o^{-(s)}$ is called the sth *approximant* of Φ_o^-. It follows from *Theorem 3* in Hallin (1982a) that $\Phi_o^{-(s)}$ is p.d., as well as $\Phi_o^- - \Phi_o^{-(s)}$, for any value of $s \in \mathbb{Z}$.

There is a close connection between ordinary continued fractions and difference equations of order two (cf., for example, Wall (1967)) : a similar connection holds for *matrix continued fractions*. Consider the homogeneous linear difference equation of order two

$$X_s = (\Sigma_{-2s}\,\Gamma_{-2(s-1)}^{-1})X_{s-1} - (\Gamma'_{-2(s-1)}\,\Gamma_{-2(s-2)}^{-1})X_{s-2} \qquad s \in \mathbb{Z}, \qquad (20)$$

(X_s being a 2x2 matrix). Denote by G(s,o) and G(s,1) the (matrix) solutions of (20) taking on initial values

$$G(-1,0) = 0 \qquad G(0,1) = 0$$
$$G(0,0) = I \qquad G(1,1) = I \quad .$$

G(s,0) and G(s,1) are *Green's matrices*, and constitute a *fundamental set of solutions* of (20); any solution X_s of (20) can therefore be expressed as a linear combination

$$X_s = G(s,0)M + G(s,1)N \qquad s \in \mathbb{Z} \ ,$$

M and N being unequivocally determined by X_s's initial values. The approximants $\Phi_o^{-(s)}$ of Φ_o^- are related to (20) in the following way :

$$\Phi_o^{-(s)} = \Gamma_o\, G'(s,1)G'^{-1}(s,0) \quad ,$$

and (19) can be written as

$$\Phi_o^- = \Gamma_o \lim_{s=\infty} G'(s,1)\, G'^{-1}(s,0) \ . \qquad (21)$$

Before going to the proof of *Theorem 3*, let us establish the following lemma .

Lemma. Let Φ_o be such that $\Phi_o - \Phi_o^-$ and $\Phi_o^+ - \Phi_o$ are both p.s.d., and denote by Π_s the product

$$\Pi_s = \Phi_o \Gamma_o'^{-1} \Phi_{-2} \Gamma_{-2}'^{-1} \cdots \Gamma_{-2(s-1)}'^{-1} \Phi_{-2s} \quad . \tag{22}$$

The model associated with Φ_o (according to (8) or (11)) is an invertible one iff $\lim_{s=\infty} \Pi_s = 0$.

Proof. According to *Theorems 1* and *2*, Φ_o defines an adequate MA(2) model (2) for z_t. (12) associates with this univariate MA(2) model (2) a bivariate MA(1) model (13) for $\underline{z}_t$. Denote by A_{tj} its coefficients : we have

$$A_{t1} A_{t1}' = \Phi_t \qquad A_{to} = \Gamma_{t+2}' A_{t+2,1}'^{-1} \quad , \tag{23}$$

with Φ_t satisfying the same recursions as Φ_t^-

$$\Phi_{t+2} = \Gamma_{t+2}(\Sigma_t - \Phi_t)^{-1} \Gamma_{t+2}' \qquad \Phi_{t-2} = \Sigma_{t-2} - \Gamma_t' \Phi_t^{-1} \Gamma_t \ . \tag{24}$$

The univariate MA(2) model (2) is invertible iff the corresponding bivariate one (13) is invertible. Denote by $\|M\|$ the norm $\|M\| = \Sigma_{i,j} M_{ij}^2$: (13) is invertible iff

$$\lim_{s=-\infty} \|A_{t0}^{-1} A_{t1} A_{t-2,0}^{-1} A_{t-2,1} \cdots A_{s0}^{-1}\| = 0 \qquad t \in \mathbb{Z} \ .$$

Indeed, if $\underline{z}_t = A_{t0}\, \underline{\varepsilon}_t + A_{t1}\, \underline{\varepsilon}_{t-2}$, we have

$$\underline{\varepsilon}_t - A_{t0}^{-1} \underline{z}_t + A_{t0}^{-1} A_{t1} A_{t-2,0}^{-1} \underline{z}_{t-2} - \cdots \pm A_{t0}^{-1} A_{t1} \cdots A_{s0}^{-1} \underline{z}_s$$

$$= A_{t0}^{-1} A_{t1} \cdots A_{s0}^{-1} A_{s1}\, \underline{\varepsilon}_s \ . \tag{25}$$

The left hand side of (25) thus converges (in quadratic mean) to 0 iff (15) vanishes. Actually, the products $\pm A_{t0}^{-1} A_{t1} \cdots A_{s0}^{-1}$ are the *Green's matrices* G(t,s) associated with (13) (cf. Miller (1968) or Hallin (1982c)). But

$$A_{to}^{-1} A_{t1} A_{t-2,0}^{-1} A_{t-2,1} \cdots A_{s0}^{-1}$$

$$= A_{t+2,1}' \Gamma_{t+2}'^{-1} A_{t1} A_{t1}' \Gamma_t'^{-1} \cdots A_{s+2,1}' \Gamma_{s+2}'^{-1}$$

$$= A_{t+2,1}' \Gamma_{t+2}'^{-1} \Phi_t \Gamma_t'^{-1} \Phi_{t-2} \cdots \Phi_{s+2} \Gamma_{s+2}'^{-1} \qquad t > s \in \mathbb{Z}$$

(recall that $A_{to} = \Gamma_{t+2}' A_{t+2,1}'^{-1}$). Hence (2) is invertible iff the infinite product

$$\Phi_t \Gamma_t'^{-1} \Phi_{t-2} \Gamma_{t-2}'^{-1} \Phi_{t-4} \Gamma_{t-4}'^{-1} \cdots$$

vanishes for any $t \in \mathbb{Z}$, which happens as soon as it vanishes for some arbitrary $t_o \in \mathbb{Z}$. □

We may now turn to the proof of *Theorem 3*.

Proof of Theorem 3. Let $t_o = 0$, and denote by Π_s^- the product

$$\Pi_s^- = \Phi_o^- \Gamma_o'^{-1} \Phi_{-2}^- \Gamma_{-2}'^{-1} \cdots \Phi_{-2s}^- \ .$$

In view of the above lemma, it will be sufficient to show that $\lim_{s=\infty} \Pi_s = 0$ implies $\lim_{s=\infty} \Pi_s^- = 0$. The invertibility of (2) then implies that of (16), and these two models therefore cannot be distinct. (17) and (18) immediately follow from *Theorems 1* and *2*.

Taking (24) into account, we have

$$\Pi_s = \Phi_0 \Gamma_0'^{-1} \cdots \Phi_{-2(s-1)} \Gamma_{-2(s-1)}'^{-1} (\Sigma_{-2s} - \Gamma_{-2(s-1)}'^{-1} \Phi_{-2(s-1)}^{-1} \Gamma_{-2(s-1)})$$

$$= \Pi_{s-1} \Gamma_{-2(s-1)}'^{-1} \Sigma_{-2s} - \Pi_{s-2} \Gamma_{-2(s-2)}'^{-1} \Gamma_{-2(s-1)} ,$$

or, equivalently,

$$\Pi_s' = \Sigma_{-2s} \Gamma_{-2(s-1)}^{-1} \Pi_{s-1}' - \Gamma_{-2(s-1)}' \Gamma_{-2(s-2)}^{-1} \Pi_{s-2}' .$$

Π_s' is thus a solution of equation (20), and can be expressed in terms of the Green's matrices G(s,0) and G(s,1) : it is easy to see that

$$\Pi_s = \Phi_0 \, G'(s,0) - \Gamma_0 \, G'(s,1) . \tag{26}$$

Similarly,

$$\Pi_s^- = \Phi_0^- \, G'(s,0) - \Gamma_0 \, G'(s,1) . \tag{27}$$

If we assume that G(s,0) is uniformly bounded (i.e. there exists a constant G such that $[G(s,0)]_{ij} \leqslant G \;\; \forall \; s, i, j$), it follows from (21) that

$$\lim_{s=\infty} \Pi_s^- = \lim_{s=\infty} \{[\Phi_0^- - \Gamma_0 \, G'(s,1) G'^{-1}(s,0)] \, G'(s,0)\} = 0 .$$

$\lim_{s=\infty} \Pi_s$ thus cannot be zero unless $\Pi_s = \Pi_s^-$ and $\Phi_0 = \Phi_0^-$.

Now consider the case when G(s,0) is not bounded and $\lim_{s=\infty} \Pi_s = 0$. If $\Phi_0 - \Phi_0^-$ is strictly p.d., (21) and (26) imply that

$$\lim_{s=\infty} \Pi_s \, G'^{-1}(s,0) = \Phi_0 - \Phi_0^- \; ; \tag{28}$$

Π_s and G(s,0) being non singular, we have

$$\lim_{s=\infty} [\Pi_s \, G'^{-1}(s,0)]^{-1} = (\Phi_0 - \Phi_0^-)^{-1} ,$$

hence

$$\lim_{s=\infty} [\Pi_s \, G'^{-1}(s,0)]^{-1} \, \Pi_s = 0 .$$

But this latter expression, of course, is nothing else than $\lim_{s=\infty} G(s,0)$, which was supposed not to exist. Hence $\Phi_0 - \Phi_0^-$ cannot be of full rank : assume therefore that it is of rank 1. Since it is p.s.d., there exists an orthogonal matrix P such that

$$\Phi_0 - \Phi_0^- = P' \begin{pmatrix} \lambda^2 & 0 \\ 0 & 0 \end{pmatrix} P .$$

Π_s can be written as

$$\Pi_s = (\Phi_o - \Phi_o^- + \Phi_o^- - \Phi_o^{-(s)})G'(s,0)$$

$$= P'(\ \Lambda + P(\Phi_o^- - \Phi_o^{-(s)})P')P\ G'(s,0) \quad ,$$

with

$$\Lambda = \begin{vmatrix} \lambda^2 & 0 \\ 0 & 0 \end{vmatrix} \ .$$

Put $g_s = P\ G'(s,0) = (g_{ij}^s)$ and

$$\Delta_s = \begin{pmatrix} \delta_1^s & \delta^s \\ & \\ \delta^s & \delta_2^s \end{pmatrix} = P(\Phi_o^- - \Phi_o^{-(s)})P' \quad :$$

we know that Δ_s is p.d. (the difference $\Phi_o^- - \Phi_o^{-(s)}$ is p.d.).
If $\lim_{s=\infty} \Pi_s = 0$, we also have

$$\lim_{s=\infty} (\Lambda + \Delta_s)g_s = 0 \quad ,$$

hence

$$\begin{cases} (\lambda^2 + \delta_1^s)g_{11}^s + \delta^s\ g_{21}^s \to 0 \\ \delta^s\ g_{11}^s + \delta_2^s\ g_{21}^s \to 0 \end{cases}$$

It is easy to check that this is impossible unless $\lambda^2 = 0$. It follows that $\Phi_o^- = \Phi_o$ and $\Pi_s^- = \Pi_s$, which completes the proof. □

REFERENCES

Anderson, O.D. (1977) : The time series concept of invertibility. *Math. Operationsforsch., Ser. Statistics, 8,* 399-406.

Hallin, M. (1980) : Invertibility and generalized invertibility of time series models. *J.R.S.S.(B), 42,* 210-212.

Hallin, M. (1982a) : Nonstationary second-order moving average processes. In *Applied Time Series Analysis* (Proceedings of International Time Series Meeting held at Houston, Texas, August 1981). Eds : O.D. Anderson and M.R. Perryman, North-Holland, Amsterdam & New York, 75-84.

Hallin, M. (1982b) : Nonstationary moving average processes, *submitted for publication.*

Hallin, M. (1982c) : Une propriété des opérateurs moyenne mobile, *Cahiers du C.E.R.O., 24,* 229-236.

Miller, K.S. (1968) : *Linear difference equations,* Benjamin, New York.

Wall, H.S. (1948) : *Analytic Theory of Continued Fractions,* Chelsea Publishing Company, Bronx, New York.

TIME SERIES ANALYSIS: Theory and Practice 4
O.D. Anderson (editor)

RANK TESTS FOR RANDOMNESS AGAINST AUTOCORRELATED ALTERNATIVES

Z. Govindarajulu

Department of Statistics
University of Kentucky
Lexington, Kentucky 40506, USA

The test procedures for randomness against a certain dependency are surveyed and some new procedures are developed. The optimal properties of these test procedures are established. Asymptotic normality and efficiency properties are studied.

1. INTRODUCTION

The assumption of randomness is essential for almost all statistical problems. In order to ensure randomness, either samples are drawn randomly or treatments are randomly assigned to experimental units; that, however, is not always possible. The assumption of randomness is crucial when observations are collected in a sequence, in which case we might suspect some kind of trend or dependency between successive observations. Nonrandomness could arise from several sources. Thus there exists a large variety of tests for randomness. Significant controbutions have been made by Anderson (1942), Wald and Wolfowitz (1943), Mann (1945), Stuart (1954, 1956), Foster and Stuart (1954), Savage (1957), and Aiyar *et al* (1979). Most of these test procedures are sensitive to "trend" alternatives in which there is a shift in the location parameter.

Let $\underline{X} = (X_1,\ldots,X_n)$ have a continuous multivariate distribution with unknown common mean $E(X_i) = \mu$, unknown common variance σ^2 and correlation matrix $\Sigma(\rho)$ which depends on an unknown parameter θ. We wish to test H_0: $\rho = 0$ against H_1: $\rho > 0$. Let R_i denote the rank of X_i in the ordered $X_1,\ldots,X_n$ and let $R = (R_1,\ldots,R_n)$ denote the rank vector. Without loss of generality we can set $\mu = 0$ since the rank tests are invariant under location changes.

2. TEST PROCEDURES

The best known test procedure (under the joint normality assumption) is based on the (first order) serial correlation coefficient which, for large n, is equivalent to

$$S_n = (\sum_{i=1}^{n-1} X_i X_{i+1} - n\bar{X}^2)/\{\sum_{i=1}^{n} (X_i - \bar{X})^2\}. \tag{2.1}$$

Anderson (1942) and others have extensively studied the distribution of S_n under the null hypothesis. Knoke (1975) has shown that the likelihood derivative test for autocorrelated normal alternatives is equivalent to the test based on S_n. Dufour (1981) proposes a family of linear rank statistics in order to test the independence of a time series, assuming that the random variables are symmetric with zero medians.

Wald and Wolfowitz (1943) proposed a distribution-free test based on K_n where

$$K_n = n^{-1/2} \sum_{i=1}^{n-1} \{R_i/(n+1)\}\{R_{i+1}/(n+1)\}. \tag{2.2}$$

Gupta and Govindarajulu (1980) point out that if the joint density function $f(\underline{x};\rho)$ of $X_1,\ldots,X_n$ is regular with respect to ρ and permutation invariant under H_0, then the locally most powerful rank test (LMP test) of H_0: $\rho = 0$ versus H_1: $\rho > 0$ is to reject H_0 for large values of

$$E_0\{f'(\underline{X}_R;0)/f(\underline{X}_R;0)\} \tag{2.3}$$

where $\underline{X}_R = (X_{R_1},\ldots,X_{R_n})$, E_0 denotes expectation under H_0 and f' is taken with respect to ρ. The first order autoregressive normal alternative is given by

$$X_i = \rho X_{i-1} + Z_i, \; (\rho > 0), \; X_0 = 0 \tag{2.4}$$

$i = 1,2,\ldots$ where Z_i are independent and normal $(0,\sigma^2)$. Also, the first order moving average normal process is given by

$$X_i = Z_i + \rho Z_{i-1}, \; (\rho > 0), \; Z_0 = 0 \tag{2.5}$$

where the Z_i are independent and normal $(0,\sigma^2)$. In the case of (2.4), the joint density of $\underline{X} = (X_1,\ldots,X_n)$ is

$$f(\underline{x};\rho,\sigma^2) = (2\pi\sigma^2)^{-n/2} \exp[-(2\sigma^2)^{-1} \sum_{i=1}^{n} \{x_i - \rho x_{i-1}\}^2]. \tag{2.6}$$

In the case of (2.5)

$$Z_i = X_i - \rho Z_{i-1} = \sum_{k=0}^{i-1} (-1)^k \rho^k X_{i-k}$$

and

$$f(\underline{x};\rho,\sigma^2) = (2\pi\sigma^2)^{-n/2} \exp[-(2\sigma^2)^{-1} \sum_{i=1}^{n} \{ \sum_{k=0}^{i-1} \rho^{2k} x_{i-1}^2 + \sum_{k\neq k'=0}^{i-1} (-1)^{k+k'} \rho^{k+k'} x_{i-k} x_{i-k'} \}].$$

Alternatively, suppose that X is multivariate normal with mean 0 and variance-covariance matrix $\sigma^2\Sigma$ where Σ has ones in the main diagonal, ρ's in the first off-diagonals on either side of the main diagonal, ρ^2's in the second off-diagonals, etc. Then, its inverse is given by

$$\Sigma = \begin{pmatrix} 1 & -\rho & \cdots\cdots & & 0 \\ -\rho & 1+\rho^2 & -\rho & & 0 \\ 0 & \cdots\cdots & & 1+\rho^2 & -\rho \\ 0 & \cdots\cdots & & -\rho & 1 \end{pmatrix} (1-\rho^2). \tag{2.7}$$

If (2.4), (2.5) or (2.7) holds then the LMP test criterion is given by

$$T_n = n^{-1/2} \sum_{i=1}^{n-1} E_0(Y_{R_{i,n}} Y_{R_{i+1,n}}) \tag{2.8}$$

where $Y_{u,n}$ is the u-th normal score for sample size n. Aiyar (1981) has proposed a heuristic test procedure based on the statistic

$$\tau_n = n^{-1/2} \sum_{i=1}^{n-1} \Phi^{-1}\left(\frac{R_i}{n+1}\right) \Phi^{-1}\left(\frac{R_{i+1}}{n+1}\right). \tag{2.9}$$

Govindarajulu and Dwass (1979) also propose a heuristic test procedure based on

$$\mathcal{R} = n^{-1/2} \sum_{i=1}^{n-1} U_i U_{i+1} \tag{2.10}$$

where $U_i = \text{sgn}\, X_i = 1$ if $X_i > 0$ and -1 if $X_i < 0$ $(i = 1,\ldots,n)$. Notice that U_i is not defined here when $X_i = 0$ since zero observations can occur with probability zero. The statistic $\mathcal{R}$ is linearly related to the number of runs R in the "sign" sequence $U_1,\ldots,U_n$ which is given by

$$\mathcal{R} = \sum_{U_iU_{i+1}<0} U_iU_{i+1} + \sum_{U_iU_{i+1}>0} U_iU_{i+1}$$

$$= -(R-1) + [n-1 - (R-1)] = n+1-2R,$$

where we note that a "new run" starts every time $U_iU_{i+1} < 0$. Tests of randomness based on R were also studied by David (1947), Goodman (1958) and Dufour (1981). They also indicate a special kind of dependence under which the LMP test when the underlying variables are double exponential coincides with the heuristic test (2.10) provided the sample size is even. Let $Z_1,\ldots,Z_{2n}$ be a sample and let

$$\begin{aligned} X_i &= Z_{2i-1},\ Y_i = Z_{2i},\ i = 1,\ldots,n, \\ C_i &= 1(-1) \text{ if } X_i > 0(X_i < 0) \\ D_i &= 1(-1) \text{ if } Y_i > 0(Y_i < 0) \\ \text{and } U_i &= 1(-1) \text{ if } Z_i > 0(Z_i < 0). \end{aligned} \tag{2.11}$$

Hence $C_i = U_{2i-1}$ and $D_i = U_{2i}$. Combine and order C_iX_i, D_iY_i $(i = 1,\ldots,n)$. Let

$R_1^*,\ldots,R_n^*$ denote the ranks of $C_1X_1,\ldots,C_nX_n$ respectively, and $S_1^*,\ldots,S_n^*$ denote the ranks of $D_1Y_1,\ldots,D_nY_n$. Also, let $R_1,\ldots,R_{2n}$ denote the ranks of U_iZ_i $(i=1,\ldots,2n)$. Note that $R_i^* = R_{2i-1}$ and $S_i^* = R_{2i}$. We wish to test H_0: the Z_i are i.i.d. against the alternative H_1: that the Z_i are not i.i.d. Consider the sub-alternative hypothesis

$$H_\Delta:\quad X_i = X_i^* + \Delta V_i,\qquad i=1,\ldots,n \tag{2.12}$$

$$Y_i = Y_i^* + \Delta V_i,\qquad i=1,\ldots,n,$$

where X_i^*, Y_i^* and V_i are mutually independent and their distributions are free of i, X_i^*,Y_i^* have a known common density f and V_i have an unknown distribution G with $EV_1^2 < \infty$. Now, proceeding as in Section 7 of Govindarajulu and Dwass (1979), the LMP test is to reject H_0 for large values of

$$\sum_{i=1}^{n} E_0\left\{\frac{f'(U_{2i-1}W_{R_{2i-1}})}{f(U_{2i-1}W_{R_{2i-1}})}\cdot\frac{f'(U_{2i}W_{R_{2i}})}{f(U_{2i}W_{R_{2i}})}\right\} \tag{2.13}$$

where $W_1 < W_2 < \ldots < W_{2n}$ denote the ordered U_iZ_i $(i=1,\ldots,2n)$. Now by interchanging the roles of X's and Y's we can obtain for the test criterion, the expression

$$\sum_{i=1}^{n} E_0\left\{\frac{f'(U_{2i}W_{R_{2i}})}{f(U_iW_{R_{2i}})}\cdot\frac{f'(U_{2i+1}W_{R_{2i+1}})}{f(U_{2i+1}W_{R_{2i+1}})}\right\}, \tag{2.14}$$

with $U_{2n+1} = U_1$, $R_{2n+1} = R_1$. Since, under H_0, the expressions given by (2.13) and (2.14) are identically distributed, we can take the sum as the test criterion given by

$$\sum_{i=1}^{2n} E_0\left\{\frac{f'(U_iW_{R_i})}{f(U_iW_{R_i})}\cdot\frac{f'(U_{i+1}W_{R_{i+1}})}{f(U_{i+1}W_{R_{i+1}})}\right\}. \tag{2.15}$$

Now letting f be the double exponential density, we obtain the heuristic test given by (2.10).

3. A NEW AUTO-REGRESSIVE MODEL

In this section we shall propose a new auto-regressive model and show that a LMP test of H_0: $\rho = 0$ coincides with the test criterion K_n given by (2.2). Let

$$X_i = (1+e^{-X_{i-1}})^{-1}\rho + Z_i,\qquad i=1,2,\ldots,\qquad X_0 = 0 \tag{3.1}$$

where Z_i are i.i.d. having density f(x). If x is large and ρ is small, $(1+e^{-x})^{-1}\rho \doteq (1-e^{-x})\rho \doteq x\rho$. Hence (3.1) approximately corresponds to the first order moving average process. Then the LMP test of H_0 is to reject the hypothesis for large values of

$$-\sum_{i=1}^{n} E_0\left\{\frac{f'(X_{R_i})}{f(X_{R_i})}(1+e^{-X_{R_{i-1}}})^{-1}\right\}. \tag{3.2}$$

If f is the logistic density, and F denotes the logistic c.d.f. given by $(1+e^{-x})^{-1}$, then the test criterion becomes (the U_i^* denoting the uniform order statistics)

$$\sum_{i=1}^{n} E_0\{(2F(X_{R_i})-1)F(X_{R_{i-1}})\} = \sum_{i=1}^{n} E_0\{(2U^*_{R_i}-1)U^*_{R_{i-1}}\}$$

$$= 2\sum_{i=1}^{n} E_0(U^*_{R_i}U^*_{R_{i-1}}) - \sum_{i=1}^{n} E_0U^*_{R_{i-1}}. \tag{3.3}$$

Notice that

$$\sum_{i=1}^{n} E_0U^*_{R_{i-1}} = \sum_{j=1}^{n} E_0(U_j^*) - E_0U^*_{R_n},$$

since $U_0 = 0$ and $(R_1,\ldots,R_n)$ is a permutation of $(1,\ldots,n)$. Now since $E_0U^*_{R_n} \le 1$, and $\sum_1^n E_0(U_j^*) = n/2$, we surmise that for large n, $n^{-1/2}\sum_1^n E_0U^*_{R_{i-1}}$ is a non-stochastic constant. Next

$$E_0(U_j^*U_k^*) = \begin{cases} jk/(n+1)(n+2) + j/(n+1)(n+2) & \text{for } j<k \\ jk/(n+1)(n+2) + k/(n+1)(n+2) & \text{for } k<j \end{cases},$$

and the term linear in j or k will again lead to a non-stochastic constant for large n. Thus, for large n the test criterion given by (3.3) is a linear function of

$$\sum_{i=1}^{n} R_iR_{i-1}/(n+1)(n+2) \tag{3.4}$$

which except for a constant multiplier is K_n. Thus the test criterion of Wald and Wolfowitz (1943) is LMP with respect to this model when the underlying random variables have a logistic density.

4. ASYMPTOTIC EQUIVALENCE OF T_n (2.8) AND τ_n (2.9)

Gupta and Govindarajulu (1980) have shown that T_n under H_0 is asymptotically equivalent to

$$T_n^* = n^{-1/2} \sum_{i=1}^{n-1} E(Y_{R_{i,n}}) E(Y_{R_{i-1,n}}) \tag{4.1}$$

where for the sake of simplicity the subscript 0 of E is dropped. Next, one can write $T_n^* - \tau_n$ as

$$T_n^* - \tau_n = n^{-1/2} \sum_{i=1}^{n-1} E(Y_{R_{i,n}}) \{E(Y_{R_{i+1,n}}) - \Phi^{-1}(\frac{R_{i+1}}{n+1})\}$$

$$+ n^{-1/2} \sum_{i=1}^{n-1} \Phi^{-1}(\frac{R_{i+1}}{n+1}) \{E(Y_{R_{i,n}}) - \Phi^{-1}(\frac{R_i}{n+1})\}. \tag{4.2}$$

Then

$$E_0|T_n^* - \tau_n| \leq n^{-1/2} \sum_{i=1}^{n-1} \sum_{k=1}^{n} \sum_{\ell=1}^{n} |E(Y_{k,n})| \, |E(Y_{\ell,n}) - \Phi^{-1}(\frac{\ell}{n+1})| \{n(n-1)\}^{-1}$$

$$+ n^{-1/2} \sum_{i=1}^{n-1} \sum_{k=1}^{n} \sum_{\ell=1}^{n} |\Phi^{-1}(\frac{\ell}{n+1})| \, |E(Y_{k,n}) - \Phi^{-1}(\frac{k}{n+1})| \{n(n-1)\}^{-1}$$

$$= n^{-1} \sum_{k=1}^{n} [|E(Y_{k,n})| + |\Phi^{-1}(\frac{k}{n+1})|] \{n^{-1/2} \sum_{\ell=1}^{n} |E(Y_{\ell,n}) - \Phi^{-1}(\frac{\ell}{n+1})|\}. \tag{4.3}$$

Chernoff and Savage (1958), and Govindarajulu, LeCam and Raghavachari (1967) have shown that

$$n^{-1/2} \sum_{\ell=1}^{n} |E(Y_{\ell,n}) - \Phi^{-1}(\frac{\ell}{n+1})| \to 0 \quad \text{as } n \to \infty. \tag{4.4}$$

Using the Cauchy-Schwarz inequality, we have

$$\sum_{k=1}^{n} n^{-1} |E(Y_{k,n})| \leq \{n^{-1} \Sigma E^2(Y_{k,n})\}^{1/2} \sim 1, \tag{4.5}$$

by Hoeffding's (1953) theorem. Also

$$n^{-1} \sum_{i=1}^{n} |\Phi^{-1}(\frac{k}{n+1})| \sim \int_{-\infty}^{\infty} |x| d\Phi(x) = (2/\pi)^{1/2}. \tag{4.6}$$

Hence from (4.4)-(4.6) we have

$$E_0|T_n^* - \tau_n| \to 0 \text{ as } n \to \infty.$$

Now, using the Markov inequality we assert that

$$P(|T_n^* - \tau_n| > \varepsilon | H_0) \to 0 \text{ as } n \to \infty. \tag{4.7}$$

Thus, under H_0, T_n^* and τ_n and hence T_n and τ_n are asymptotically equivalent. However, Aiyar (1981) has shown that τ_n and the likelihood function under the first order auto-regressive scheme given by (2.4) have a bivariate normal distribution with specified means (see also Hajek and Sidak (1967, p. 208) for the specifications) and hence by LeCam's third lemma τ_n is asymptotically normal under contiguous alternatives. Thus, because of asymptotic equivalence, T_n and T_n^* are also asymptotically normal under contiguous alternatives.

5. ASYMPTOTIC EFFICIENCY AND POWER COMPARISONS

Asymptotic efficiency of a test procedure can be defined as the ratio of Pitman efficacies (see Noether (1955) for the definition of Pitman efficiency). Efficacy is a measure of the performance of a test. One can easily show (see Govindarajulu and Dwass (1979, pp 183-184)) that the Pitman efficacy of S_n given by (2.1) is unity. Also the Pitman efficacy of the heuristic test given by (2.10), as obtained by Govdindarajulu and Dwass (1979, p.p. 185-186) is $4f(0)J$ where

$$J = \int_0^\infty x dF(x).$$

Thus the Pitman efficiency of the heuristic test relative to the parametric test is

$$e_{R,S} = \{4f(0)[\int_0^\infty x dF(x)]\}^2. \tag{5.1}$$

Aiyar (1981) has obtained the Pitman efficiency of K_n given by (2.2) and of τ_n given by (2.9) relative to the parametric test S_n and are given by

$$e_{K,S} = 144[\int_{-\infty}^\infty xF(x)dF(x)]^2[\int_{-\infty}^\infty f^2(x)dx]^2 \tag{5.2}$$

and

$$e_{\tau,S} = [\int_{-\infty}^\infty x\Phi^{-1}(F(x))dF(x)]^2[\int_{-\infty}^\infty \frac{f^2(x)}{\phi[\Phi^{-1}(F(x))]} dx]^2 \tag{5.3}$$

for the first order auto-regressive model. Aiyar (1981) also shows that

$$\min_F e_{K,S} = 9\pi^4/1024 = .8561 \quad \text{and} \quad \min_F e_{\tau,S} = 1$$

and the equality is achieved in the latter if and only if $F = \Phi$. In view of LMP property and asymptotic equivalence under H_0, T and τ should have the same Pitman efficiency.

Gupta and Govindarajulu (1980) tabulate the exact critical values of T_n for small sample sizes and compute simulated power results of S_n, K_n and T_n based on 3000 normal samples at the 0.05 level of significance. The test T_n appears to have higher power than both S_n and K_n for almost all values of ρ for both the first order auto-regressive and the first order moving average models. Considering the achieved level of significance which corresponds to the power at $\rho = 0$, S_n seems to be more powerful than K_n, at least for large values of ρ for both the alternatives.

ACKNOWLEDGEMENT

I thank the referees for their helpful comments.

REFERENCES

AIYAR, R.J. (1981). Asymptotic efficiency of rank tests of randomness against autocorrelation. Ann. Inst. Statist. Math. 33 Part A, 255-262.

AIYAR, R.J., GUILLIER, C.L. and ALBERS, W. (1979). Asymptotic relative efficiencies of rank tests for trend alternatives. J. Amer. Statist. Assoc. 74, 226-231.

ANDERSON, R.L. (1942). Distribution of the serial correlation coefficient. Ann. Math. Statist. 13, 1-33.

CHERNOFF, H. and SAVAGE, I.R. (1958). Asymptotic normlaity and efficiency of certain nonparametric test statistics. Ann. Math. Statist. 29, 972-994.

DAVID, F.N. (1947). A power function for tests of randomness in a sequence of alternatives. Biometrika 34, 335-339.

DUFOUR, J.M. (1981). Rank tests for serial dependence. J. Time Series Analysis 2 No. 3, 117-128.

FOSTER, F.G. and STUART, A. (1954). Distribution-free tests in time series based on the breaking strength of records. J. Roy. Statist. Soc. Ser. B. 16, 1-22.

GOODMAN, L.A. (1958). Simplified runs tests and likelihood ratio tests for Markov chains. Biometrika 45, 181-197.

GOVINDARAJULU, Z. and DWASS, M. (1979). Simultaneous tests for randomness and location of symmetry. Proceedings of the Second Prague Conference on Asymptotic Statistics held in Hradec Karlove (Czechoslavakia) during August 21-25, 1978. Society of Czech. Mathematicians and Physicists, Prague, Charles University, Prague, pp. 181-197.

GOVINDARAJULU, Z., LECAM, L. and RAGHAVACHARI, M. (1967). Generalizations of theorems of Chernoff and Savage on the asymptotic normality of test statistics. Fifth Berkeley Symposium, Math. Statist. and Prob., University of California Press, Berkeley and Loss Angeles 1, 609-638.

GUPTA, G.D. and GOVINDARAJULU, Z. (1980). Nonparametric tests of randomness against autocorrelated normal alternatives. Biometrika 67 2, 375-379.

HAJEK, J. and SIDAK, Z. (1969). Theory of Rank Tests. Academic Press, New York.

HOEFFDING, W. (1953). On the distribution of expected values of the order statistics. Ann. Math. Statist. 24, 93-100.

KNOKE, J.D. (1975). Testing for randomness against autocorrelated alternatives: The parametric case. Biometrika 62, 571-575.

MANN, H.B. (1945). Nonparametric tests against trend. Biometrika 13, 245-259.

NOETHER, G.E. (1955). On a theorem of Pitman. Ann. Math. Statist. 26, 64-68.

SAVAGE, I.R. (1957). Contributions to the theory of rank order statistics: the "trend" case. Ann. Math. Statist. 14, 968-977.

STUART, A. (1954). The asymptotic relative efficiencies of distribution-free tests of randomness against normal alternatives. J. Amer. Statist. Assoc. 49, 147-157.

STUART, A. (1956). The efficiencies of tests of randomness against normal regression. J. Amer. Statist. Assoc. 51, 285-287.

WALD, A. and WOLFOWITZ, J. (1943). An exact test for randomness in nonparametric case based on serial correlation. Ann. Math. Statist. 14, 378-388.

TIME SERIES ANALYSIS: Theory and Practice 4
O.D. Anderson (editor)

PROGRESS IN FORECASTING PRICE CHANGES FROM SPECULATIVE SUPPLY AND DEMAND FUNCTIONS

Purnell H. Benson

Graduate School of Management
Rutgers University
Newark, New Jersey 07102

OVERVIEW

Benson (1981a) reported the identification of separate U-shaped supply and demand functions, which contain time-variant parameters that are themselves functions of prior market conditions, including behavior of prices and volumes transacted on the New York Stock Exchange. These functions may be useful for predicting prices and quantities by solving them for their points of intersection.

Benson (1981b) pursued a solution as follows: the points are found by searching for intersections over short intervals of change in the price variable. The combination of the multiple points for a single estimate of the price change proceeded simplistically as follows. The points of intersection in the just-prior time period are weighted by means of proximity to the observed price change and quantity in the just-prior time period. The proximity weights from the just-prior time period are then used to calculate a weighted average of the points of intersection estimated for the current time period. Since the number of intersections may change from one hourly time period to the next, the larger set is truncated by dropping the more extreme intersections to form a subset equal in size to the smaller set.

With this rationale for estimating price change during the current time period, small but significant correlations between estimated and observed price changes were found, ranging from .08 to .17 for different sets of 700 observations of the NYSE hourly data.

Benson (1982), reported the use of a moving centroid in which the points from one time period to the next are matched by both proximity and congruence with adjustment made in the centroid estimate for biases persisting from one time period to the next.

In the present paper the multiple points of intersection are also evaluated as to their validity by whether the direction of error descent from parameter change puts them in a group with higher correlation between actual and estimated price changes, or in a group with lower or minus correlation between actual and estimated price changes. A composite validity-index is constructed to weight each point in the centroid.

Opportunity is then taken by means of a direct centroid to modify the parameters in the supply and demand functions to minimize price error, rather than quantity error used in stage one of fitting parameters of the supply and demand functions. Then consideration is given in the moving centroid to matching points in successive sets of intersections by the relative slopes of the intersecting lines. This matching is needed to find the points in the current set which correspond to those in the prior set so that the proper centroid can be constructed as the basis for the current estimate of price and quantity.

THE SUPPLY AND DEMAND MODEL

Conceptualization of the speculative demand function is grounded in the bipolarity of stock market traders who buy more on a falling price in pursuit of a bargain but who also buy more on a rising price in anticipation of gain from a further price rise. At the intermediate point of little or no price change, less buying goes on. This is illustrated in Figure 1 by a U-shaped curve with price change on the horizontal axis and quantity bought on the vertical axis. The branches of the U are flattened at their extremities to express the limits to a quantity bought at extreme price changes. In Figure 1, the left hand branch depicts the forward or utilitarian buyers and the right hand the reverse or the speculative buyers. A floor for the curve is provided by those who enter the market to buy at whatever the market price is, regardless of price changes occurring during the time period of their entrance.

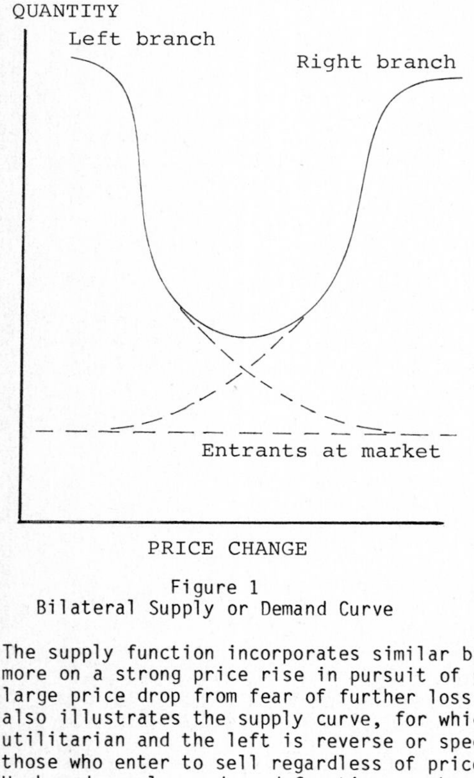

Figure 1
Bilateral Supply or Demand Curve

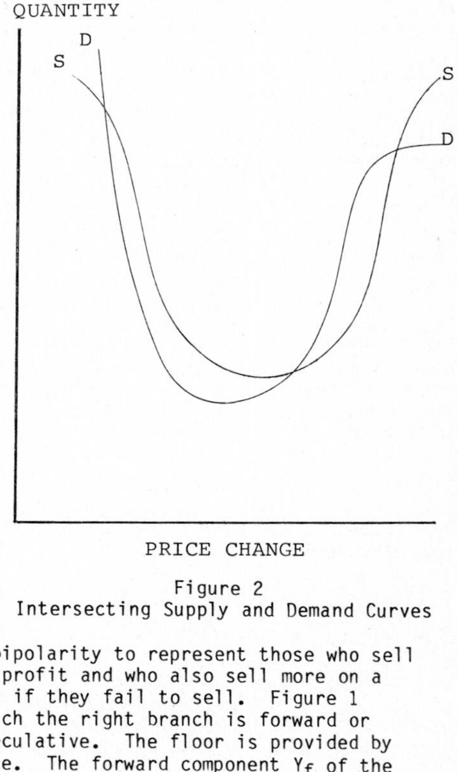

Figure 2
Intersecting Supply and Demand Curves

The supply function incorporates similar bipolarity to represent those who sell more on a strong price rise in pursuit of profit and who also sell more on a large price drop from fear of further loss if they fail to sell. Figure 1 also illustrates the supply curve, for which the right branch is forward or utilitarian and the left is reverse or speculative. The floor is provided by those who enter to sell regardless of price. The forward component Y_f of the U-shaped supply or demand function can be described by

$$Y_f = \{\arctan (G_1 + G_2 P)/\pi + .5\}^{G_7}, \qquad (1)$$

and the reverse component by

$$Y_r = \{\arctan (G_{20} + G_{21} P)/\pi + .5\}^{G_{28}}, \qquad (2)$$

which range from 0.0 to 1.0 and where P is the price change and G_1, G_2, G_{20} and G_{21} are parameters to modify P. In the full development of this function, there are additional G-parameters omitted from the analysis now reported. The subscripting used matches that in the computer programs.

The location of the center of the arctangent function for the forward branch is given by the intercept with a minus sign, that is, by $-G_1$. Parameter G_2 is the general multiplier of P. Adding 19 to the subscript defines the subscripted parameters for the reverse branch.

The arctangent function is introduced as a substitute in exploratory analysis for the normal ogive accumulation because the extremes of the function permit more gradual levelling of the arms of the bilateral supply or demand function. It may eventually appear that the lognormal accumulation can be satisfactorily used and may be preferable to the arctangent approximation to the normal.

In the development of the model, it is assumed that the number of sellers or buyers who enter with the pattern of response to price change provided by the arctangent function varies, depending upon prior market conditions. A multiple regression of prior time-series variables is introduced. The number of sellers or buyers is a functional quantity to be multiplied as a moderator of the arctangent function. For each branch there will be a separate multiple regression multiplier. For the forward and reverse branches for Y_f and Y_r we write:

$$X_f = \sum_{j=1}^{j=n} A_j S_i , \quad X_r = \sum_{j=1}^{j=n} A_{j+n} S_i \tag{3}$$

where there are $\underline{n}$ time series variables S_i and $2\underline{n}$ linear coefficients A_j.

A second multiple regression multiplier is provided by dummy constants for each of the six hours of the day's trading. Even though some standardization is accomplished by adjusting the hourly volumes to their means, there remains an hourly influence on the separate branches of the supply and demand functions. With 6 hourly periods, the non-zero values for the hourly parameters define multipliers H_f and H_r.

$$H_f = G_h, \ h=3...8; \quad H_r = G_h, \ h=22...27 \tag{4}$$

To accomodate irregular changes over time, five harmonic regression parameters are introduced. The sum of these provides a third multiplier of the basic arctangent function for each branch. Opportunity is provided for six sine, six cosine, linear, and squared time terms. If this multiplier for the forward branch is designated by Z_f, and $\underline{t}$ is the time period we have with parameters S_k and 1500 the maximum duration of a one-quarter harmonic cycle,

$$Z_f = \sum_{k=1}^{k=6} [S_k \text{ cosine}(k\ t\ /1500) + S_{k+6} \text{ sine}(k\ t\ /1500)] + S_{13}t + S_{14}t^2 + S_{15} \tag{5}$$

The other components require similar sets of terms.

For a time there remain in the market remnants of buyers or sellers whose price change requirements were not met by the prior time period, (Benson, 1981a, pp. 36-37). First for remnants is the shift to a new origin, given by U_i for the $\underline{i\text{th}}$ remnant, $\underline{i}$ extending to the number of time periods since the remnant entered. $\underline{P_{-1}}$ and P_{-2} are just prior price changes assumed to influence this cumulative shift U_i. The quantity U_i is initially G_1. For the forward branch we write

$$U_i = U_{i-1} + G_{10} + G_{11}P_{-1} + G_{12}P_{-2} \tag{6}$$

Next is the arctangent response of the remnant to the current price change, W_i.

$$W_i = \{\arctan(U_i + G_2\ G_{18}\ P_0)/\pi + .5\}^{G_{19}} \tag{7}$$

Next is introduced an expression to determine whether the quantitative response in the arctangent function for remnants is high enough to bring the buyers or sellers into action. The invasion V_i of unsatisfied remnant W_i involves Z_i, the level of invasion already reached for remnant _i_.

$$V_i = (1.0-Z_i)^{G_{13}} G_{15} \{W_i - Z_i + G_{14} + ((W_i - Z_i + G_{14})^2 + G_{16})^{.5}\} \tag{8}$$

Then Z_i, the highest level reached by invasion of the remnant is defined by the sum of prior invasions $V_{i(t)}$ over the life _i_ of the remnant.

$$Z_i = \sum_{t=1}^{t=i} V_{i(t)} \tag{9}$$

If the proportion remaining after each time period is G_{17}, the ultimate remainder is G_{17} raised to the number of time periods the remnant has been in existence, _i_. We may write for Y_{fm}, the quantity sum of the invasions of forward remnants at the current time,

$$Y_{fm} = \sum_{i=1}^{i=D} V_i (G_{17})^i, \tag{10}$$

where D is the maximum appreciable life of a remnant. Corresponding expressions for remnants in the reverse branch, Y_{rm}, are written with the subscript of G advanced by 19.

The remaining component of transactors consists of those who enter the market ready to buy (or sell) regardless of price. The functions for these involve only hourly, time series, and harmonic parameters.

$$H_e = G_h, \; h=39...44 \tag{11}$$

$$X_e = \sum_{j=1}^{j=n} A_{j+2n} S_{j+2n} \tag{12}$$

The quantities of buyers (or sellers) engaging in transactions within the hourly intervals are given by the cross-products of Y's, or proportionate responses to price changes, the H's, or hourly variations, the X's, or moderators induced by past market behavior, and the Z's as harmonic functions. We then have, letting the subscripts _f_ and _r_ designate forward and reverse, and _fc_, _fm_, _rc_, _rm_, and _e_ designate for supply (or demand) the forward current sellers (or buyers), the forward remnant invasion, the reverse current, the reverse remnant invasion, and the entrants regardless of price,

$$O = (Y_{fc} + Y_{fm})(X_f)(H_f)(Z_f) + (Y_{rc} + Y_{rm})(X_r)(H_r)(Z_r) + (X_e)(H_e)(Z_e) \tag{13}$$

The quantity O is the total number of sellers (or buyers) active during the hourly time period of price change. The same structural function is assumed to prevail for either supply or demand, but with corresponding sets of parameters.

The iteration to minimize quantity error successively goes through the non-linear routine and the three separate regressions. At each of the four stages, the aggregate results from the other three stages are used as multipliers multiplied together.

IMPROVEMENT IN SUPPLY AND DEMAND PARAMETERS TO MINIMIZE PRICE ERROR

In the first stage of effort to determine the time-variant parameters, it is necessary to use quantity as the dependent variable, the error of which is minimized. The U-shaped parameters have only a single quantity corresponding to a price value, but have multiple (and therefore ambiguous) price values for a particular quantity transacted.

The fit achieved for parameters of the supply and demand functions contains least squares biases from assuming that all of the errors of the fit are due to measurement or modelling of the quantity variable. With the complexity of the functions, maximum likelihood analysis to fit the parameters is not a feasible option.

With the approximate fit found by minimizing quantity error, it then becomes possible to use the points of intersection generated by the approximate supply and demand functions as the basis for minimizing price error by parameter improvement. With multiple points of intersection, a direct centroid of these forms a single estimate of the price-change variable, the mean-squared error of which then becomes the minimand for improving upon the parameters already found with quantity as the dependent variable. The analytical approach with finite-difference derivatives already requires the use of trial values before error descent can be pursued. With price error as the minimand, the starting trial values for parameters are those found from minimizing quantity error.

A centroid of points of intersection weighted by proximity of intersections to actual price change was first tried to minimize price error by parameter improvement. Instead of looking at each of 100 or more parameters, multipliers of groups of similar parameters are considered. Derivatives for descent were calculated by introducing tiny finite increments into the supply and demand parameters. This effort at first was unsuccessful. It became apparent that invalid points of intersection were causing large distortions or reversals of the correct changes in parameters for error descent.

Which are valid and which are invalid points of intersection? It occurred to the author that the points could be grouped by the direction of change in the error minimand, plus or minus, occasioned by introducing a positive increment into each parameter. That is, the finite-difference derivative is calculated for the price error at each point of intersection for each observation with respect to the parameter incremented. One group of points is defined by plus derivatives, and the other by minus derivatives. Then the correlation between observed price change and the price change estimated from a centroid of the points (usually a single point) is calculated for each group. The group with the larger positive correlation is taken as consisting predominantly of valid points. For some of the 15 parameters tried the group of invalid points displayed a small negative correlation, while the group of valid points showed a correlation of from .2 to .5.

The direction of the error derivative, whether plus or minus, for the valid group defines the sign to be attached (with reversal since minimization, not maximization is involved) to the validity weight for each point. The validity weight is simply the derivative for that point with the correct sign affixed. The scanning of 15 parameters permits the construction of an aggregate validity index for each of one or more points of intersection in each observed time period.

This validity index is introduced into the centroid calculation by using e raised to the validity index as its exponent with a minus sign attached. A heuristic designation of the exponent weight at -1.0 and the exponent for the distance weight at -1.0 was made. The correlation then for the entire sample between actual and estimated price change, within the current time period is .197. If distance alone provided the weighting in the centroid, this would be in excess of .5, since the correlation calculated is then merely that between the observed price change and whichever of 1 to 4 points is nearest the actual price change. If the points of intersection of which the nearest is selected are random, one would expect such a spurious correlation.

When the correlation between the actual and estimated price changes is calculated from using the points with high validity indexes, a striking result emerges. The top 150 observations (out of 660) from the standpoint of validity scores of points have a correlation between observed and estimated price of .549, while the correlation for the entire sample of 660 is .197. There seems little question that the U-shaped supply and demand functions do generate valid points, and that the detection of these points by a validity procedure independent of the correlation calculated is feasible. However, this still does not resolve the prediction problem since the calculation of the derivative as a validity measure for each point depends upon availability of the observed price change, as well as the estimated price change.

With the select group of price changes estimated from valid points, the search for parameter improvement was renewed. As a result of this search, changes in parameters were disclosed, and the correlation between estimated and actual price change for the top 150 observations rose to .724 from .549. The correlations for cumulative groups down to lower validity scores are given in Table 1. The frequency distribution of validity scores is given in the same table. The values obtained for changes in selected groups of supply and demand parameters are given in Table 2. These are given as adjustments in the relative magnitudes of supply and demand parameters, as well as relative changes in the sizes of the parameters. The unadjusted values for parameters from fitting by minimizing quantity error are listed in Table 3, 4, 5 and 6. In stage two of the parameter calculation, the original 720 observations were reduced to the initial 660 consecutive observations in order to fit practical computer priorities on Rutgers computing machinery. Owing to the need for calculating many supply and demand points at intervals for interpolating points of intersection, the drain on processing time is quite large.

USE OF A MOVING CENTROID OF POINTS OF INTERSECTION

The strategy for estimating the price change taking place during each current time period includes: (1) forecasting the unique supply and demand functions for which the time-variant parameters in the current time period are determined by prior time series values, and (2) using the points of intersection of these functions to estimate the current price change.

The time-variant parameters of the supply and demand functions are found by minimizing quantity error during stage one of the analysis and by improving them during stage two in minimizing price-change errors. The time-variant parameters are entirely a function of variables prior to the current period when the prediction is to be made.

The problem which remains is how to use the current multiple points of intersection created by the criss-crossing of the U-shaped functions (Figure 2). If there were only one point of intersection, this would specify the predicted price change, as well as the quantity transacted.

Owing to the multiplicity of points of intersection, a procedure is needed to select which of the points or combination of points averaged is to be considered the basis for estimating the price change and quantity transacted. The validity index developed for each point is not of direct use in the prediction task since it can be compiled for each point only with the current price change which is not yet available.

However, another insight suggests basis for point selection (Benson, 1982). From time period to time period, the valid point of intersection must be presumed to move continuously in its location from one time period to the next. The valid point of intersection (or composite average of points) is known in the prior time period. What is needed is to track the system of points from the prior to the current time period, and then to select whichever point or points in the current period have continuity with a point or points in the prior period. The details for tracking the system of points have been developed in the paper referred to

(Benson, 1982) and are briefly explained here.

If the number of points from the prior to the current time period remains the same, the procedure for maintaining continuity is straight-forward. A weighted average of points provides the centroid of points for the just-prior time period. The weights in this centroid are used as weights for the corresponding points in calculating the new centroid in the current time period.

The situation is more complicated when the number of points changes from one time period to the next. Then the points require examination to ascertain which ones provide continuity for a fixed number of points from set to set. The extraneous point or points are then eliminated from the set so that the number of points is unchanged from one time period to the next. When the sets of points are the same in number, the procedure already described for calculating the moving centroid is followed.

In comparing points to reduce them to equal sets, useful factors to consider are , (1) proximity of each point in one set to the corresponding point in the next set (2) the congruence of interpoint distances in one set with interpoint distances in the other set, and (3) matching of points of intersection so that the relative slopes of the intersecting supply and demand lines are compatible. If the demand line descends across the supply line, the same direction of intersection should prevail for a corresponding point in the succeeding time period. These rules are incorporated into the matching of points in successive sets for the moving centroid.

RESULTS

The values of parameters obtained from the first stage of minimizing least-squared quantity error and contained in Tables 3, 4, 5 and 6. The stationary points for these were found in stage one by Box's method of evolutionary operation (Box, 1957). Table 2 contains the proportionate changes in parameters or groups of parameters found by minimizing least-squared error in price change during stage two. During this stage error descent is pursued by calculating finite-difference derivatives from introducing tiny increments, of the order of .00001, into the parameters or parameter groups (Murray, 1972, p. 116).

While project work continues, it may now be reported that, with the procedures described, the percentage of right guesses for the direction of price change is 57 percent. This increases to 58 percent with the assist provided by including prior price-change as an estimator combined with the estimate from points of intersection. This corresponds to a tetrachoric coefficient of correlation of .24. Here, the tetrachoric coefficient is more indicative than the product-moment coefficient because of the disturbances from extreme values for observed and estimated price changes.

Starting with a more selected group of 265 prior observations with higher validity points, the level of correlation goes up, and the estimated proportion of right guesses as to direction of price change exceeds 3 out of 5 correct guesses. This shows that predictive efficiency is increased if the forecaster is permitted to select the times when a forecast is to be made.

Many problems and opportunities for improvement remain. The use of the second derivative of price error offers added opportunity for current point selecting. The extension of the lines of analysis to scrutiny of individual stocks is another opportunity to be pursued.

REFERENCES

Benson, P.H., Defining latent supply and demand functions by cumulative arc-tangent distributions with time-variant parameters, in O.D. Anderson and M.R. Perryman, Eds., Time Series Analysis, North-Holland Publishing Company 1981, 33-42. (a)

Benson, P.H., Two-stage fitting of parameters for speculative supply and demand functions, Proceedings of the American Statistical Association, Business and Economics Statistics Section, 1981, 229-234. (b)

Benson, P.H., Using a moving centroid of points to solve supply and demand functions with multiple intersections, American Statistical Association, Business and Economics Statistics Section, Cincinnati, Ohio, August 19, 1982.

Box, G.E.P., Evolutionary operation: A method for increasing industrial productivity. Applied Statistics, 5, 1957, 81-101.

Murray, W., Numerical Methods for Unconstrained Optimization, Academic Press, 1972.

Table 1

DISTRIBUTION OF VALIDITIES AND CUMULATIVE CORRELATIONS BETWEEN OBSERVED AND ESTIMATED PRICE CHANGES FOR POINTS OF INTERSECTION OF SUPPLY AND DEMAND

Class interval for validity index	Class Frequency	Cumulative frequency of this and remaining classes	Correlation for this and remaining classes
Up to -.48	199	654*	.20
-.47 to -.36	88	455	.23
-.35 to -.24	68	367	.27
-.23 to -.12	21	299	.31
-.11 to 0.0	6	278	.31
0.0 to .12	12	272	.31
.13 to .24	24	260	.34
.25 to .36	43	236	.37
.37 to .48	43	193	.43
.49 or more	150	150	.55

*Of 660 observations, there are: 0 solutions, 6; near-approaches, 50; 1 solution, 313; 2 solutions, 252; 3 solutions, 37; 4 solutions, 2.

Table 2

PROPORTIONATE CHANGES IN PARAMETERS FROM MINIMIZING PRICE ERROR

Functional Component or Parameter Group	Supply	Demand
Overall multiplier of function	1.023	.977
Forward current component	.980	1.000
Forward remnant component	.990	1.000
Reverse current component	.980	1.000
Reverse remnant component	.995	.995
Entrants at market component	1.030	1.000
Time series intercept	1.030	1.012
Day before quantity	1.021	1.041
Price change since opening	.980	.940
Hour before quantity	.985	.997
Ratio stock earnings/bond yield	.974	.992
Hourly price change	.975	1.035
Cosine, 1/4 cycle=500	1.000	1.010
Cosine, 1/4 cycle=250	.990	1.000
Sine, 1/4 cycle=500	1.010	1.000
Sine, 1/4 cycle=250	1.000	.990

Table 3

NON-LINEAR PARAMETERS FOR BILATERAL SUPPLY AND DEMAND FUNCTIONS FITTED TO NYSE HOURLY DATA

		Parameter			
		Supply function		Demand function	
Parameter Description*	Symbol	Forward or right branch	Reverse or left branch	Forward or left branch	Reverse or right branch
Price intercept	G_1, G_{20}	-.025	.241	.030	-.193
Price multiplier	G_2, G_{21}	.918	-1.500	-1.761	2.011
Overall multiplier	G_3', G_{22}'	1.040	1.040	.738	1.079
Remnant intercept increment	G_{10}, G_{29}	.274	.342	.015	.110
Remnant prior price multiplier	G_{11}, G_{30}	.086	.180	-.036	.411
Remainder exponent	G_{13}, G_{32}	1.234	1.027	2.338	2.489
Remnant multiplier	G_{15}, G_{34}	.530	.490	.392	.457
Remnant proportion retained	G_{17}, G_{36}	.534	.695	.564	.934
Remnant price moderator	G_{18}, G_{37}	1.238	.935	.727	1.044

*Parameters not listed were not iterated and were set at .0 and 1.0, as required.

Table 4

MARKET TIME-VARIANT PARAMETERS FOR BILATERAL SUPPLY AND DEMAND FUNCTIONS FITTED TO NYSE HOURLY DATA

		Value					
		Supply function			Demand function		
Parameter Description*	Symbol	Forward or right branch	Reverse or left branch	Base	Forward or left branch	Reverse or right branch	Base
Intercept for the branch or base	A_1, A_{13}, A_{25}	9.189	45.978	-23.602	-5.644	2.4892	-1.496
5-day price change	A_2, A_{14}, A_{26}	.118	-.345	.016	-.410	-.071	.285
5-day price turnaround	A_3, A_{15}, A_{27}	-.081	-.021	.101	.459	-.154	-.319
Day-before price change	A_4, A_{16}, A_{28}	-3.011	-5.454	3.348	1.962	.753	-1.684
Day-before quantity	A_5, A_{17}, A_{29}	.340	.711	-.396	.019	.031	-.023
Price change since open	A_6, A_{18}, A_{30}	3.283	.845	-2.470	-2.818	-.243	1.800
Price change hour prev.	A_7, A_{19}, A_{31}	-4.726	-23.437	8.866	.534	1.063	-.922
Hour-before quantity	A_8, A_{20}, A_{32}	-1.285	-3.503	2.360	.078	-.009	.634
5-day bond price change	A_9, A_{21}, A_{33}	-.061	-.090	.068	-.729	-.049	.432
Ratio Stock earnings/bond yield	A_{10}, A_{22}, A_{34}	-15.792	46.822	24.431	-1.215	.846	2.455
Current price as a moderator	A_{11}, A_{23}	4.555	-1.378	**	-5.028	.191	**

*Price changes are calculated as a percentage from the prior price as a Base of 100. Quantity is in units of one million shares.

**Parameter not applicable to entrants at market.

Table 5

HOURLY TIME PERIOD PARAMETERS FOR BILATERAL SUPPLY AND DEMAND FUNCTIONS FITTED TO NYSE HOURLY DATA

		Value					
		Supply function			Demand function		
Parameter Description*	Symbol	Forward or right branch	Reverse or left branch	Base	Forward or left branch	Reverse or right branch	Base
Hourly multipliers:							
10:00-11:00 AM	G_3, G_{22}, G_{39}	.928	.362	.719	.804	1.116	.783
11:00-12:00 Noon	G_4, G_{23}, G_{40}	1.677	.660	1.222	1.173	1.855	1.221
12:00- 1:00 PM	G_5, G_{24}, G_{41}	1.621	.671	1.239	.981	1.749	1.026
1:00- 2:00 PM	G_6, G_{25}, G_{42}	1.448	.655	1.140	1.105	1.497	.997
2:00- 3:00 PM	G_7, G_{26}, G_{43}	1.435	.572	1.007	1.209	.954	1.143
3:00- 4:00 PM	G_8, G_{27}, G_{44}	.856	.360	.673	.838	.873	.830
Intercept for the system	A_0			6.233*			5.552*

*The first four hourly dummy constants are -.349, .298, .465 and .098 for supply and -.596, .217, .204 and .362 for demand. The last two hourly dummy constants are not defined, since the base intercept and the overall intercept for the system use up two determinants. The hourly quantities were divided by hourly means before processing.

Table 6

HARMONIC SERIES PARAMETERS FOR BILATERAL SUPPLY AND DEMAND FUNCTIONS FITTED TO NYSE HOURLY DATA

		Value					
		Supply function			Demand function		
Parameter Description*	Symbol	Forward or right branch	Reverse or left branch	Base	Forward or left branch	Reverse or right branch	Base
Cosine, 1/4 cycle=500	S_3	.327	-.252	.545	1.616	-.425	1.095
Cosine, 1/4 cycle=250	S_6	.214	-.068	.247	.690	.590	.555
Sine, 1/4 cycle=500	S_7	.920	.623	1.276	3.378	.888	2.522
Sine, 1/4 cycle=250	S_{12}	-.139	.191	.235	-.950	.289	-.642
Intercept for series	S_{15}	.288	1.273	-.022	-1.668	.568	-.462

TIME SERIES ANALYSIS: Theory and Practice 4
O.D. Anderson (editor)

AN ALTERNATIVE APPROACH TO THE SPECIFICATION OF MULTIPLE TIME-SERIES MODELS

J. Keith Ord

Division of Management Science
The Pennsylvania State University
University Park, Pennsylvania 16802
USA

The different approaches to modeling structural form in multiple time-series are reviewed and their relative strengths and weaknesses outlined. A different formulation is suggested which makes direct use of the conditional expectation for each dependent variable, given all other variables in the model. Although this method requires some constraints upon the parameters to ensure a valid joint distribution for all the dependent variates, it offers the advantages that structural, rather than reduced, forms may be specified and that the model may be consistently estimated by least squares. The maximum likelihood estimators for the complete system are derived, and it is shown that they may be computed iteratively from the least squares estimates. Their large sample variances are also given. The differences between the conditional and existing econometric structural approaches are then illustrated by means of two examples.

1. INTRODUCTION

The specification of a linear model (whether or not in a time-series setting) has long been recognized as something of an art form. Nevertheless, within the statistical realm of model specification there are two important questions to be addressed:

model identification (in the sense of Box and Jenkins, 1970) which refers to the selection of variables and appropriate lag structures and the choice of a suitable (moving average) formulation for the error process;

model identifiability (or estimability) which refers to the feasibility of estimating the parameters in the formulated model (cf. Fisher, 1966). We will sometimes use the term estimability to avoid semantic confusion, although identifiability is the standard term in the econometric literature.

In a single equation model, the identification stage may proceed by use of some or all of stepwise regression, distributed lag methods (Griliches, 1967), transfer functions (Box and Jenkins, 1970, Chapters 10 and 11) or prior theoretical considerations. Once the model has been identified, estimability typically requires certain conditions on the parameters; see section 2. Other difficulties may arise if we have an errors-in-variables model (cf. Kendall and Stuart, 1979, Chapter 29).

Leaving aside the question of errors-in-variables, we may write the single equation model as

$$E(Y_t | \underset{\sim}{x}_t) = \underset{\sim}{\beta}' \underset{\sim}{x}_t \tag{1}$$

$$Y_t = E(Y_t|\underset{\sim}{x}_t) + u_t, \tag{2}$$

where $\underset{\sim}{u} \sim D(\underset{\sim}{0},\underset{\sim}{\Sigma})$ is the error process with some distribution D and the vector x of independent variables may include both past Y values (termed lagged endogenous) and current or past values of other variables (exogenous).

Although both the time-series analysis and econometric research traditions start from specification (1) - (2), they have tended to diverge with the time-series analyst relying upon stepwise regression and transfer functions, whereas the econometrician has depended more upon distributed lags and prior theory. At the univariate level, this divergence represents primarily a difference in practice, and one would expect a synthesis to emerge to the benefit of both disciplines. Indeed, considerable progress has been made in blending the two approaches, as noted by Granger (1980).

Turning to multiple equation models, the early work of Sargan (1961) on models with autoregressive structure has been amplified and extended by Hendry (1971) and Hendry and Tremayne (1976). Zellner and Palm (1974), Prothero and Wallis (1976), and Wallis (1977) give examples of how transfer function methods may be used to identify individual equations which are then combined into multiple equation econometric models. For a recent discussion, see Zellner (1979).

It is at this point that questions of model specification become critical, as we show in the next section.

The plan of the remainder of this paper is as follows. In section 2, we review the different approaches to the formulation of multiple series models, including both the reduced form schemes usually considered in time-series and the structural forms more common in econometrics. Then, in section 3, we present a different approach which makes use of the conditional expectation of each dependent variable given <u>all</u> other variables in the system, including other current dependent variables. The conditions for these conditional distributions to generate a valid joint distribution are derived.

In section 4, the maximum likelihood estimators for the entire system are developed, together with their large sample properties. It is shown that the ML estimators may be computed in iterative fashion from the initial least squares estimators.

In section 5, the differences between this and other approaches are illustrated by means of two examples. Finally, in section 6, the relative advantages and weaknesses of the conditional approach are summarized.

2. MULTIPLE TIME-SERIES MODELS

A general linear model for m series $\underset{\sim}{Y}_t$ may be written as

$$\sum_{s=0}^{p} A_s B^s \underset{\sim}{Y}_t = \sum_{i=0}^{r} G_i B^i \underset{\sim}{x}_t + \sum_{j=0}^{q} D_j B^j \underset{\sim}{\varepsilon}_t \tag{3}$$

where $\underset{\sim}{Y}_t$ and $\underset{\sim}{\varepsilon}_t$ are (m x 1) vectors

$\underset{\sim}{x}_t$ is a (k x 1) vector of exogenous variables

A_s and D_j are (m x m) matrices

G_i are (m x k) matrices, and

B is the backward shift operator.

Further, it is assumed that the model is not degenerate in any way (A_0^{-1} and D_0^{-1} both exist) and that

$$E(\underset{\sim}{\varepsilon}_t) = \underset{\sim}{0} \quad \text{and} \tag{4}$$

$$E(\underset{\sim}{\varepsilon}_s \underset{\sim}{\varepsilon}_t') = \delta_{st}\Sigma, \tag{5}$$

where $\delta_{st} = 1$ if $s = t$ and $\delta_{st} = 0$ otherwise. Also, we assume that Σ is non-singular.

To avoid redundancy in the parameters, we assume that $D_0 = I$ and either

$$\text{(a)} \quad A_0 = I \quad \text{or} \quad \text{(b)} \quad \Sigma = I. \tag{6}$$

Version (b) may also be expressed equivalently as the requirements that Σ is diagonal and that the elements on the main diagonal of A_0 are unity.

2.1 The reduced form. The choice of conditions in (6) may appear unimportant, but it has a crucial effect upon model specification. The time-series literature has usually taken $A_0 = I$ so that the model may be expressed as

$$\underset{\sim}{Y}_t = C\underset{\sim}{z}_t + \underset{\sim}{u}_t, \tag{7}$$

where $\underset{\sim}{u}_t = D(B)\underset{\sim}{\varepsilon}_t = \Sigma D_s B^s \underset{\sim}{\varepsilon}_t$ and $\underset{\sim}{z}_t$ represents the complete set of lagged y and x and current x values, with corresponding parameter matrix C. This is the reduced form of the model. Hannan (1969) develops conditions under which the representation is uniquely determined. In addition to the absence of common factors and conditions on the roots of the operator polynomials to ensure stationarity of the error process, it is required that

$$\text{rank } [A_p, D_q] = m. \tag{8}$$

We shall assume the conditions hold. It then follows (Hannan, 1975) that the sample cross-correlations are consistent estimators for their population counterparts. Thus, once the model is identified, we know that it is estimable and that the (ordinary) least squares estimators are consistent.

The reduced form model as we have described it is sometimes known as a canonical form for the multiple series model. This and other canonical forms and their properties are discussed in Deistler (1980).

2.2 The structural form. In purely statistical terms, the reduced form provides a complete formulation, but econometricians have preferred to work with the structural form

$$A\underset{\sim}{Y}_t = G\underset{\sim}{x}_t + \underset{\sim}{u}_t^*, \tag{9}$$

where $A = A_0 \neq I$ and condition 6b) is imposed. (Other variants are possible, but we use (9) for conceptual simplicity.) The econometric formulation could be completed by assuming that Σ is diagonal. However, a more general form for Σ is often assumed and parameter redundancy then removed by imposing exclusion and other linear restrictions upon the parameters. The structural form is preferred because it allows several current y variables to be incorporated into each equation.

Since A^{-1} exists, (9) could be written in the form of (7) with

$$C = A^{-1}G \quad \text{and} \quad \underset{\sim}{u}_t = A^{-1}\underset{\sim}{u}_t^*. \tag{10}$$

If there is an isomorphism between the two sets of parameters (for the structural and reduced forms), the parameters in (9) are estimable; this is known as the <u>exactly identified</u> case. If there is at least one distinct function of the reduced form parameters for every structural parameter, the model is <u>over-identified</u> (but remains estimable). On the other hand, if every structural parameter cannot be related to a different function of the reduced form parameters, the model is <u>under-identified</u> (and some parameters are non-identifiable). A full account of the necessary and sufficient conditions for identifiability appears in Fisher (1966).

In addition to the question of identifiability just discussed, a further problem with the structural form is that the (ordinary) least squares estimators are inconsistent, a point first noted by Haavelmo (1943). This has led to a series of other estimators being proposed. See Fisk (1967) and Kendall, Stuart and Ord (1983, Chapter 51) for further details (or any standard econometrics text).

A further implication of this point is that any addition of new dependent variables to the set of structural equations requires that the whole system must be re-estimated. For example, suppose that $\underset{\sim}{y}_t = (\underset{\sim}{y}_{1t}', \underset{\sim}{y}_{2t}')$ and we estimate the parameters of the two structural subsystems, for $\underset{\sim}{Y}_{1t}$ given $(\underset{\sim}{y}_{2t}, \underset{\sim}{x}_t)$ and for $\underset{\sim}{Y}_{2t}$ given $(\underset{\sim}{y}_{1t}, \underset{\sim}{x}_t)$. That is, each subsystem regards the other subset of y values as exogenous. Any of the standard econometric estimators (not least squares) would yield consistent estimators for each subset. However, if the two subsets are now combined to form $\underset{\sim}{y}_t$ as the complete set of dependent variables, the subset estimators become inconsistent in general. This is most easily seen for the two equation model when ordinary least squares is consistent for each single equation but inconsistent for the two equation system. The result extends to the more general case.

An often overlooked implication of this result is a set of single equations estimated by least squares cannot be validly combined into a multiple equation model; the complete set must be re-estimated. No such problem exists for the reduced form. A further problem, noted by Strotz (1960) is that if there is a time lag (however small) in the effect of Y_1 on Y_2, then the appropriate model is the reduced, not the structural, form.

2.3 <u>Interdependent models</u>. Several proposals have been made which try to retain the modeling attractions of the structural form and yet avoid its associated difficulties. If we denote the m current endogenous variables by $Y_{1t}, \ldots, Y_{mt}$, their joint density function may be written, given $\underset{\sim}{x}_t$, as

$$f(y_{1t}, \ldots, y_{mt} | \underset{\sim}{x}_t) = f_1(y_{1t}|\underset{\sim}{x}_t) f_2(y_{2t}|y_{1t}, \underset{\sim}{x}_t) \ldots f_m(y_{mt}|y_{mt}^*, \underset{\sim}{x}_t), \tag{11}$$

where $\underset{\sim}{y}_{jt}^* = y_{it}: \ i \neq j$. For the linear model, (11) implies a recursive scheme such as

$$\left.\begin{aligned} y_{1t} &= \underset{\sim}{\gamma}_1'\underset{\sim}{x}_t + \varepsilon_{1t} \\ a_{21}y_{1t} + y_{2t} &= \underset{\sim}{\gamma}_2'\underset{\sim}{x}_t + \varepsilon_{2t} \\ a_{m1}y_{1t} + a_{m2}y_{2t} + \dots + y_{mt} &= \underset{\sim}{\gamma}_m'\underset{\sim}{x}_t + \varepsilon_{mt}. \end{aligned}\right\} \qquad (12)$$

Note that only y_{it} in the ith equation is a random variable; this follows from (11) The recursive scheme (12) is known as a causal chain (Wold, 1960, 1965, 1981; Lyttkens, 1973). The causal chain scheme admits consistent estimation by least squares and achieves identifiability by assumption. As noted by Strotz and Wold (1960, p. 426), "if the laboratory notion of causality is to be sustained [the underlying dynamic model] will be recursive in character."

Despite these advantages, most econometricians have been reluctant to use the causal chain approach, primarily because of the identification difficulties; many economic aggregates are perceived to be interrelated, and a causal chain may be very difficult to specify.

In response to this difficulty, Wold (1965) formulated a more general reduced form interdependent scheme (REID) as

$$\underset{\sim}{y}_t = A_1\underset{\sim}{\eta}_t + G\underset{\sim}{x}_t + \underset{\sim}{u}_t \qquad (13)$$

where $$\underset{\sim}{\eta}_t = (I - A_1)^{-1}G\underset{\sim}{x}_t = E(\underset{\sim}{y}_t|\underset{\sim}{x}_t) \qquad (14)$$

and $$\underset{\sim}{u}_t = (I - A_1)^{-1}\underset{\sim}{\varepsilon}_t. \qquad (15)$$

Later work (cf. Lyttkens, 1973) uses a generalization of the REID scheme called GEID which has a more flexible correlation structure for the error processes. The advantages of the GEID scheme are that it allows direct modeling of interdependence and avoids identifiability problems. Parameter estimation is performed by the fix-point method; a comprehensive account is given in the volume edited by Wold (1981). The only drawback, if it is to be viewed as such, is the need to estimate the conditional expectations, $\underset{\sim}{\eta}_t$.

2.4 Other approaches. Two other important approaches to modeling multiple time-series deserve mention, although they are not directly relevant to the structural versus reduced form question being addressed here.

The first is the state-space approach of Akaike (1974), which is a different formulation of the reduced form with the advantage that an appropriate model may be identified (selected) by use of an information criterion. Further details and an example are given in Kendall, Stuart, and Ord (1983, sections 51.24-26).

The other method is, of course, that developed by Box and Jenkins (1970) and more recently extended to ARMA models for multiple series by Tiao and Box (1981). This approach also concentrates upon the reduced form and offers much better possibilities for accurate modeling of the error process.

3. THE CONDITIONAL FORMULATION

From the discussion in the previous section, it is apparent that some or all of the following criteria are desirable for multiple time-series models for econometric and similar work:

i) a structural form should be available;
ii) models must be identifiable (estimable);
iii) least squares estimators should be consistent;
and iv) minimal distributional assumptions on the error terms are required.

As noted by one of the referees, there is a sense in which requirement ii) is trivial in that identifiability can always be achieved by imposing enough restrictions. However, requirement ii) should be interpreted in the following way. Suppose that a linear model for Y_{it} is written down in terms of $\underset{\sim}{y}_{it}^*$ and $\underset{\sim}{x}_t$. Assuming no linear dependencies among these variables and a sufficiently large sample size, can all the coefficients be estimated? For the reduced form models (and, as we shall see, for the conditional scheme), the answer is yes. For the structural scheme, we may have to impose further conditions, as in Fisher (1966).

We now examine the question of extending the model to include additional variables, or to combine sub-models for different sectors of the economy. Specifically, suppose we decide to treat as dependent a variable which was previously exogenous. This may be achieved by adding a new equation to the model, which is otherwise unchanged. An appealing requirement is that

v) the creation of a new dependent variable from an existing exogenous variable, which does not change the structure of any previously existing equations, should not render previously consistent estimators of the regression parameters inconsistent.

The standard econometric models achieve i) and iv) by construction and ii) with some difficulty. However, iii) cannot hold (Haavelmo, 1943) and v) typically does not. Given that no model is ever all-embracing, the failure of v) to hold seems important. Although this failure is well-known, at least implicitly, it does not seem to have caused great concern. Since

$$f(\underset{\sim}{y}_t, y_{m+1,t}|\underset{\sim}{x}_t) = f(\underset{\sim}{y}_t|y_{m+1,t},\underset{\sim}{x}_t)f(y_{m+1,t}|\underset{\sim}{x}_t) = f(y_{m+1,t}|\underset{\sim}{y}_t,\underset{\sim}{x}_t)f(\underset{\sim}{y}_t|\underset{\sim}{x}_t),$$

it follows that the reduced form satisfies v). Also, the various reduced form models discussed in section 2 drop requirement i) and thereby achieve ii)-iv). For some purposes, this may seem a modest price to pay, but we cannot take that option since we assume i) is of prime importance. Thus, we are left with the interdependent approach which satisfies i)-iv) but makes use of constructed conditional expectations which must be estimated along the way.

The formulation which we now explore has its roots in the work of Besag (1974) on purely spatial models and involves the specification of conditional expectations directly. For further details of spatial and spatio-temporal models, see Cliff and Ord (1981, Chapter 6).

Referring back to the original single equation model in equations (1) and (2), let us now consider an m-equation system with current endogenous variables $Y_{1t}, \ldots, Y_{mt}$ and

$$\mu_{it} = E(Y_{it}|\underset{\sim}{y}_{it}^*, \underset{\sim}{x}_t) = \underset{\sim}{\gamma}_i' \underset{\sim}{y}_{it}^* + \underset{\sim}{\beta}_i' \underset{\sim}{x}_t, \quad i = 1, \ldots, m, \tag{16}$$

where $\underset{\sim}{y}_{it}^*$ denotes the vector of current y values excluding y_{it} and $\underset{\sim}{x}_t$ denotes the complete set of predetermined variables, a$(q \times 1)$ vector, say. If x_j does not appear in the ith relation, we specify $\beta_{ij} = 0$. Thus, the conditional distribution for Y_{it} is specified with mean μ_{it} and variance, say, ω_i which is constant over time. Equation (16) is best interpreted in the sense "if $\underset{\sim}{Y}^*_{it}$ were to have realized value $\underset{\sim}{y}^*_{it}$, then Y_{it} would have expectation μ_{it}."

Thus, (16) may be regarded as a schedule of possibilities rather than requiring an ordering in time. This is entirely consistent with notions such as supply and demand schedules where, say, quantities are specified for given prices in advance of any transactions taking place. Other contemporaneous relations may be fashioned in the same way.

For a single equation, (16) is the standard regression form. The current proposal is that each of the m equations in the model be specified in this way, so that the scheme is both structural (in this conditional sense) and non-recursive. The error terms for each equation may be autocorrelated, and the full range of single equation techniques may be utilized to formulate the model. However, for the present, we restrict attention to the simplest case where no autocorrelation exists.

Because we are formulating the equations one at a time in a structural fashion, it follows that requirements i) - iii) are satisfied. Also, it follows from our earlier discussion that v) is satisfied. The question that remains is whether such a formulation is internally consistent and what conditions are necessary under iv) to achieve this. We shall refer to an internally consistent model as a valid model, otherwise invalid, to avoid over-working the term consistency. However, before examining this, we consider the causal structure inherent in the model.

3.1 Causality. An important feature of any econometric model is the causal structure implied by the specification. In this regard, we follow Granger (1969) and assume that X "causes" Y if given data D_t up to time t

$$\sigma^2(Y_t|\underset{\sim}{X},D_t) < \sigma^2(Y_t|D_t).$$

Although not essential in principle (cf. Simon, 1957, Chapter 1), it is usually assumed that the $\underset{\sim}{X}$ vector includes only current and past values of that variable. Given this definition, the test developed by Sims (1972) is equally applicable in the present setting. Sims noted that the test could fail when the causality was purely contemporaneous (Y_t depends on X_t only). In the conditional scheme, either X_t would be exogenous by specification when its effect could be examined by testing the β coefficient in (16), or both would be endogenous when they would be treated symmetrically (Y appears in the equation for X and vice versa).

A further flexibility may be imparted to the conditional scheme by the following factorization. Suppose that $\underset{\sim}{Y}_1$ denotes a set of "key" endogenous variables (such as GNP) which "drive" the overall system. Then

$$f(\underset{\sim}{Y}_1,\underset{\sim}{Y}_2|X) = f(\underset{\sim}{Y}_1|\underset{\sim}{X})f(\underset{\sim}{Y}_2|\underset{\sim}{Y}_1,\underset{\sim}{X})$$

so that the initial subsystem for Y_1 may be developed and then the subsystem for $\underset{\sim}{Y}_2$ may be formulated treating the $\underset{\sim}{Y}_1$ variables as predetermined. Such a factorization allows the model to be factored into sub-models which lock together in a valid fashion and is in accord with Simon's (1957, Chapter 1) ideas on ordered causal subsets. This observation is, of course, not new and may be considered almost trivial. Nevertheless, it is important from an operational viewpoint since it allows valid sub-models to be developed and consistently estimated. Indeed, when the parameter sets for the different factors are disjoint, the estimators will be fully efficient.

3.2 The joint distribution. If the conditional distributions are assumed to be normal, it is well known (cf. Besag, 1974) that the joint distribution for $\underset{\sim}{Y}_t$ is

$$\underset{\sim}{Y}_t | \underset{\sim}{x}_t \quad \text{is MVN } (G\underset{\sim}{x}_t, \Sigma). \tag{17}$$

From (17) it is apparent that the joint model is a form of the multivariate regression model (cf. Kendall, Stuart, and Ord, 1983, Chapter 42). However, it will almost invariably be the case that many elements of $\underset{\sim}{\gamma}_i$ and $\underset{\sim}{\beta}_i$ will be set to zero a priori. It is this prior information which sets the model apart from multivariate regression and requires that a different approach be used to obtain efficient estimators.

Suppose we reorder the equations in (17) so that the ith original equation now appears first. Then we may partition G and Σ with respect to the ith variable as

$$\Sigma = \begin{pmatrix} \sigma_{ii} & \underset{\sim}{\sigma}_i' \\ \underset{\sim}{\sigma}_i & \Sigma_i \end{pmatrix} \text{ and } G = \begin{pmatrix} \underset{\sim}{g}_i' \\ G_i \end{pmatrix}.$$

We can relate the parameters in (16) to those in (17) by the expressions

$$\underset{\sim}{\gamma}_i = \Sigma_i^{-1} \underset{\sim}{\sigma}_i \tag{18}$$

$$\text{Var } (Y_{it} | \underset{\sim}{y}_{it}^*, \underset{\sim}{x}_t) = \omega_i = \sigma_{ii} - \underset{\sim}{\sigma}_i' \Sigma_i^{-1} \underset{\sim}{\sigma}_i \tag{19}$$

$$\text{and} \quad \underset{\sim}{\beta}_i = \underset{\sim}{g}_i - G_i' \underset{\sim}{\gamma}_i. \tag{20}$$

Equation (19) may also be written as

$$\text{Var } (Y_{it}) = \sigma_{ii} = \omega_i + \underset{\sim}{\gamma}_i' \underset{\sim}{\sigma}_i. \tag{21}$$

Since Σ is symmetric, the model is valid if and only if

$$\sigma_{ij} = \sigma_{ji}, \quad \text{for all } i \neq j. \tag{22}$$

To express (G,Σ) in terms of the β,γ and ω parameters, we note that

$$\begin{pmatrix} \sigma_{ii} & \underset{\sim}{\sigma}_i' \\ \underset{\sim}{\sigma}_i & \Sigma \end{pmatrix} \begin{pmatrix} -1 \\ \underset{\sim}{\gamma}_i \end{pmatrix} = \begin{pmatrix} -\omega_i \\ \underset{\sim}{0} \end{pmatrix} \tag{23}$$

If the rows of Σ are reshuffled back into order and expressions (23) are combined for all i, we have

$$-\Sigma C = \Omega, \tag{24}$$

where $C = (\underset{\sim}{c}_1, \ldots, \underset{\sim}{c}_m)$ has ith column given by reordering $\begin{pmatrix} -1 \\ \underset{\sim}{\gamma}_i \end{pmatrix}$ and $\Omega = \text{diag}$ $(\omega_1, \ldots, \omega_m)$.

It follows from (24) that

$$\Sigma^{-1} = -C\Omega^{-1} \tag{25}$$

and so the symmetry conditions may be rewritten as the requirement that

$$\gamma_{ij}\omega_j = \gamma_{ji}\omega_i, \quad \text{all } i \neq j, \tag{26}$$

where γ_{ij} is the jth element of $\underset{\sim}{\gamma}_i$.

By similar arguments, it may be shown that

$$B = \begin{pmatrix} \underset{\sim}{\beta}_1' \\ \cdot \\ \cdot \\ \cdot \\ \underset{\sim}{\beta}_m' \end{pmatrix} = -GC. \tag{27}$$

Since Σ is an (m x m) matrix, (26) imposes a total of $\frac{1}{2}$ m(m - 1) constraints. There are two implications from these requirements:

i) the least squares estimators, taken one equation at a time, are consistent but will be inefficient in that they do not exploit conditions (26). Also, this implies that the estimated Σ, $\hat{\Sigma}_0$ say, derived from the single equation estimates, may not be symmetric. However, a revised estimator such as

$$\hat{\Sigma}_1 = \frac{1}{2}(\hat{\Sigma}_0 + \hat{\Sigma}_0')$$

is symmetric. Alternatively, the maximum likelihood estimators may be found which do satisfy (26).

ii) the conditional structural form requires that y_{jt} appears in the equation for Y_{it} whenever y_{it} appears in the relation for Y_{jt}. At first sight, this seems an unusual requirement, but it follows directly from the conditional specification and the symmetry of $C\Omega^{-1}$ in (25).

In order to understand the form of the model, we now consider an example.

<u>Example 3.1</u> Let m = 2, q = 2 so that

$$\begin{pmatrix} Y_1 \\ Y_2 \end{pmatrix} \Bigg| \begin{pmatrix} x_1 \\ x_2 \end{pmatrix} \sim \text{MVN}\left\{ \begin{pmatrix} g_{11} & g_{12} \\ g_{21} & g_{22} \end{pmatrix}\begin{pmatrix} x_1 \\ x_2 \end{pmatrix}, \begin{pmatrix} \sigma_{11} & \sigma_{12} \\ \sigma_{21} & \sigma_{22} \end{pmatrix} \right\}.$$

(The t subscript has been dropped for simplicity.) The conditional expectation (16) for Y_1 is

$$E(Y_1|y_2,x_1,x_2) = \gamma_{12}y_2 + \beta_{11}x_1 + \beta_{12}x_2,$$

where, from (18) - (20),

$$\gamma_{12} = \sigma_{12}/\sigma_{22}, \tag{28}$$

$$\beta_{1j} = (\sigma_{22}g_{1j} - \sigma_{12}g_{2j})/\sigma_{22}, \tag{29}$$

and $\omega_1 = \sigma_{11} - \sigma_{12}^2/\sigma_{22}.$

The conditional specification will be valid if and only if

$$\gamma_{12}\sigma_{22} = \gamma_{21}\sigma_{11} \; (= \sigma_{12}).$$

A check on the two parametrisations reveals that (16) has 8 parameters (but one restriction), where (17) has 7 parameters (unrestricted). Any prior restrictions upon the β_{ij} or γ_{ij} translate, through (28) or (20), into an equal number of restrictions upon G and Σ. Thus, there is no identifiability problem.

3.3 Uniqueness of the normal distribution. Thus far, we have concentrated exclusively upon the multivariate normal distribution. This might be justified by force of habit alone, but there is a more compelling reason, namely the following result:

If each of the linear regressions of a set of m variates on the remaining (m - 1) has constant variance, the m variates are jointly normally distributed unless they are (a) independent or (b) functionally related (see Kendall and Stuart, 1979, sections 28.8 - 10 for a proof). Technically, the condition for multivariate normality depends upon the pattern of non-zero γ_{ij} coefficients, but it seems fair to state that the assumption of normality becomes a practical necessity for the conditional approach. Given the assumptions about linear, additive models made already, this additional assumption is almost inevitable but, arguably, less critical to the modeling process.

4. PARAMETER ESTIMATION

If we start with the conditional expectation in (16), it is apparent that the standard least squares estimators will be consistent. Further, when a lot of parameters are specified a priori, it is likely that the least squares estimators will be more efficient than those derived from a multivariate regression without constraints.

In order to develop the estimation procedure more fully, we shall rewrite (16) as

$$E(Y_{it}|\underset{\sim}{y}^*_{it},\underset{\sim}{x}_t) = \underset{\sim}{\delta}'_i\underset{\sim}{z}_{it}, \tag{30}$$

where

$$\underset{\sim}{\delta}_i = \begin{pmatrix} \underset{\sim}{\gamma}_i \\ \underset{\sim}{\beta}^*_i \end{pmatrix} \quad \text{and } z_{it} = \begin{pmatrix} \underset{\sim}{y}^*_{it} \\ \underset{\sim}{x}_{it} \end{pmatrix}$$

x_{it} and β_i^* refer only to those x contained in the ith expectation. Given observations for t = 1, 2, ..., T time periods, we may write the model in vector form as

$$y_i = Z_i \delta_i + u_i, \tag{31}$$

where the nature of the conditional expectations is to be understood from (30). Clearly, the least squares estimators are

$$\hat{\delta}_{i0} = (Z_i'Z_i)^{-1}Z_i'y_i, \quad i = 1, \ldots, m. \tag{32}$$

A standard assumption in least squares is that

$$E(u_{it}|z_{it}) = 0,$$

and this ensures that the LS estimators are consistent. It is well known that this condition is not satisfied by the structural model, but it is a natural assumption in the present context.

4.1 Maximum likelihood. The log-likelihood function may be written as

$$\ell = \text{const} - \frac{1}{2} T\ell n\omega_i - \frac{1}{2\omega_i}(y_i - Z_i\delta_i)'(y_i - Z_i\delta_i) + \ell(\delta_i^*, \omega_i^*) \tag{33}$$

where

$$\omega_i^* = \{\omega_j, j \neq i\}$$

and

$$\delta_i^* = \{\delta_j : j \neq i \text{ and excluding all } \gamma_{ji} \in \delta_i\}.$$

Expression (33) may be verified from (17) - (20). Thus, the ML estimators for $\hat{\delta}_i$ depend on $\hat{\delta}_i^*$ and ω_i^* only through the symmetry conditions given in (26). Let $\{\lambda_{ij}, i < j\}$ denote a set of Lagrange multipliers and set $\lambda_i' = \{\lambda_{i1}, \ldots, \lambda_{im}\}$, i = 1, ..., m with the conventions that $\lambda_{ii} = 0$ and $\lambda_{ij} = -\lambda_{ji}$. Also, $\lambda_{ij} = 0$ whenever $\gamma_{ij} = 0$ a priori, and we shall write $\mu_i' = (\lambda_i', 0')$ for convenience. The ML estimators are obtained by maximizing the augmented function

$$\ell - \sum_{i<j}\sum \lambda_{ij}\left(\frac{\gamma_{ij}}{\omega_i} - \frac{\gamma_{ij}}{\omega_j}\right)$$

$$= \ell - \sum_{ij}\sum(\lambda_{ij}\gamma_{ij}/\omega_i) \tag{34}$$

The first order conditions for a maximum yield the estimating equations (for i = 1, ..., m)

$$\hat{\delta}_i = (Z_i'Z_i)^{-1}(Z_i'y_i - \hat{\mu}_i) \tag{35}$$

$$\text{and} \quad T\hat{\omega}_i = S_{ii} + 2\hat{\mu}_i'\hat{\delta}_i, \tag{36}$$

where $S_{ii} = (y_i - Z_i\hat{\delta}_i)'(y_i - Z_i\hat{\delta}_i)$, and we also impose the conditions

$$\hat{\gamma}_{ij}\hat{\omega}_j = \hat{\gamma}_{ji}\hat{\omega}_i, \quad \text{all } i \neq j.$$

In order to solve these equations, we rewrite (35) and (37) as

$$\begin{pmatrix} Z_1'Z_1 & 0 & \cdots & 0 & W_1 \\ 0 & Z_2'Z_2 & \cdots & 0 & W_2 \\ \vdots & & & & \\ 0 & 0 & \cdots & Z_m'Z_m & W_m \\ U_1 & U_2 & \cdots & U_m & 0 \end{pmatrix} \begin{pmatrix} \underset{\sim}{\delta}_1 \\ \underset{\sim}{\delta}_2 \\ \vdots \\ \underset{\sim}{\delta}_m \\ \underset{\sim}{\Lambda} \end{pmatrix} = \begin{pmatrix} Z_1'\underset{\sim}{y}_1 \\ Z_2'\underset{\sim}{y}_2 \\ \vdots \\ Z_m'\underset{\sim}{y}_m \\ \underset{\sim}{0} \end{pmatrix}.$$

where $\underset{\sim}{\Lambda}$ is the vector of all distinct Lagrange multipliers. We assume that the ith equation contains m_i endogenous variables other than Y_i, so that $\underset{\sim}{\Lambda}$ consists of $M = \frac{1}{2}\Sigma m_i$ elements in all. If the ith equation contains q_i exogenous/lagged endogenous variables, the matrices T_i and U_i' will be order $(m_i + q_i) \times M$; their elements are chosen so that $W_i\underset{\sim}{\Lambda} = \underset{\sim}{\mu}_i$ and $\sum_i U_i\underset{\sim}{\delta}_i = \underset{\sim}{0}$ corresponding to the constraints in (37). The form of these may be seen from the following example.

Example 4.1 Let $m = 3$ and assume that $\gamma_{12} \neq 0$, $\gamma_{13} \neq 0$, $\gamma_{23} = 0$, so that $M = 2$. The set of estimating equations is given by (38) with

$$\underset{\sim}{\delta}_1 = \begin{pmatrix} \gamma_{12} \\ \gamma_{13} \\ \underset{\sim}{\beta}_1 \end{pmatrix}, \quad \underset{\sim}{\delta}_2 = \begin{pmatrix} \gamma_{21} \\ \underset{\sim}{\beta}_2 \end{pmatrix}, \quad \underset{\sim}{\delta}_3 = \begin{pmatrix} \gamma_{31} \\ \underset{\sim}{\beta}_3 \end{pmatrix} \quad \text{and } \underset{\sim}{\Lambda} = \begin{pmatrix} \lambda_{12} \\ \lambda_{13} \end{pmatrix}.$$

The W_i matrices are

$$W_1 = \begin{pmatrix} 1 & 0 \\ 0 & 1 \\ \underset{\sim}{0} & \underset{\sim}{0} \end{pmatrix}, \quad W_2 = \begin{pmatrix} -1 & 0 \\ \underset{\sim}{0} & \underset{\sim}{0} \end{pmatrix}, \quad W_3 = \begin{pmatrix} 0 & -1 \\ \underset{\sim}{0} & \underset{\sim}{0} \end{pmatrix},$$

where the U_i matrices are

$$U_1 = \begin{pmatrix} \omega_2 & 0 & \underset{\sim}{0}' \\ 0 & \omega_3 & \underset{\sim}{0}' \end{pmatrix}, \quad U_2 = \begin{pmatrix} -\omega_1 & \underset{\sim}{0}' \\ 0 & \underset{\sim}{0}' \end{pmatrix}, \quad U_3 = \begin{pmatrix} 0 & \underset{\sim}{0}' \\ -\omega_1 & \underset{\sim}{0}' \end{pmatrix}$$

corresponding to the two conditions on γ_{12} and γ_{13}.

This appears rather cumbersome, but the special structure of (38) allows explicit solution. Let the usual least squares estimators be $\hat{\underset{\sim}{\delta}}_{i0} = (Z_i'Z_i)^{-1}Z_i'\underset{\sim}{y}_i$ and let $C_i = (Z_i'Z_i)^{-1}W_i$. If we multiply (38) by

$$\begin{pmatrix} I & & & 0 \\ & \ddots & & \vdots \\ & & I & 0 \\ -U_1 & \cdots & -U_m & I \end{pmatrix} \begin{pmatrix} (Z_1'Z_1)^{-1} & & & \\ & \ddots & & \\ & & (Z_m'Z_m)^{-1} & \\ & & & I \end{pmatrix},$$

we obtain

$$\begin{pmatrix} I & & & C_1 \\ & \ddots & & \vdots \\ & & I & C_m \\ 0 & \cdots & 0 & -\Sigma U_i C_i \end{pmatrix} \begin{pmatrix} \hat{\delta}_1 \\ \vdots \\ \hat{\delta}_m \\ \hat{\Lambda} \end{pmatrix} = \begin{pmatrix} \hat{\delta}_{10} \\ \vdots \\ \hat{\delta}_{m0} \\ -\Sigma U_i \hat{\delta}_{i0} \end{pmatrix}. \tag{39}$$

From (39) we arrive at

$$\hat{\Lambda} = (\Sigma U_i C_i)^{-1} (\Sigma U_i \hat{\delta}_{i0}) \tag{40}$$

$$\text{and} \quad \hat{\delta}_i = \hat{\delta}_{i0} - C_i \hat{\Lambda}, \quad i = 1, \ldots, m. \tag{41}$$

If the ω_i were known (40) and (41) would represent the final solution. When the ω_i are unknown, we may solve, iteratively,as follows:

i) estimate the ω_i by least squares;

ii) use these estimates to solve (40) and (41);

iii) use (40) to compute

$$\hat{\mu}_i = W_i \hat{\Lambda} \text{ and hence } \hat{\omega}_i \text{ from (36);}$$

iv) repeat steps ii) and iii) until the solution converges.

Although there is no guarantee of convergence, the experience in similar problems is encouraging. Further, it is conjectured that a single run through i) - iii) may often suffice.

4.2 Large sample variances. In order to obtain the large sample covariance matrix, we compute the matrix of second derivatives and invert the negative of its large sample expectation in the usual way. Although we are dealing with a parameter set augmented by the Lagrange multipliers, the fact that the constraints serve to pool distinct estimates ensures that the relevant components of the inverse matrix can be interpreted in the usual way (Silvey, 1970, section 4.7). Given the vector of parameters $[\Lambda \;\; \delta_1 \cdots \delta_m \;\; \omega_1 \cdots \omega_m]$, the large sample covariance matrix turns out to be the inverse of the matrix:

$$\begin{pmatrix} 0 & D_1 & \cdots & D_m & \underset{\sim}{f}_1 & \cdots & \underset{\sim}{f}_m \\ D_1' & \Sigma_1^{-1} & & & & & \\ \vdots & & \ddots & & & & \\ D_m' & & & \Sigma_m^{-1} & & & \\ \underset{\sim}{f}_1' & & & & v_1^{-1} & & \\ \vdots & & & & & \ddots & \\ \underset{\sim}{f}_m' & & & & & & v_m^{-1} \end{pmatrix} \tag{42}$$

where $\Sigma_i = \omega_i[E(Z_i'Z_i)]^{-1}$, the unconditional covariance matrix for the LS estimators

$v_i = 2\omega_i^2 T$, the variance of $\hat{\omega}_{i0}$

D_i has elements that are zero unless $D_{(i)r,j}$ refers to the rth element of $\underset{\sim}{\Lambda}$ which relates to a constraint on γ_{ij} when

$$\begin{aligned} D_{(i)r,j} &= -\omega_i \quad \text{if } i < j \\ &= +\omega_i \quad \text{if } i > j \end{aligned}$$

f_i has elements that are zero unless f_{ir} refers to the rth element of $\underset{\sim}{\Lambda}$ which relates to a constraint on γ_{ij} when

$$\begin{aligned} f_{ir} &= \gamma_{ij}/\omega_i^2 \quad \text{if } i < j \\ &= -\gamma_{ij}/\omega_i^2 \quad \text{if } i > j. \end{aligned}$$

Matrix (42) is of a similar form to (38) and inversion yields

$$\text{cov}\,(\hat{\underset{\sim}{\delta}}_i) = \Sigma_i - \Sigma_i D_i' H D_i \Sigma_i \tag{43}$$

$$\text{var}\,(\hat{\omega}_i) = v_i - v_i^2 \underset{\sim}{f}_i' H \underset{\sim}{f}_i \tag{44}$$

with cross-covariance matrices

$$\text{cov}\,(\underset{\sim}{\delta}_i, \underset{\sim}{\delta}_j) = -\Sigma_i D_i' H D_j \Sigma_j \tag{45}$$

$$\text{cov}\,(\underset{\sim}{\delta}_i, \omega_j) = -\Sigma_i D_i' H \underset{\sim}{f}_j \omega_j \tag{46}$$

$$\text{and} \quad \text{cov}\,(\omega_i, \omega_j) = -v_i v_j \underset{\sim}{f}_i' H \underset{\sim}{f}_j, \tag{47}$$

where

$$H = [\Sigma(D_j \Sigma_j D_j' + v_j \underset{\sim}{f}_j \underset{\sim}{f}_j')]^{-1}. \tag{48}$$

The form of (43) - (47) indicates that the ML estimators will generally be more efficient, as we would expect.

The advantage of the LS estimators is that they are not affected by misspecifications in other equations in the system. This makes preliminary model identification by least squares a natural procedure prior to estimating the final model by maximum likelihood.

4.3 Significance of estimates. The statement that $\gamma_{12} \neq 0$ implies $\gamma_{21} \neq 0$ sometimes evokes the criticism that one coefficient may be significant whereas the other is not. This may indeed happen as the following example shows.

Example 4.3 Consider the model

$$E(Y_1|Y_2,X) = \gamma_{12}Y_2 + \beta_1 X$$

$$E(Y_2|Y_1,X) = \gamma_{21}Y_1 + \beta_2 X,$$

where the error variances are ω_1 and ω_2. Writing s_{12}, S_{1x} for the sample covariances of (Y_1,Y_2) and of (Y_1,X) and so on, the ratios $t_{ij} = \hat{\gamma}_{ij}/SE(\hat{\gamma}_{ij})$ yield

$$(t_{12}/t_{21})^2 = \left(\frac{\hat{\gamma}_{12}^2 s_{22}}{\hat{\gamma}_{21}^2 s_{11}}\right)\left(\frac{s_{xx}s_{22} - s_{2x}^2}{s_{xx}s_{11} - s_{1x}^2}\right).$$

The first term converges to 1 as n increases, but the second term does not. Thus, two significance tests on the least squares estimators may well yield different results.

To recognize that this does not upset the conditional model, we need to recognize that the inclusion of both γ_{12} and γ_{21} (or neither) is a question of specification, not significance. Including (or excluding) both yields a valid model for which, in turn, ML will give as good or better estimates. It may still be the case that the effect of Y_2 on Y_1, or vice-versa, is slight even when it is stronger in the other direction.

5. SOME EXAMPLES

To illustrate some of the differences between the conditional and existing structural formulations, we present two examples, relating to a simple macroeconomic model and to a pair of supply and demand equations.

5.1 A simple macroeconomic model. Consider a simple closed economy with no government sector which could be represented by the structural model

$$C_t = a_1 Y_t + g_{11} C_{t-1} + \varepsilon_{1t}$$

$$I_t = a_2 Y_t + g_{22} K_{t-1} + \varepsilon_{2t}$$

$$Y_t \equiv C_t + I_t,$$

where I_t = investment, C_t = national income, and K_t = stock of capital. In this form, the zero values of coefficients g_{12} and g_{21} are necessary for identifiability. After elimination of Y_t through the identity, we obtain the structural form

$$\left.\begin{aligned} C_t &= a_1^* I_t + g_{11}^* C_{t-1} + \qquad\qquad + \varepsilon_{1t}^* \\ I_t &= a_2^* C_t + \qquad\qquad + g_{22}^* K_{t-1} + \varepsilon_{1t}^* \end{aligned}\right\}, \tag{49}$$

where $a_1^* = a_1/(1 - a_1)$, $g_{11}^* = g_{11}/(1 - a_1)$, and so on.

The reduced form corresponding to (49) is

$$\left.\begin{aligned} C_t &= h_{11} C_{t-1} + h_{12} K_{t-1} + \varepsilon_{1t}^{**} \\ I_t &= h_{21} C_{t-1} + h_{22} K_{t-1} + \varepsilon_{2t}^{**} \end{aligned}\right\}, \tag{50}$$

where

$$h_{11} = g_{11}^*/(1 - a_1^* a_2^*) = g_{11}(1 - a_2)/(1 - a_1 - a_2),$$

$$h_{12} = g_{22}^*/(1 - a_1^* a_2^*) = g_{22}(1 - a_1)/(1 - a_1 - a_2),$$

and so on. If the ε_{it} are independent with variance σ_i^2 $(i = 1, 2)$, it follows that $\mathrm{var}(\varepsilon_{1t}^{**}) = ((1 - a_2)^2\sigma_1^2 + a_1^2\sigma_2^2)/(1 - a_1 - a_2)^2$ and $\mathrm{cov}(\varepsilon_{1t}^{**}, \varepsilon_{2t}^{**})$ $= \{a_1(1 - a_1)\sigma_2^2 + a_2(1 - a_2)\sigma_1^2\}/(1 - a_1 - a_2)^2$. The conditional approach starts from the conditional expectations

$$\left.\begin{aligned} E(C_t | I_t) &= \gamma_{12} I_t + \beta_{11} C_{t-1} + \beta_{12} K_{t-1} \\ E(C_t | I_t) &= \gamma_{21} C_t + \beta_{21} C_{t-1} + \beta_{22} K_{t-1} \end{aligned}\right\}, \tag{51}$$

which, by Example 3.1, yields a reduced form like (50) but with

$$h_{11} = (\beta_{11} + \beta_{12}\gamma_{21})/(1 - \gamma_{12}\gamma_{21})$$

$$h_{12} = (\beta_{12} + \beta_{11}\gamma_{12})/(1 - \gamma_{12}\gamma_{21})$$

and covariance matrix

$$(1 - \gamma_{12}\gamma_{21})^{-1} \begin{bmatrix} \omega_1 & \omega_1\gamma_{12} \\ \omega_2\gamma_{21} & \omega_2 \end{bmatrix}. \tag{52}$$

In each case, there are seven parameters (after allowing for constraints). The theme of this paper is that the specification of (51) involves a more natural "if, then" approach than does the structural approach which requires both constraints for identifiability and a simultaneous dependence structure which may be difficult to explain. Further, the a priori specification of parameters in (51) is not restricted by identifiability considerations.

5.2 Supply and demand. Let (Q_t^S, Q_t^D), (P_t^S, P_t^D) denote, respectively, the current supply and demand quantities and prices. The demand schedule is obtained by fixing prices and asking for the quantity demanded given that price--a natural lead-in to a conditional model such as

$$E(Q_t^D|P_t^D) = \gamma_{12}P_t^D + \beta_{11}Y_t + \beta_{12}Z_t. \tag{53}$$

The suppliers may fix prices in light of the potential quantities by

$$E(P_t^S|Q_t^S) = \gamma_{21}Q_t^S + \beta_{21}Y_t + \beta_{22}Z_t. \tag{54}$$

If we now impose the market equilibrium conditions that, a posteriori,

$$P_t^D = P_t^S = P_t \text{ and } Q_t^D = Q_t^S = Q_t, \tag{55}$$

(53) and (54) give a conditional specification for the two endogenous variables P_t and Q_t. This model is of the same form as (51); no conditions are necessary for identifiability.

If the suppliers' relationship is of the form

$$E(Q_t^S|P_t^S) = \gamma_{21}P_t^S + \beta_{21}Y_t + \beta_{22}Z_t, \tag{56}$$

the imposition of (55) leads to a reduced form model where conditions such as $\beta_{12} = \beta_{21} = 0$ are required if the underlying schedules (53) and (56) are to be identified. Thus, the conditional framework is able to give an appropriate stochastic interpretation whether consumers and/or suppliers are price fixers or price takers.

6. CONCLUSION

After a review of existing approaches to the formulation of multiple time-series models, a different method is suggested which uses the conditional distribution for each dependent variable, given all other variables in the model. Under appropriate conditions, these conditional distributions may be combined to form the joint distribution.

The advantages of the "conditional approach" are as follows:

(i) the model may be developed on a one equation at a time basis so that the wealth of techniques available for single equation models can be exploited, yet these single equations are logically consistent with the complete model;

(ii) the single equation least squares estimators are consistent for the whole system;

(iii) there are no problems of parameter identifiability.

The only restriction which must be imposed to gain these advantages is that we must assume multivariate normality for the errors in the linear model. This is entirely in accord with custom, even if structural models do not formally require such an assumption.

In later sections of the paper, the maximum likelihood estimators are derived and so is their large sample covariance matrix. The ML estimators are derivable from the least squares estimators and do not appear to hold any numerical difficulties, although this remains to be explored more fully.

At the present stage, the conditional scheme operates with the simplest possible error structure, and further research on this and many other aspects of the approach is necessary.

ACKNOWLEDGEMENT

I would like to thank both referees for their helpful comments.

REFERENCES

AKAIKE, H. (1974). Markovian representation of stochastic processes and its application to the analysis of autoregressive-moving average series. Ann. Inst. Statistical Math. 26, 363-387.

BESAG, J. E. (1974). Spatial interaction and the statistical analysis of lattice systems. J. Royal Statistical Soc. B36, 192-256.

BOX, G. E. P. and JENKINS, G. M. (1970). Time Series Analysis, Forecasting and Control. Holden-day, San Francisco. (Revised 1976)

CLIFF, A. D. and ORD, J. K. (1981). Spatial Processes: Models and Applications. Pion, London.

DEISTLER, M. (1981). Parameterization and consistent estimation of ARMA systems. In Time Series, O. D. Anderson (ed.), North-Holland, Amsterdam and New York, 373-385.

FISHER, F. M. (1977). The Identification Problem. McGraw-Hill, New York.

FISK, P. R. (1967). Stochastically Dependent Equations. Griffin, London.

GRANGER, C. W. J. (1969). Investigating causal relations by econometric models and cross-spectral methods. Econometrica 37, 424-438.

GRANGER, C. W. J. (1980). On the systhesis of time-series and econometric models. In Directions in Time Series, D. R. Brillinger and G. C. Tiao (eds.), Inst. of Math. Statistics, 149-167.

GRILICHES, Z. (1967). Distributed lags: a survey. Econometrica 35, 16-49.

HAAVELMO, T. (1943). The statistical implications of a system of simultaneous equations. Econometrica 11, 1-12.

HANNAN, E. J. (1969). The identification of vector mixed autoregressive-moving average systems. Biometrika 56, 223-225.

HANNAN, E. J. (1975). The estimation of autoregressive-moving average models. Ann. Statistics 3, 975-981.

HENDRY, D. F. (1971). Maximum likelihood estimation of systems of simultaneous regression equations with errors generated by a vector autoregressive process. Int. Economic Review 22, 257-272.

HENDRY, D. F. (1974). Stochastic specification in an aggregate demand model of the United Kingdom. Econometrica 42, 559-578.

HENDRY, D. F. and TREMAYNE, A. R. (1976). Estimating systems of dynamic reduced form equations with vector autoregressive errors. Int. Economic Review 17, 463-471.

KENDALL, M. G. and STUART, A. (1979). The Advanced Theory of Statistics, Vol. 2. Griffin, London. 4th edition.

KENDALL, M. G., STUART, A. and ORD, J. K. (1983). The Advanced Theory of Statistics, Vol. 3. Griffin, London. 4th edition.

LYTTKENS, E. (1973). The fix-point method for estimating interdependent systems with the underlying model specification. J. Royal Statistical Soc. A 136, 353-394.

PROTHERO, D. L. and WALLIS, K. F. (1976). Modelling macroeconomic time series. J. Royal Statistical Soc. A 139, 469-500.

SARGAN, J. D. (1961). The maximum likelihood estimation of economic relationships with autoregressive residuals. Econometrica 29, 414-426.

SILVEY, S. D. (1970). Statistical Inference. Chapman and Hall, London.

SIMON, H. A. (1957). Models of Man. Wiley, New York.

SIMS, C. A. (1972). Money, income and causality. Amer. Economic Review 62, 540-552.

STROTZ, R. H. and WOLD, H. O. A. (1960). Recursive versus non-recursive systems: an attempt at synthesis. Econometrica 28, 417-427.

TIAO, G. C. and BOX, G. E. P. (1981). Modeling multiple time series with applications. J. Amer. Statist. Ass. 76, 802-816.

WALLIS, K. F. (1977). Multiple time series analysis and the final form of econometric models. Econometrica 45, 1481-1497.

WOLD, H. O. A. (1960). A generalization of causal chain models. Econometrica 28, 443-463.

WOLD, H. O. A. (1965). A fix-point theorem with econometric background I, II. Arkiv f. Matematik 6, 209-240.

WOLD, H. O. A. (1981). The Fix-Point Approach to Interdependent Systems. North-Holland, Amsterdam and New York.

ZELLNER, A. (1979). Statistical analysis of econometric models (with discussion). J. Amer. Statist. Ass. 74, 628-651.

ZELLNER, A. and PALM, F. (1974). Time-series analysis and simultaneous econometric models. J. Econometrics 2, 17-54.

TIME SERIES ANALYSIS: Theory and Practice 4
O.D. Anderson (editor)

DETERRENCE: A RATIONAL EXPECTATIONS FORMULATION†

Llad Phillips and Subhash Ray

Department of Economics
University of California,
Santa Barbara
USA

There is increasing political pressure in the United States to pass punitive "per se" legislation, making it a crime to drive if alcohol in the blood is above a given level. However, a review of the experience in Great Britain suggests skepticism about the wisdom of such legislation. The British Road Safety Act of 1967 did have an immediate impact in reducing the number of drivers killed upon the highways and in reducing blood alcohol levels in driver fatalities. However, in a few years blood alcohol levels again rose and exceeded their previous highs. This paper attempts to explain this pattern of behavior by constructing and estimating a rational expectations model of deterrence.

INTRODUCTION

There is increasing political pressure in the United States to pass punitive drunk driving laws. California introduced legislation effective January 1, 1982 making it a crime to drive with a blood alcohol content of 100 micrograms per milliliter or more. Per se legislation makes drunken driving a crime on the basis of potential injury to others rather than actual injury. To justify such a policy, at least it should be effective. Perhaps some lessons can be learned from the British Road Safety Act of 1967.

Using interrupted time series methods, Ross detected a 24.7 per cent decline in fatalities, a 14.6 per cent decline in serious injuries and a 9.8 per cent decline in slight injuries following the introduction of the Act in October 1967. Using Box-Jenkins time series techniques as well as regression, Phillips, Votey, and Ray constructed a forecasting model for seriously injured which supported Ross' findings about the impact of the law on seriously injured. However, they also found that arrests for drunken driving were not significantly correlated with seriously injured, but were correlated with road fatalities. This raised a question of why the per se legislation was effective in reducing seriously injured and how enforcement affected road fatalities.

In this paper, some dynamic extensions of the standard economic model of rational agency behavior and the supply of offenses are explored. The empirical findings suggest that arrests for drunken driving have a transitory impact on road fatalities. This result seems more consistent with arrests having a temporary incapacitation effect rather than being a deterrent. Whether the passage of per se legislation, as distinct from its enforcement, has a permanent effect in reducing road casualties remains an open question.

---

† This research was supported under grants #79-NI-AX-0069 and #81-IJ-CX-0019 from the National Institute of Law Enforcement and Criminal Justice. Points of view are those of the authors and do not necessarily reflect the position of the U.S. Department of Justice.

Theory

Rational Agency behavior is presumed to follow the objective criterion of minimizing the sum of damages to the victims of crime plus the costs of control achieved by enforcement, adjudication and corrections. This is the objective criterion suggested by Stigler [1970]. Minimize

$$\sum_{i=1}^{m} r_i OF_i + C \qquad (1)$$

where r_i is the loss or damage rate per offense for crime i, OF_i is the number of offenses for crime i, and C is the cost of control. The offense rate and cost could be expressed in per capita terms.

The cost function for the criminal justice system can be expressed as a function of a vector of system outputs, $\vec{q}$, and a vector of system input prices $\vec{w}$,

$$C = C(\vec{q},\vec{w}) \qquad (2)$$

where the vector of system outputs can include for various crimes a vector of arrests $\vec{A}$, felony charges, $\vec{F}$, convictions $\vec{V}$, imprisonments $\vec{I}$, time served $\vec{TS}$, and other control measures such as probation $\vec{R}$,

$$C = C(\vec{A},\vec{F},\vec{V},\vec{I},\vec{TS},\vec{R},\vec{w}) \ . \qquad (3)$$

The offense rate is determined by the sum of behaviors by individuals who allocate their time between leisure and legal and illegal activities in response to incentives and disincentives. The supply of offenses, as developed by Becker and Ehrlich, will vary with a vector of probabilities $\vec{P}$, such as apprehension, charge given apprehension, conviction given charge, imprisonment given conviction, a vector of sanctions $\vec{S}$, such as time served TS or probation R, and a vector of socioeconomic and demographic variables $\vec{X}$,

$$OF_i = f_i(\vec{P}_i,\vec{S}_i,\vec{X}_i) \ . \qquad (4)$$

The probability vector $\vec{P}$, can be expressed in terms of the offense rate and the system outputs, for example the probability of arrest, P_A, as the ratio of arrests to offenses,

$$P_A^{(i)} = \frac{A(i)}{OF(i)} \quad i=1,m \ . \qquad (5)$$

or the probability of imprisonment given conviction

$$P_{I/V}^{(i)} = \frac{I(i)}{V(i)} \quad i=1,m \ . \qquad (6)$$

We presume that the supply of offenses can be expressed in terms of system outputs by combining equations such as (4) with equations such as (5) and (6),

$$OF_i = g_i(A_i,F_i,V_i,R_i,I_i,TS_i,X_1,X_2,\ldots X_n), \ i=1,m \qquad (7)$$

where, if a crime is controllable, the derivative with respect to a system output, for example arrests, is negative,

$$\frac{\partial g_i}{\partial A_i} < 0 \; . \qquad (8)$$

If the agencies of the criminal justice system act to minimize damages and control costs, Eq. (1), subject to the criminal justice system technology, Eq. (3), and the supply of offenses, Eq. (4), the optimal level of arrests will be determined by the first order condition

$$r_i \frac{\partial g_i}{\partial A_i} + \frac{\partial C}{\partial A_i} = 0 \qquad i=1,m \qquad (9)$$

where $\frac{\partial C}{\partial A_i}$ is the increment in criminal justice system costs of an additional arrest, or equivalently the supply curve of arrests, $(-r_i \frac{\partial g_i}{\partial A_i})$ is the decrement in damages to victims of an additional arrest, or equivalently the demand curve for arrests. The demand and supply curve for arrests is illustrated in Fig. 1. The demand curve will be downward sloping if the partial of the demand curve $(-r \frac{\partial g_i}{\partial A_i})$ with respect to arrests, A_i , is negative, which will be the case if

$$\frac{\partial^2 g_i}{\partial A_i^2} > 0 \; . \qquad (10)$$

As socioeconomic and demographic factors, $\vec{X}$, change to increase crime, this will shift the demand curve for arrests up and to the right if

$$\frac{\partial^2 g_i}{\partial X \partial A_i} < 0 \qquad (11)$$

which will certainly be the case if the supply of offenses is separable in system outputs and in exogenous factors X ,

$$g(\vec{q},\vec{X}) = h(\vec{q})k(\vec{X}) \qquad (12)$$

Under these conditions, we will observe the arrest rate, A , increasing as the offense rate, OF , increases.

Criminal justice budgets and outputs are determined through a political process rather than market exchange, and hence the price or marginal cost of an arrest is not observed, limiting the value of the demand and supply perspective of Fig. 1.

Another perspective on the first order condition, Eq. 9, is illustrated in Fig. 2, where the optimal offense rate, OF , and arrest rate, A , are determined by the tangency of the supply of offenses (Eq. 7), plotted with the offense rate on the abscissa, and the total social cost curve (damages plus control), with slope equal to the ratio of the price of offenses (damage rate, r) to price of arrests (marginal cost, $\frac{\partial C}{\partial A}$) .

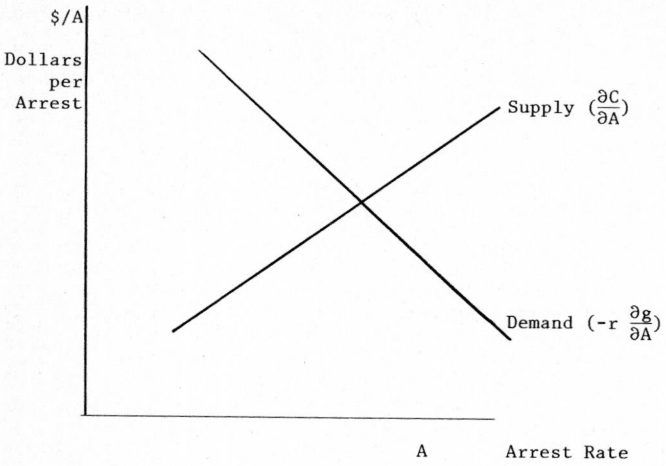

Figure 1: DEMAND AND SUPPLY OF ARRESTS

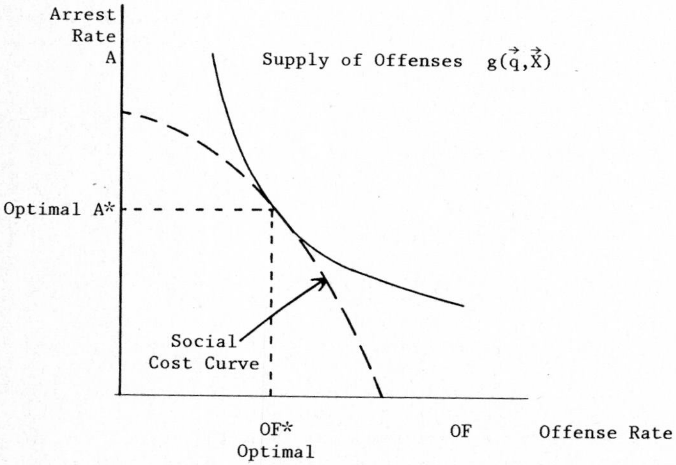

Figure 2: SUPPLY OF OFFENSES AND THE SOCIAL COST CURVE

$$1/\frac{\partial g_i}{\partial A_i} = -\frac{r_i}{\frac{\partial C}{\partial A_i}} \quad (13)$$

The slope of the total social cost curve is illustated in Fig. 2 for the case of increasing marginal cost of arrest.

Changes in socioeconomic and demographic factors, X , that increase crime will shift the supply of offenses function to the right in Fig. 2, and the optimal arrest rate, A , will increase if Eq. (11) holds.

STATIC MODEL

As variations in socioeconomic and demographic factors increase the offense rate, the locus of optimal arrest and offense rates, or expansion path, will depend upon the functional structure of the supply of offenses and the social cost curve. For simplicity, presume the marginal cost of arrest is constant, i.e. Eq. (2) can be expressed as

$$C = A\ m(\vec{q}^*,\vec{w}) \quad (14)$$

where $\vec{q}^*$ is the vector of other system outputs. Then the slope of the social cost curve will be constant, as illustrated in Fig. 3.

(i) Homothetic Supply of Offenses Function

If the supply of offenses function is homothetic such that the family of iso-offense curves parametric on $\vec{q}^*$ and $\vec{X}$ is nested, then the slope of the supply of offenses functions will be the same along a ray from the origin, as illustrated in Fig. 3. An example of such a function is:

$$OF = A^{-\beta}\ h(\vec{q}^*)k(\vec{X})\ ,\ \beta > 0 \quad (15)$$

The slope of the social cost curve is $-r/m(\vec{q}^*,\vec{w})$ where $m(\vec{q}^*,\vec{w})$ is the marginal cost of an arrest, and if this slope stays constant as the supply of offenses shifts to the right, then the expansion path will be a straight line and the probability of arrest, $P_A = A/OF$, will remain constant, as given by the tangent of the angle α in Fig. 3. This follows from using Eq.'s (14) and (15) to evaluate (13),

$$-\frac{1}{\beta}\frac{A}{OF} = -\frac{r}{m(\vec{q}^*,\vec{w})} \quad (16)$$

Of course the slope of the social cost curve may not remain constant if the damage rate changes, or the marginal cost of an arrest changes because of changes in the prices, w , of criminal justice inputs or changes in other system outputs, q^* . In that case, the probability of arrest could either increase or decrease.

(ii) Nonhomothetic Supply of Offenses Function

If the supply of offenses function is non-homothetic, then the expansion path will not be a ray from the origin. For example, the expansion path could be a straight line from a point on the ordinate, as illustrated in Fig. 4, implying a declining probability of arrest expansion path.

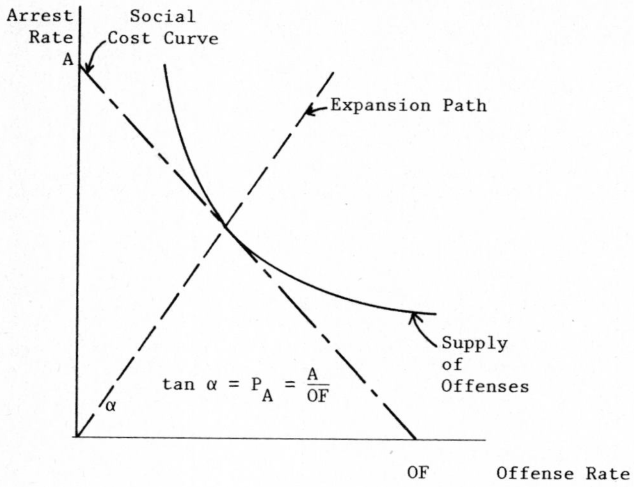

Figure 3: CONSTANT PROBABILITY OF ARREST EXPANSION PATH

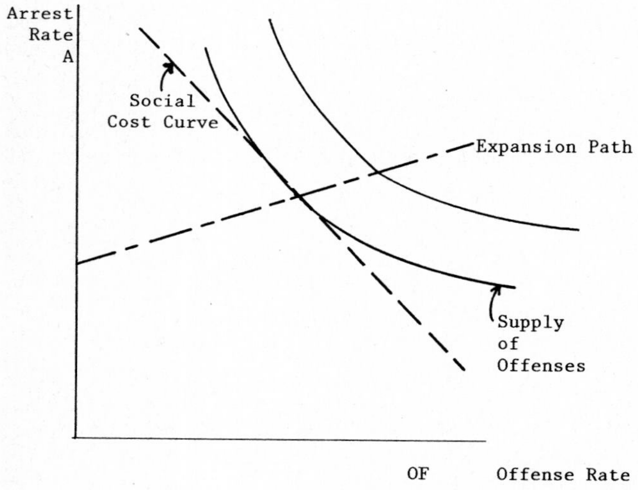

Figure. 4: DECLINING PROBABILITY OF ARREST EXPANSION PATH

DYNAMIC ADJUSTMENT PROCESSES

(i) The Supply of Offenses

Equation (15) can be written as

$$\ln OF(t) = \ln h(q^*) - \beta \ln A(t) + \gamma \ln X(t) + e_1(t) \qquad (17)$$

assuming $X(t)$ to index exogenous socioeconomic factors or to be a vector of variables and that $h(q^*)$ can be presumed constant, abstracting from changes in other system outputs. The latter could be controlled for by explicitly including them in Eq. (17). The stochastic residual is $e_1(t)$.

There is some question, of course, about whether individuals have accurate knowledge of the arrest rate and it can be replaced by the value they expect $\tilde{A}$. If individuals have knowledge of the behaviour of the system and form their expectations of $\ln A(t)$ on the basis of the information available to them at time $t-1$, denoted as the information set Ω_{t-1} , then their expectations of the logarithm of the arrest rate in period t , $\widetilde{\ln A(t)}$, will be:

$$\ln\tilde{A}(t) = E[\ln A(t)/\Omega_{t-1}] \qquad (18)$$

i.e. the expected value of $\ln A(t)$ conditional on the information set Ω_{t-1} , and the supply of offenses will be

$$\ln OF(t) = \ln h(q^*) - \beta \ln\tilde{A}(t) + \gamma \ln X(t) + e_1(t) \qquad (19)$$

(ii) The Budget

In the event of changing socioeconomic conditions and demographic variables leading to an increase in crime, the supply of offenses will shift to the right, as illustrated in Fig. 3. Under the assumption of a homothetic supply of offenses function, Eq. (15), and costs proportional to the arrest rate, Eq. (14), the cost minimizing expansion path would keep the probability of arrest constant as illustrated in Fig. 3. However, the production of arrests is limited by the budget, and the budget is determined by a political process. We presume that criminal justice agencies in period $t-1$ justify their desired budget $C^*(t)$ for period t on the funds needed to produce arrests in period $t-1$,

$$C^*(t) = A^*(t-1)m(\vec{q}^*,\vec{w}) \qquad (20)$$

and arrests in period $t-1$ are those needed to stay on the expansion path, Eq. (16),

$$A^*(t-1) = \frac{\beta r}{m(\vec{q}^*,\vec{w})} OF(t-1) \qquad (21)$$

In turn, we assume that the political process works so as to allocate budgets in period t only on the basis of demonstrated realized need from the previous period, and then only meets some fraction λ of the difference between the desired percentage change and actual percentage change in the budget,

$$\Delta \ln C(t) = \Delta \ln C(t-1) + \lambda[\Delta \ln C^*(t) - \Delta \ln C(t-1)] \qquad (22)$$

noting that the percentage change in the budget is approximately $\Delta \ln C(t)$,

$$\Delta \ln C(t) = \ln C(t) - \ln C(t-1) \simeq \frac{\Delta C(t)}{C(t)} \qquad (23)$$

and integrating Eq. (22), we have, equivalently,

$$\ln C(t) = \ln\delta + \ln C(t-1) + \lambda[\ln C^*(t) - \ln C(t-1)] \quad (24)$$

where $\ln\delta$ is a constant.

Assuming that the marginal cost $m(q^*,w)$ is changing negligibly over time, Eq.'s (14), (20) and (21) can be used to express Eq. (24) in terms of the arrest rate and offense rate,

$$\ln A(t)+\ln m = \ln\delta+\ln A(t-1)+\ln m+\lambda[\ln A^*(t-1)+\ln m-\ln A(t-1)+\ln m] \quad (25)$$

or

$$\ln A(t)-\ln A(t-1)+\lambda\ln A(t-1)=\ln\delta+\lambda[\ln\frac{\beta r}{m}+\ln OF(t-1)] \quad (26)$$

and defining Z as the lag operator,

$$Z^n\, x(t) = x(t-n) \quad (27)$$

Eq. (26) can be expressed as,

$$[1-(1-\lambda)Z]\ln A(t) = \ln\delta\ (\frac{\beta r}{m})^{\lambda} + \lambda\ln OF(t-1) \quad (28)$$

or

$$\ln A(t) = \frac{1}{\lambda}\ln\delta\ (\frac{\beta r}{m})^{\lambda} + \frac{\lambda}{[1-(1-\lambda)Z]}\ln OF(t-1) + e_2(t) \quad (29)$$

where any difference from $\ln A(t)$ being a geometrically distributed lag of $\ln OF(t-1)$, for example due to the slope of the social cost curve, $\frac{r}{m}$, not remaining constant, is presumed to be captured by the stochastic residual $e_2(t)$.

Implications of the Rational Expectations - Adaptive Expectations Model

The behavior of the public has been assumed to be rational, forming expectations of the behavior of the authorities on the information available to them, leading to a supply of offenses as expressed in Eq. (19). The behavior of the authorities has been assumed to reflect some inertia, responding to the known situation without extrapolation, as embodied in the system response equation (29). These two behavioral equations are the structural equations of the model. Note that the model has a certain recursive character with arrests depending on a distributed lag of past offense rates. The expectation of the variable $\ln A(t)$ is,

$$\ln\tilde{A}(t) = E[\ln A(t)] = \frac{1}{\lambda}\ln\delta(\frac{\beta r}{m})^{\lambda} + \frac{\lambda}{[1-(1-\lambda)Z]}\ln OF(t-1) \quad (30)$$

Assuming $e_2(t)$ to have mean zero. Thus,

$$\ln\tilde{A}(t) = \ln A(t) - e_2(t) \quad (31)$$

and substituting Eq. (31) in Eq. (19) we obtain,

$$\ln OF(t) = \ln h(q^*) - \beta\ln A(t) + \gamma\ln X(t) + e_1(t) - \beta e_2(t) \quad (32)$$

Note that the offense rate is negatively correlated only with the current arrest rate, while from Eq. (29) the arrest rate is a positive distributed lag of past offense rates. Hence, for these special expectations assumptions, the transfer function model between these variables appears to be one sided, with offense rates leading arrest rates.

Since lnA(t) is not independent of the error term, $e_1(t) - \beta e_2(t)$, an attempt to estimate the supply of offenses using ordinary least squares will result in inconsistent estimates with the deterrent parameter β biased towards zero. In addition, if other system outputs, q^*, are varying and not controlled for, the residuals $e_1(t)$ and $e_2(t)$ may be correlated as well.

If we substitute Eq. (30) in Eq. (19) we obtain,

$$\ln OF(t) = \ln h(q^*) - \frac{\beta}{\lambda} \ln\delta \left(\frac{\beta r}{m}\right)^{\lambda} - \frac{\beta\lambda}{[1-(1-\lambda)Z]} \ln OF(t-1) + \gamma \ln X(t) + e_1(t) \quad (33)$$

and lnOF(t) has an autoregressive structure, where if lnX(t) or $e_1(t)$ has a moving average structure, lnOF(t) will have an autoregressive-moving average structure.

<u>Some Evidence of a One-Sided Transfer Function</u>

Using annual data for aggravated assault in California for the period 1945 to 1978, Phillips and Ray report the following transfer function between the probability of imprisonment and the offense rate (with student's t-statistics in parenthesis),

$$P(t) = \underset{(t=8.1)}{0.0127} - \frac{\overset{(t=3.5)}{7.02}\,[1-Z]}{[1-\underset{(t=3.7)}{0.59}Z]} OF(t) + \frac{1}{[1-\underset{(t=4.3)}{0.551}Z-\underset{(t=2.5)}{0.319}Z^5]} e(t) \quad (34)$$

Attempts to treat the probability as an input did not yield significant results.

The supply of offenses-budget adaptations model can be modified to relate the probability of imprisonment and the offense rate. First, the analysis can be specified in terms of imprisonments, I(t), rather than arrests. Second, the definition of the probability of imprisonments in logarithms,

$$\ln P(t) = \ln I(t) - \ln OF(t) \quad (35)$$

can be combined with the supply of offenses, Eq. (17), expressed in terms of imprisonments instead of arrests, to express the supply of offenses as,

$$\ln OF(t) = \frac{1}{1+\beta} \ln h - \frac{\beta}{1+\beta} \ln P(t) + \frac{\gamma}{1+\beta} \ln X(t) + \frac{1}{1+\beta} e_1(t) \quad (36)$$

where the public's supply of offenses is assumed to depend on knowledge of the current probability of imprisonment. Third, Eq. (35) can be combined with Eq. (29), expressed in terms of imprisonments rather than arrests,

$$\ln I(t) = \frac{1}{\lambda} \ln\delta\left(\frac{\beta r}{m}\right)^{\lambda} + \frac{\lambda}{[1-(1-\lambda)Z]} \ln OF(t-1) + e_2(t) \quad (37)$$

to re-express it in terms of the probability of imprisonment,

$$\ln P(t) = \frac{1}{\lambda} \ln\delta\left(\frac{\beta r}{m}\right)^{\lambda} - \frac{(1-Z)}{[1-(1-\lambda)Z]} \ln OF(t) + e_2(t) \quad (38)$$

Note that this two equation model, (36) and (38), is not recursive but simultaneous, and it is one-sided with the probability being a negative distributed lag of the offense rate. Also note that, differencing Eq. (38),

$$\Delta \ln P(t) = - \frac{(1-Z)}{[1-(1-\lambda)Z]} \Delta \ln OF(t) \quad (39)$$

or

$$\Delta \ln P(t) = -1[\Delta \ln OF(t) - \lambda\Delta \ln OF(t-1) - \lambda(1-\lambda)\Delta \ln OF(t-2)\ldots] \quad (40)$$

i.e. the percentage change in the probability of imprisonment is a weighted average of the percentage change in the offense rate where the weights in brackets in Eq. (40) sum to zero:

$$1-\lambda[1+(1-\lambda)+(1-\lambda)^2\ldots] = 1 - \frac{\lambda}{1-(1-\lambda)} \quad (41)$$

Consequently, even if the offense rate is growing at a constant percentage rate, the probability of imprisonment has a homeostatic tendency to return to the expansion path, but first decreases in response to increases in the offense rate, as illustrated in Fig. 5. In the estimate for aggravated assault, the probability of imprisonment tends toward .0127 if the assault rate is constant. The average probability of imprisonment for the period 1945-1978 was .0105. The coefficient -7.02 in Eq. (34) implies an elasticity at the means of -1.08, close to the theoretical value of -1 in Eq. (38).

Phillips and Ray (1982a,b) also find one sided transfer functions for homicide, robbery, burglary and auto theft using the California data but the offense rate is a negative distributed lag of the probability of imprisonment, suggesting that public expectations of the probability are adaptive. The contrasting results for aggravated assault and the other crimes is puzzling.

Drunken Driving Casualties and Per Se Legislation

California introduced legislation effective January 1, 1982 making it a crime (per se) to drive with a blood alcohol content of 100 micrograms per milliliter or greater. Fifteen years ago, the British Road Safety Act of 1967 became effective on October 9th of that year, making it illegal to drive with a blood alcohol content of 80 micrograms per milliliter or greater. In a classic article, H. Laurence Ross detected a 24.7 per cent decline in fatalities, a 14.6 per cent decline in serious injuries and a 9.8 per cent decline in slight injuries following the introduction of the Act in October 1967. Ross standardized for driving activity by expressing casualties per 100 million vehicle miles. He deseasonalized the monthly data.

Using Box-Jenkins time series techniques and regression in parallel, Phillips, Votey, and Ray constructed forecasting models for seriously injured using monthly data for the period January 1960 through December 1974, and showed that an index of miles driven explained 20% of the variance in seriously injured, the addition of rainfall explained another 29% of the variance, and the addition of a measure of alcohol consumption explained another 4% of the variance. Adding an intervention or dummy variable coinciding with the introduction of the British Road Safety Act only added 2.5% to the explained variance. However, the reduction in injuries was approximately 15.7 per cent, comparable to Ross' estimate of 14.6 per cent.

The significant impact of the intervention variable is consistent with deterrence. Drunken drivers should be deterred not only by arrests, but also by convictions given arrest. One of the motivations for introducing per se legislation is to increase the probability of conviction given arrest. The British Road Safety Act of 1967 increased the fraction of convictions compared to the Traffic Safety Act of 1962, as detailed in Ross. The probability of conviction given arrest under section 1 of the 1967 Road Safety Act was 93 per cent in 1968 and 91 per cent in 1969 as reported by Beaumont and Newby. The comparable figure was 93 per cent in 1973, as reported by the Home Office (1973). Thus for six years following the per se legislation, there was a large and permanent increase in the probability of conviction given arrest.

It is, of course, more difficult to follow the course of the probability of arrest for drunken driving because the denominator, the offense rate for drivers exceeding a blood alcohol level of 80 micrograms per milliliter is unobservable. It is possible to relate the arrest rate for drunken driving to casualties. As reported in Phillips, Votey and Ray, there was no significant relationship between seriously injured and the arrest rate, either in the regression model, or using the Box-Jenkins techniques. Hence it appears the intervention variable is measuring some impact other than the deterrent effect of conviction given arrest.

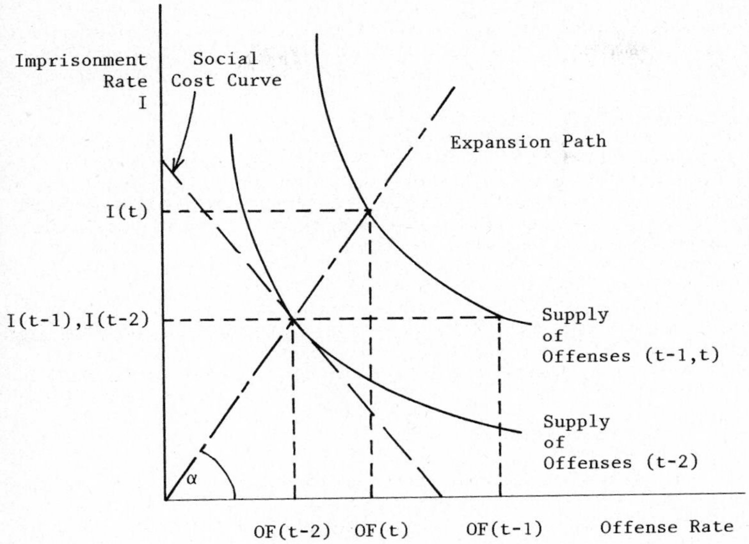

Figure 5: HOMEOSTATIC TENDENCY FOR THE PROBABILITY OF IMPRISONMENT, $P = I/OF$.

However, as Ross' analysis indicated, fatalities declined more in percentage terms than seriously injured, following the introduction of the British Road Safety Act, and fatalities may be a more sensitive index of drunken driving than seriously injured, although both will be affected by other factors such as miles driven and variations in risk, if driving sober, caused by conditions such as weather.

Phillips, Votey and Ray used Box-Jenkins techniques to check for a relationship between arrests for drunken driving and road fatalities using monthly data. For the period October 1967 to December 1972 there appeared to be a significant correlation with both leads and lags. However, for the extended period October 1967 to April 1975, the correlations were smaller and insignificant at the 5% level. Thus the evidence is mixed and further interpretation and analysis is needed to understand the impact of per se legislation.

Probability Model of Road Fatalities

The probability of being killed while driving, $P(K \wedge D)$, measured for example in fatalities per capita, will consist of those killed while driving and below the per se blood alcohol level or legally sober, $P(K \wedge D \wedge S)$ and those killed while driving at and above the per se blood alcohol limit of 80 micrograms per milliliter or .08 percent, $P(K \wedge D \wedge \bar{S})$,

$$P(K \wedge D) = P(K \wedge D \wedge S) + P(K \wedge D \wedge \bar{S}) \qquad (42)$$

or using the definition for conditional probabilities

$$P(K \wedge D) = P(K/D \wedge S)P(S/D)P(D) + P(K/D \wedge \bar{S})P(\bar{S}/D)P(D) \qquad (43)$$

where, for the distribution of drivers by blood alcohol content, $f(x)$, the probability of being sober given you are driving, $P(S/D)$ is

$$P(S/D) = \int_0^{.08} f(x)dx = F(.08) \qquad (44)$$

and the probability of being drunk given you are driving, or the offense rate for drunken driving, $OF(t)$, is

$$OF(t) = P(\bar{S}/D) = 1-F\ (.08) \qquad (45)$$

and the probability of driving is $P(D)$, measured for example, in road miles per capita.

Using Eq. (45), Eq. (43) can be expressed as

$$P(K \wedge D) = P(K/D \wedge S)P(D)[1-OF(t) + \frac{P(K/D \wedge \bar{S})}{P(K/D \wedge S)} OF(t)] \qquad (46)$$

or

$$P(K \wedge D) = P(K/D \wedge S)P(D) + P(K/D \wedge S)P(D)\ OF(t)\ (\frac{P(K/D \wedge \bar{S})}{P(K/D \wedge S)} -1) \qquad (47)$$

The time series for road fatalities or killed, $K(t)$, is a measure of $P(K \wedge D)$ multiplied by the population, and the risk of being killed given that you are driving while sober, $P(K/D \wedge S)$, will vary with rainfall, $R(t)$, the vehicle mix and the quality of roads, for which we do not have monthly measures. The probability of driving P(D) multiplied by the population can be measured by an index of motor vehicle traffic MV(t) . The ratio $\frac{P(K/D\wedge\bar{S})}{P(K/D\wedge S)}$, the probability of being killed if driving while drunk relative to the probability of being killed while driving while sober is a measure of relative risk, r , and as investigated in Phillips, appears to be approximately constant. Thus Eq. (47) can be expressed as,

$$K(t) = aR(t)MV(t)+aR(t)MV(t)OF(t)(r-1)+e(t) \qquad (48)$$

where a is a constant of proportionality and the additive residual e(t) captures the influence of the omitted variables. It is clear from Eq. (48), that the time series for killed can vary, even if the offense rate for drunken driving is constant, since varying weather conditions and road traffic will affect fatalities. Eq. (48) can be expressed as

$$K(t) = a\ R(t)\ MV(t)\ [1-(r-1)OF(t)]e^{e_3(t)} \qquad (49)$$

where $e^{e_3(t)}$ is a multiplicative residual, and taking logarithms, we have

$$\ln K(t) = \ln a+\ln R(t)+\ln MV(t)+\ln[1-(r-1)OF(t)] + e_3(t) \qquad (50)$$

We could approximate the logarithm of the linear function of OF(t) , [1+(r-1)OF(t)] , by

$$\ln[1+(r-1)OF(t)] \cong (r-1)OF(t) \qquad (51)$$

but choose

$$\ln[1+(r-1)OF(t)] \cong b(r-1)\ln OF(t) \qquad (52)$$

where b = E OF(t) and hence

$$\ln K(t) = \ln a+\ln R(t)+\ln MV(t)+b(r-1)\ln OF(t)+e_3(t). \qquad (53)$$

Equation (53), combined with a supply of offenses function like Eq. (19),

$$\ln OF(t) = \ln h - \beta\ln\tilde{A}(t) + e_1(t) \qquad (54)$$

and a system response equation like Eq. (28),

$$\ln A(t) = \ln d + \frac{\lambda}{[1-(1-\lambda)Z]} \ln OF(t-1) + e_2(t) \qquad (55)$$

defines a three equation system in two observable variables, road fatalities K(t) , and drunken driving arrests A(t) , and one unobservable variable, the offense rate for drunken driving OF(t) . For a model similar in structure, but different in specification, see Votey (1982).

<u>Unobservable Variable Problem</u>

As discussed in Judge et al. (1980), the elimination of an unobservable variable from a system of equations leads to an errors in variables problem. The expected value of the variable lnA(t) is

$$\ln\tilde{A}(t) = E[\ln A(t)/\Omega_{t-1}] = \ln d + \frac{\lambda}{[1-(1-\lambda)Z]} \ln OF(t-1) \qquad (56)$$

or

$$\ln\tilde{A}(t) = \ln A(t) - e_2(t) \qquad (57)$$

and substituting in Eq. (54) we obtain,

$$\ln OF(t) = \ln h - \beta \ln A(t) + e_1(t) + \beta e_2(t) . \qquad (58)$$

Substituting Eq. (58) into Eq. (53) we have,

$$\ln K(t) = \ln a + \ln R(t) + \ln MV(t) + b(r-1)[\ln h - \beta \ln A(t) + e_1(t) + \beta e_2(t)] + e_3(t) \qquad (59)$$

Note that the variable logarithm of arrests for drunken driving, $\ln A(t)$, is not independent of the error term,

$$e_3(t) + b(r-1)e_1(t) + b(r-1)\beta e_2(t) ,$$

which will lead to inconsistent ordinary least squares estimates and a coefficient on $\ln A(t)$ biased towards zero, indicating no deterrence.

Substituting Eq. (58) into Eq. (55) we obtain

$$\ln A(t) = \ln d + \ln h - \frac{\lambda\beta}{[1-(1-\lambda)Z]} A(t-1) + \frac{\lambda}{[1-(1-\lambda)Z]} [e_1(t) + \beta e_2(t)] + e_2(t) \qquad (60)$$

or

$$\ln A(t) = \lambda(\ln d + \ln h) + [1-\lambda(1+\beta)]\ln A(t-1) + \lambda e_1(t-1) + e_2(t) - [1-\lambda(1+\beta)]e_2(t-1) \qquad (61)$$

Thus $\ln A(t)$ will have an autoregressive-moving average (ARMA) structure. Equations (59) and (61) are the two structural equations of the model, after eliminating the unobservable variable. The variable $\ln A(t-1)$ is the only predetermined variable excluded from Eq. (59), and will be a satisfactory instrumental variable if the three residual components of the error structure of Eq. (59), $e_1(t)$, $e_2(t)$ and $e_3(t)$, are not autocorrelated.

If this model is valid, from Eq. (59), road fatalities will be negatively correlated with current arrests for drunken driving, and since road fatalities are positively correlated with the unobservable drunken driving offenses, from Eq. (55), the arrest rate will be a positive distributed lag of road fatalities. The actual lead-lag pattern was estimated using Box-Jenkins techniques.

EMPIRICAL PROCEDURES

The relationship between the road fatalities, $K(t)$, and arrests for drunken driving, $A(t)$, was investigated by fitting univariate ARIMA models to each to obtain the white noise innovations $\varepsilon_1(t)$ and $\varepsilon_2(t)$, respectively,

$$A(Z)\,\Delta^{d}\,K(t) = B(Z)\varepsilon_1(t) \qquad (62)$$

and

$$C(Z)\,\Delta^{d^*}\,A(t) = D(Z)\,\varepsilon_2\,(t) \qquad (63)$$

The innovations were then cross-correlated to determine whether the relationship was one-sided, with $\varepsilon_2(t)$ leading $\varepsilon_1(t)$ or vice versa, or two sided. This determined the model complexity, i.e. the choice of

$$\varepsilon_1(t) = \alpha_0\varepsilon_2(t) + \alpha_1\varepsilon_2(t-1) + \ldots + \mu_1(t) \qquad (64)$$

or

$$\varepsilon_2(t) = \beta_0\varepsilon_1(t) + \beta_1\varepsilon_1(t-1) + \ldots + \mu_2(t) \qquad (65)$$

or both Eq.'s (64) and (65). With the selection of the appropriate model, for example, a particular realization of Eq. 64,

$$\varepsilon_1(t) = \hat{\alpha}_0\ \varepsilon_2(t) + \mu_1(t) \tag{66}$$

the estimates from Eq.'s (62) and (63) could be substituted into Eq. (66) to infer the parameters of a transfer function, $\Lambda(Z)$,

$$\Lambda(Z) = 1+\lambda_1 Z+\lambda_2 Z^2+\ldots = \frac{\hat{B}(Z)\ \Delta^{\hat{d}*}\ \hat{C}(Z)}{\hat{A}(Z)\ \Delta^{\hat{d}}\ \hat{D}(Z)}$$

where from Eq.'s (62), (63), and (66)

$$K(t) = \frac{\hat{\alpha}_0\ \hat{B}(Z)\ \Delta^{\hat{d}*}\ \hat{C}(Z)}{\hat{A}(Z)\ \Delta^{\hat{d}}\ \hat{D}(Z)}\ A(t) + \mu_1(t) \tag{67}$$

In the case of a one-sided model, such as Eq. (66) the transfer function was estimated by first filtering both $K(t)$ and $A(t)$ by the estimated filter,

$$\frac{\hat{C}(Z)\Delta^{\hat{d}*}}{\hat{D}(Z)} \tag{68}$$

Cross-correlating the filtered series, and using the estimates to determine initial estimates of the transfer function between $K(t)$ and $A(t)$.

Data

Monthly data for road fatalities is published in the Monthly Digest of Statistics for Great Britain. The monthly series on arrests for driving under the influence of alcohol, beginning October 1967, was obtained by Professor Harold L. Votey Jr. from the British Transport and Road Research Laboratory.

Empirical Results

The monthly time series for road fatalities, $K(t)$, for the period October 1967 through December 1972, was seasonally differenced, $(1-Z^{12})$, and a moving average structure was estimated:

$$(1-Z^{12})K(t) = \underset{(t=1.3)}{(1-.172Z^{12})}\ \varepsilon_1(t) \tag{69}$$

with a residual mean square of 2,750. The residuals $\varepsilon_1(t)$ are white noise.

The monthly time series for drunken driving arrests, $A(t)$, for the same period, was seasonably differenced and first differenced and also had a moving average structure:

$$(1-Z)(1-Z^{12})A(t) = (1-\underset{(t=4.2)}{.445Z} + \underset{(t=2.6)}{.296Z^6} - \underset{(t=2.5)}{.304Z^{12}})\ \varepsilon_2(t) \tag{70}$$

with a residual mean square of 118,355. The residuals $\varepsilon_2(t)$ are white noise.

The residuals $\varepsilon_1(t)$ and $\varepsilon_2(t)$ were cross-correlated. There were

FIGURE 6

CROSS CORRELATIONS OF INNOVATIONS FROM UNIVARIATE MODELS OF
ARRESTS AND KILLED
OCTOBER 1967 - DECEMBER 1972

```
                     CROSS-CORRELATION
  -1.0 -0.8 -0.6 -0.4 -0.2  0.0  0.2  0.4  0.6  0.8  1.0
    +----+----+----+----+----+----+----+----+----+----+
LAG
                             I
-24  -0.018         +        I       +
-23   0.045         +        IX      +
-22   0.064         +        IXX     +
-21  -0.093         +      XXI       +
-20  -0.079           +    XXI      +
-19   0.050           +      IX     +
-18  -0.079           +    XXI      +
-17   0.092           +      IXX    +
-16   0.007           +      I      +
-15   0.092           +      IXX    +
-14   0.024           +      IX     +
-13   0.083           +      IXX    +
-12   0.338           +      IXXXXXX+X
-11   0.045           +      IX     +
-10   0.189           +      IXXXXX +
 -9   0.346           +      IXXXXXX+XX
 -8   0.006           +      I      +
 -7   0.040           +      IX     +
 -6  -0.020            +     I     +
 -5  -0.006            +     I     +
 -4   0.216            +     IXXXXX+
 -3  -0.049            +     I     +
 -2  -0.080            +     IXX   +
 -1  -0.141            +  XXXI     +
  0  -0.056            +     I     +
  1   0.105            +     IXXX  +
  2  -0.143            + XXXXI     +
  3  -0.238            XXXXXXI     +
  4  -0.014            +     I     +
  5  -0.317          XX+XXXXXI     +
  6   0.172            +     IXXXX +
  7   0.117           +      IXXX   +
  8  -0.089           +    XXI      +
  9   0.121           +      IXXX   +
 10  -0.113           +   XXXI      +
 11  -0.104           +   XXXI      +
 12  -0.110           +   XXXI      +
 13  -0.053           +      I      +
 14   0.029           +      IX     +
 15   0.061           +      IXX    +
 16  -0.016           +      I      +
 17   0.130           +      IXXX   +
 18   0.038           +      IX     +
 19   0.072           +      IXX    +
 20  -0.100           +   XXXI      +
 21  -0.201          +  XXXXXI       +
 22  -0.107          +    XXXI       +
 23  -0.051          +    XXXI       +
```

\+ 95% confidence intervals around zero.

significant negative correlations at lags 3 and 5 with $\varepsilon_2(t)$ (arrests) leading $\varepsilon_1(t)$ (fatalities) and significant positive correlations at lags 9 and 12 with $\varepsilon_1(t)$ (fatalities) leading $\varepsilon_2(t)$ (arrests). This is illustrated in Fig. 6 and suggests the model:

$$\varepsilon_1(t) = -.041\ \varepsilon_2(t-3) - .038\ \varepsilon_2(t-5) + \mu_1(t) \qquad (71)$$

and

$$\varepsilon_2(t) = +2.03\ \varepsilon_1(t-9) - 2.09\ \varepsilon_1(t-12) + \mu_2(t) \qquad (72)$$

This finding of a two-way model suggests adaptive expectations on the part of both the public and the authorities is more realistic than rational expectations for the public and adaptive expectations for the authorities, at least where the reaction time is measured in months rather than years.

For Eq. (71), the inferred transfer function can be calculated by substituting for $\varepsilon_1(t)$ and $\varepsilon_2(t)$ from Eq.'s (69) and (70):

$$\frac{(1-Z^{12})}{(1-.172Z^{12})} K(t) = \frac{-.041(1+.927Z^2)(1-Z)(1-Z^{12})A(t-3)}{(1-.445Z+.296Z^6-.394Z^{12})} + u_1(t) \qquad (73)$$

with a transfer function

$$\Lambda(Z) = (1+\lambda_1 Z+\lambda_2 Z^2\cdots) = \frac{(1+.927Z^2)(1-.172Z^{12})}{(1-.445Z+.296Z\ -.304Z\ \ )}$$

with lag weights

λ_0	=	1.0	λ_6	=	-.272	λ_{12}	=	.178
$\lambda 1$	=	1.38	λ_7	=	-.529	λ_{13}	=	.495
λ_2	=	.615	λ_8	=	-.418	λ_{14}	=	.531
$\lambda 3$	=	.273	λ_9	=	-.267	λ_{15}	=	.348
λ_4	=	.121	λ_{10}	=	-.154			
$\lambda 5$	=	.054	λ_{11}	=	-.085			

as illustrated in Figure 7. This suggests using a declining geometric lag as an approximation, ignoring the residual seasonal structure at lags 7 and 14.

The imputed model is:

$$K(t) = .041(1.0+1.38Z+0.615Z^2\ldots)(1-z)A(t-3) + \frac{(1-.172Z^{12})}{(1-Z^{12}} \mu_1(t)$$

Surprisingly, it indicates that road fatalities is a geometric distributed lag of the change in arrests, lagged one quarter.

A transfer function with this form was estimated, relating seasonally differenced road fatalities to an index of motor vehicle traffic, MV(t), differenced and seasonally differenced, and the difference between monthly rainfall and its average value:

FIGURE 7

INFERRED TRANSFER FUNCTION WEIGHTS

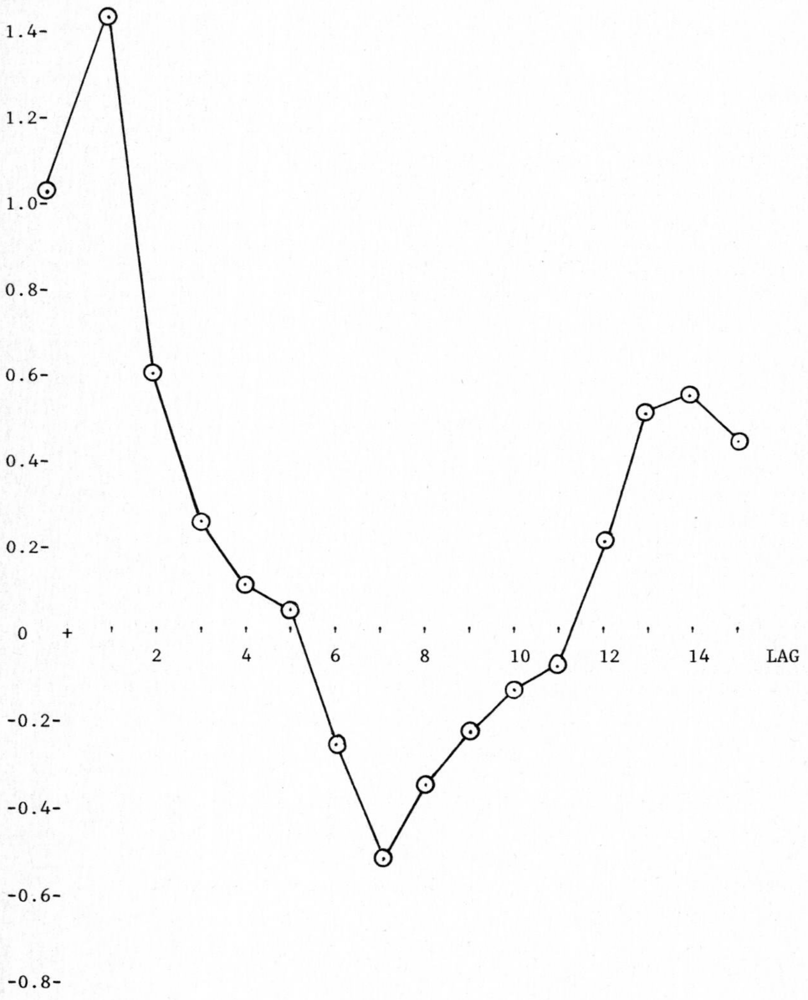

$$(1-Z^{12})K(t) = \underset{(t=2.2)}{\overset{(t=2.2)}{\frac{-0.0427(1-Z)(1-Z^{12})}{[1-0.347Z]}}}\,A(t-3) + \underset{(t=1.6)}{8.823}\,[R(t)-ER(t)]$$

$$+ \underset{(t=0.8)}{0.759}\,(1-Z)(1-Z^{12})\,MV(t) + \varepsilon(t) \qquad (75)$$

with a residual mean square error of 2,392 and student's t-statistics in parenthesis. The index of motor vehicle traffic was not significant but did have the anticipated sign. Rainfall increased accidents as expected. Arrests had a deterrent effect, but with a geometric distributed lag, suggesting adaptive expectations, and with an impact beginning after a lag of one quarter. This may be due to an information lag if drivers obtain information through the media. Lastly, it is the change in arrests activity which drivers respond to rather than the level of arrests, suggesting a more complex model of behavior than anticipated. Road fatalities have a homeostatic tendency to return to a constant level, as revealed by the estimated transfer function,

$$\frac{(1-Z)}{[1-0.347Z]},$$

(refer to the discussion of Eq.'s (39), (40) and (41) above). This suggests that arresting drunken drivers may have a transitory incapacitation effect as drunken drivers are taken off the roads for a while but then returned, and that deterrence is not operating at all. Thus it may prove possible, using dynamic models and Box-Jenkins techniques, to identify whether crime control is achieved through deterrence or incapacitation (see Blumstein et al. for a discussion of the difficulties in identifying deterrence.)

Conclusions:

The empirical estimates show that road fatalities are inversely related to a distributed lag of arrests for drunken driving. Surprisingly, the lag structure reveals a homeostatic tendency for road fatalities, suggesting that enforcement may be operating through a transitory incapacitation effect rather than through deterrence.

References

BEAUMONT, K. and NEWBY, R.F. (1972). Traffic Law and Road Safety Research in the United Kingdom - British Counter-measures. National Road Safety Symposium, Canberra, Australia.

BECKER, Gary S. (1968). Crime and Punishment: An Economic Approach. Journal of Political Economy, 526-36.

BLUMSTEIN, A., COHEN, J. and NAGIN, D., eds. (1978). Deterrence and Incapacitation. National Academy of Science, Washington, D.C.

BOX, George E.P., and JENKINS, Gwilym M. (1970). Time Series Analysis, Forecasting, and Control. Holden-Day, San Francisco.

CENTRAL STATISTICAL OFFICE, Monthly Digest of Statistics.

EHRLICH, Isaac (1973). Participation in Illegitimate Activities: A Theoretical and Empirical Investigation. Journal of Political Economy 81, 521-565.

EHRLICH, Isaac (1975). The Deterrent Effect of Capital Punishment: A Question of Life and Death, American Economic Review 63, 397-417.

HOME OFFICE (1973). Offences Relating to Motor Vehicles.

JUDGE, George G., GRIFFITHS, William, HILL, R. Carter, and LEE, Tsoung-Chao (1980). The Theory and Practice of Econometrics. John Wiley, New York.

PHILLIPS, Llad (1975). The Logistics of Driving While Drunk. Unpublished manuscript.

PHILLIPS, Llad and RAY, Subhash (1982). The Dynamics of Deterrence: Some Contrasting Evidence for Index Felonies. Unpublished manuscript.

PHILLIPS, Llad and RAY, Subhash (1982). Evidence on the Identification and Causality Dispute about the Death Penalty. In Applied Time Series Analysis (Proceedings of the International Conference at Houston, Texas, August 1981). Ed: O.D. Anderson and M.R. Perryman, North-Holland, Amsterdam & New York, 313-340.

PHILLIPS, Llad, RAY, Subhash and VOTEY, Harold L., Jr. (1981). Forecasting Highway Casualties: The British Road Safety Act and a Sense of Déja Vu. Unpublished manuscript.

ROSS, H. Laurence (1973). Law, Science, and Accidents: The British Road Safety Act of 1967. The Journal of Legal Studies 2, 1-78.

STIGLER, George J. (1970). The Optimal Enforcement of Laws. Journal of Political Economy 78, 526-539.

VOTEY, Harold L., Jr. (1982). Scandinavian Drinking-Driving Control: Myth or Intuition? Journal of Legal Studies 11, 93-116.

TIME SERIES ANALYSIS: Theory and Practice 4
O.D. Anderson (editor)

THE DYNAMIC LINKS BETWEEN INFLATION AND UNEMPLOYMENT: SOME EMPIRICAL EVIDENCE

Stephen E. Haynes
and
Joe A. Stone

University of Oregon
Eugene, Oregon

This study explores the dynamic links between inflation and unemployment for the United States in two ways: first, by exploiting the lead-lag properties between inflation and unemployment to identify aggregate demand and supply; and second by using cross-spectral analysis to discriminate among competing macroeconomic models. Time series tests of the lead-lag relationship directly demonstrate the significance of both demand and supply links, and frequency domain tests provide clear support for the monetary natural unemployment rate hypothesis.

INTRODUCTION

The relationship between unemployment and inflation, an issue central to macroeconomic policy debates, remains both a theoretical and empirical puzzle. There is substantial agreement regarding the immediate linkages induced between unemployment and inflation by various shocks to aggregate demand or supply, but widespread disagreement regarding the final, steady-state relationships. Virtually all economists agree, for example, that unemployment can be reduced at least temporarily by expansionary monetary or fiscal policies, but no such consensus exists on whether such reductions can be permanently sustained.

Conclusive empirical evidence regarding the steady-state relationship between unemployment and inflation does not exist. The evidence that does exist is typically ambiguous and highly sensitive to alternative specifications and sample periods. This paper surmounts these empirical difficulties using lead-lag information to identify aggregate demand and supply linkages between unemployment and inflation; and spectral analysis to discriminate among competing macroeconomic models. Under the assumptions that price expectations are exactly met in a steady-state and that exogenous supply shocks have little recurring long-term structure, the results provide strong support for the natural unemployment rate hypothesis.

The paper is organized as follows. Section II presents a brief survey of standard theoretical and empirical analyses of aggregate demand and supply linkages between unemployment and inflation, and Section III disentangles short-run demand and supply links using their lead-lag properties and cross-correlation methods. Section IV estimates the steady-state link using cross-spectral estimates of a quasi-reduced form, dynamic equation for unemployment and inflation that includes competing macroeconomic models as special cases. The section also conducts sensitivity tests using proxies for inflationary expectations and uncertainty. A final section provides a brief summary.

STANDARD ANALYSES AND TESTS

In this section, we survey essential elements of standard models of inflation and unemployment, describe standard empirical tests of these models, and summarize deficiencies of such tests.[1] Standard macroeconomic models distinguish two types of inflation: demand-induced (demand-pull) and supply-induced (cost-push). Demand-induced inflation can be illustrated using Fig. 1a with DD and SS as the aggregate demand and supply schedules, respectively, and with P and Q as the aggregate price level and real output, respectively. A demand shock shifts the aggregate demand schedule outward to D'D', raising both P and Q. As the economy adjusts to the higher price level, however, the aggregate supply schedule shifts inward to S'S' reducing real output and increasing price. To induce continued inflation, the demand shock must be maintained, otherwise only a one-time increase in the price level occurs. Two types of supply shifts can be distinguished: demand-induced shifts just described, and exogenous shocks, which cause one-shot cost-push inflation.

The standard Keynesian and finite Phillips-curve models suggest that output (and unemployment) can be permanently altered by management of aggregate demand. That is, under appropriate circumstances permanent demand shifts can sustain a permanent change in real output and unemployment. The standard monetary model, however, holds that as the higher inflation induced by demand expansion becomes fully anticipated, the aggregate supply schedule shifts inward and the short-run Phillips-curve shifts outward, fully offsetting real effects. The direct and induced effects are illustrated in Fig. 1b by an inward movement along the short-run Phillips curve aa and by the outward shift in the short-run Phillips curve to a'a' as the higher inflation becomes fully anticipated. The vertical curve bb depicts the natural unemployment rate hypothesis (NRH) introduced into the monetary model by Friedman (1966) and Phelps (1967). In this case when demand-induced inflation is fully anticipated, actual unemployment equals the "natural rate" and steady-state unemployment is invariant to demand-induced inflation.[2] A feature of the monetary model is the hypothesis that most instability is induced by monetary and fiscal policies, hence that a substantial proportion of supply shocks is demand-induced, as in Fig. 1a. The time dynamics are important for the empirical test in Section III. The initial demand shift identifies the supply equation, with unemployment inversely correlated with subsequent rates of inflation, but the induced supply shift identifies the demand equation, with inflation positively correlated with subsequent rates of unemployment. This follows from the standard wisdom that

Figure 1a: Aggregate Demand and Supply

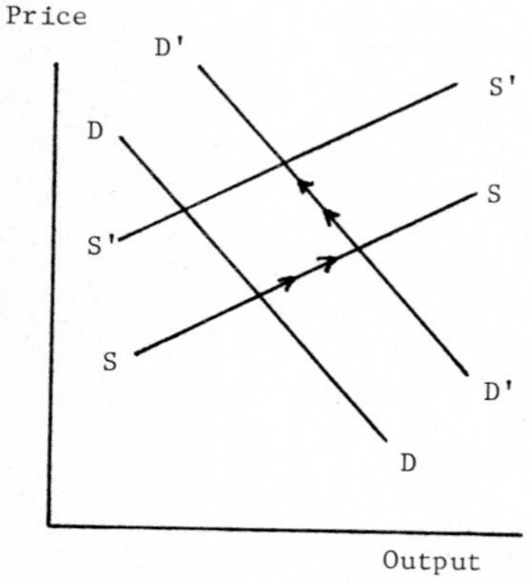

Figure 1b: Phillips Curve

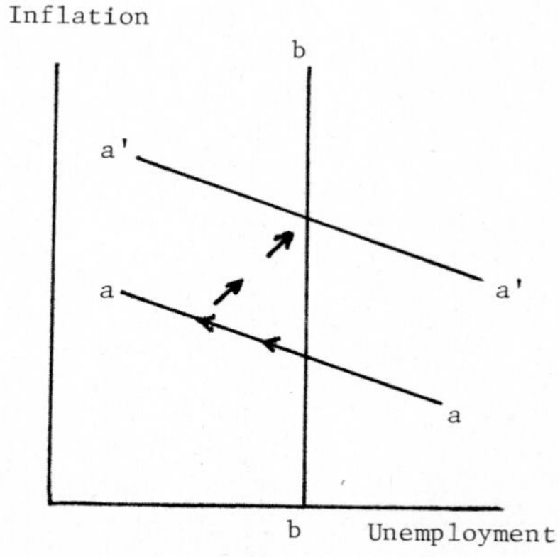

unemployment responds more rapidly to demand shocks, and that prices respond more rapidly to supply shocks (Gordon (1981), Stein (1974), and Van Order (1977)).

Empirical tests of linkages between unemployment and inflation concentrate primarily on supply (the Phillips curve), and assume supply is identified by shifts in demand.[3] Such tests are based exclusively on standard time domain techniques which require explicit measures of price expectations and which are sensitive to dynamic misspecification (Howrey (1980) and Engle (1980)). These limitations are evidenced by failures to discriminate among competing models and sensitivities to alternative specifications and sample periods. The volatile signs and magnitudes of parameter estimates over periods of varying demand and supply shocks suggest that demand and supply are inadequately identified with standard time domain techniques -- aggregate demand and supply for industrial economies are functions of the same arguments.

We address these limitations in two ways. First, we use cross-correlation analysis and lead-lag properties of the linkages between inflation and unemployment to identify both supply and demand effects. Second, we use cross-spectral analysis to discriminate competing macroeconomic models. Whereas the steady-state Phillips curve between unemployment and inflation is difficult to identify using time-domain techniques, it is directly obtained using cross-spectral techniques.[4]

CROSS-CORRELATION ESTIMATES

The expectations-augmented Phillips-curve model specifies the supply relationship between inflation and unemployment conditional on expected inflation. The major difficulties in testing the model are the absence of direct measures of expected inflation and the inability to identify the separate demand and supply links. In this section, we use cross-correlation methods and lead-lag properties to identify the aggregate demand and supply linkages between unemployment and inflation. The cross-correlation technique is used because it separates demand and supply by lead-lag properties if the dynamic model discussed above is correct. For the demand link, unemployment is predicted to lag inflation with a positive sign, but for the supply link, inflation is predicted to lag unemployment with a negative sign.[5]

Our sample is quarterly, seasonally unadjusted, data on the consumer price index (CPI) and the official unemployment rate for the United States from the first quarter 1953 to the fourth quarter 1981. The starting date is 1953 because it marks the date the Federal Reserve abandoned its policy of "pegging" government bond prices.[6] The inflation rate is computed as the log difference (annualized) of the CPI. Both series are examined after formal "prewhitening."[7] Prewhitening inflation tends to control for expected inflation since expectations are usually based on the history of inflation. Thus, the estimated supply link is short-run in nature.

Table 1 records cross-correlation estimates between inflation and unemployment for a lag of eight quarters. Seven of the eight estimates with inflation lagging unemployment are negative, the sign and dynamics supporting the supply link. Based on the appropriate chi-square test, inflation significantly lags unemployment at the one percent level, and the cumulative sum of these lagged correlation coefficients is significantly negative at the one percent level for every lag.[8] Also, seven of the eight cross-correlation estimates with unemployment lagging inflation are positive, the sign and dynamics supporting the demand link. The appropriate chi-square test indicates that unemployment significantly lags inflation at the five percent level, and the cumulative sum of these lagged correlation coefficients is significantly positive at the five percent level for lag five, and at the one percent level for every lag thereafter. Dividing the sample at the end of 1969 and reestimating by subsample does not substantively alter these findings.

Previous regression studies of the Phillips curve employ a wide variety of specifications, including both static and dynamic models, and models with inflation or unemployment as the dependent variable. The cross-correlation evidence in Table 1 indicates that the most appropriate method to identify the short-run supply link is to specify inflation as the dependent variable and permit lagged adjustment. Static models are appropriate only if one is confident that shift factors in supply are held constant, and dynamic models with unemployment as the dependent variable are appropriate only for identifying the demand relationship.

Table 1 Cross-Correlation Estimates between U.S. Inflation and Unemployment (1953I - 1981IV)

Lag in Quarters	Inflation Lagging		Unemployment Lagging	
	Correlation Coefficient	Cumulative Coefficient	Correlation Coefficient	Cumulative Coefficient
0	-.051	--	-.051	--
1	-.383**	-.383**	.029	.029
2	-.082	-.465**	.010	.039
3	-.159*	-.624**	.110	.149
4	-.158*	-.782**	-.009	.140
5	-.003	-.785**	.030	.170*
6	-.007	-.792**	.191*	.361**
7	.087	-.705**	.319**	.680**
8	-.114	-.819**	.060	.740**
χ^2	24.67		17.14	

* significant at the five percent level.
** significant at the one percent level.

Note: Inflation is the annualized percentage change in the consumer price index; unemployment is the official unemployment rate. The cumulative correlation coefficient includes only the lagged correlation coefficients.

CROSS-SPECTRAL ESTIMATES

This section presents cross-spectral estimates of the intermediate and long-run Phillips curves, explicitly discriminating competing models relating unemployment and inflation. Cross-spectral techniques are used because they describe the dynamic relationship between two variables without requiring the specification of a particular dynamic functional form. Standard regression techniques have little power against dynamic misspecification, but frequency domain methods may detect such misspecification (Engel (1980)). Steady-state parameters, often impossible to estimate in the time domain, are directly obtained in the frequency domain.

Consider the following simple supply model of Pippenger (1982) relating unemployment and inflation:

$$U(t) = -a\ I(t) - b\ \ \Delta I(t) + z(t) \tag{1}$$

where U(t) is the unemployment rate, I(t) is the inflation rate, z(t) is an error term, Δ is the difference operator, and a and b are positive constants. The term b ΔI(t) in eq. (1) is a mathematical representation of demand-induced supply shifts described by the movement from points B to C in Fig. 1a. Based on the assumption that expected and actual rates of inflation converge in the long-run, the term b ΔI(t) implies that the Phillips curve becomes steeper the longer the time horizon. To see this, solve eq. (1) for I(t):

$$I(t) = \frac{1}{(-a\ -b\ (1 - L^{-1}))}\ (U(t) - z(t)) \tag{2}$$

where L^{-1} is the lag operator. Eq. (2) implies that inflation is an exponentially-declining distributed lag of unemployment, i.e., that the Phillips curve is horizontal in the short run, but downward sloping and fixed in long-run steady state.

Eq. (2) contains standard Keynesian and monetarist models as special cases. The Keynesian special case is obtained by setting the parameter b equal to zero, which implies that the short and long-run Phillips curves are identical with a slope of -1/a. The monetarist special case is obtained by setting the parameter a equal to zero, which implies that the short-run Phillips curve is horizontal, but that the long-run curve is vertical. We assume that at intermediate and long-run cycles the variation in z(t) is small relative to exogenous shifts in demand, e.g., those stemming from monetary policy. Thus, at these cycles time series data on inflation and unemployment would tend to trace out the parameters of supply eq. (2).

To discriminate among the various cases of eq. (2), we compare estimated cross spectral statistics for intermediate and long-run relationships between inflation and unemployment to corresponding theoretical statistics implied by the competing models. Spectral analysis summarizes the dynamic link between time series in three statistics -- the coherence square, gain, and phase.[9] Each is a function of differing time horizons or cycles, and to some extent has a regression analog. The coherence square between two series measures by cycle the percentage variation of one series that can be explained by that of the other series, and is analogous to the coefficient of determination (R^2) of regression analysis decomposed by cycle. The gain is analogous to the absolute value of a regression coefficient decomposed by cycle. The phase at a particular cycle measures the delay between two series, but in the frequency domain.

The theoretical coherence square, gain, and phase implied by the various cases of eq. (2) are derived in Appendix 1 and plotted in Fig. 2. For the general case of eq. (2), the gain (from unemployment to inflation) increases as length of cycle increases, which indicates that the Phillips curve becomes steeper as the steady-state is approached. A similar pattern is predicted for the coherence square. At the longest cycle, the gain is finite, i.e., the Phillips curve is not vertical in the long run, and the coherence square is less than unity. The implied phase is 0.5 at both the shortest and longest cycles, and remains between 0.5 and 0.25 for intermediate cycles.

For the Keynesian special case (b equals zero), gain is constant over cycles (the Phillips curve does not shift), the coherence square increases with the length of cycle, and the phase is 0.5 for all cycles, reflecting the inverse link.

For the monetarist special case (a equals zero), theoretical gain and coherence square increase as the length of cycle increases, as with the general form of eq. (2). At the longest cycle, however, coherence square becomes undefined as the gain approaches infinity, i.e., the Phillips curve is vertical in the long run.

Figure 2: Theoretical Cross Spectral Statistics

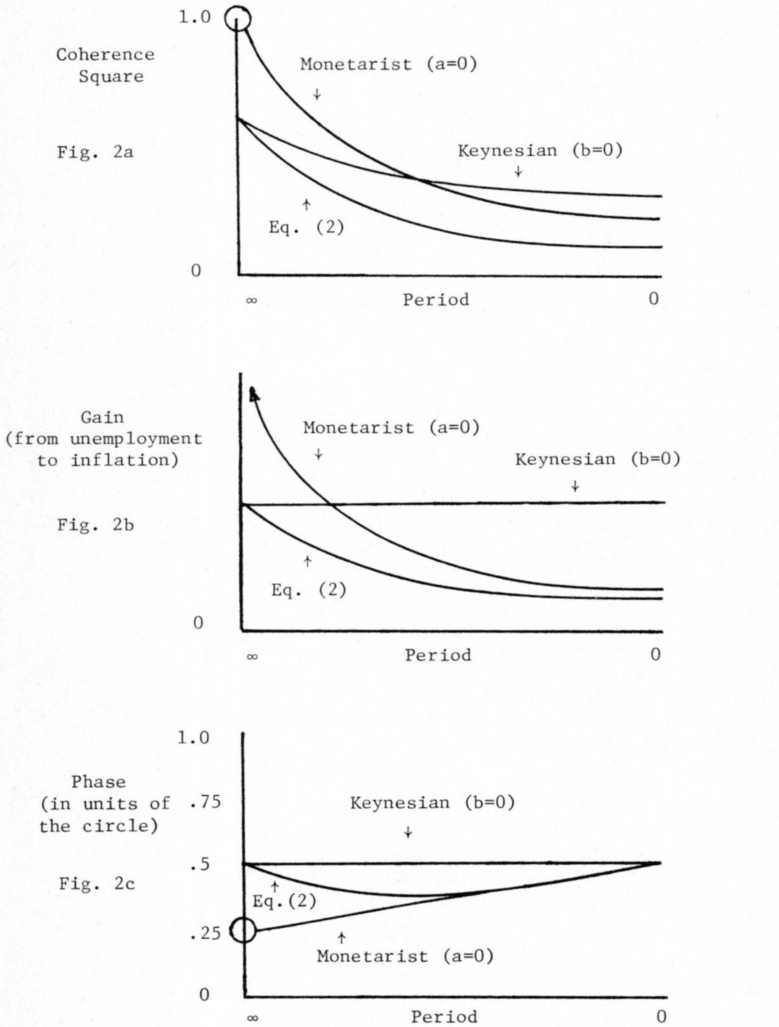

Note: See text for eq. (2) and the Keynesian and monetarist special cases, and Appendix 1 for formal derivation of the theoretical statistics.

As a consequence, estimated gain and coherence square should be zero at the longest cycle. Finally, the theoretical phase for the monetarist case moves from 0.5 to 0.25 as the length of cycle increases, and is also undefined at the longest cycle.

In summary, a phase between 0.25 and 0.5, or equal to 0.5, indicates identification of a supply model (a simple demand relationship between unemployment and inflation implies a phase of zero (and unity)). In addition, two characteristics of the theoretical statistics in Fig. 2 discriminate among the supply models. First, a constant gain over cycles supports the Keynesian special case, whereas an increasing gain is consistent with the other two supply models. Second, an insignificant coherence square and gain at the longest cycle supports only the monetarist model.

The comparison between empirical and theoretical cross-spectral statistics is made at cycles of two years and longer. The comparison is made only for these longer cycles because we assume that it is at these cycles that the supply equation is identified. This follows if (1) expected and actual inflation converge over longer cycles and (2) the variation in error term z(t) is small relative to exogenous shifts in the demand equation over longer cycles. The first assumption is an explicit part of the expectations-augmented Phillips-curve model, and is consistent with our spectral evidence regarding the link between U.S. inflation and expected inflation.[10] The second holds if exogenous supply shocks have little or no recurring long-term structure (i.e., structure at longer cycles).

Table 2 records the empirical cross-spectral statistics between inflation and unemployment based on the U.S. data.[11] The phases which are associated with significant gains are significantly less than 0.5 and numerically greater than 0.25, supporting identification of a supply equation. Other than the seasonal estimate, coherence square and gain are insignificant until the 2.5-year cycle and tend to increase in magnitude as the length of cycle increases to the 10-year estimate. This gain evidence at intermediate cycles is consistent with the predictions of the general form of eq. (2) and the monetarist special case, but not the Keynesian special case.

Coherence square and gain estimates in Table 2 at the longest cycle, i.e., the infinite cycle, clearly discriminate the monetarist model from the other two models. These statistics are insignificant, indicating no steady-state relationship between inflation and unemployment. This finding provides strong support only for the monetarist natural unemployment rate hypothesis. Unlike regression tests of the hypothesis, this evidence is direct in that it by definition estimates the steady-state link and is not based on a proxy for inflationary expectations.

The finding of significant demand and supply responses between unemployment and inflation (in the cross-correlation estimates) yet a strong supply response at cycles of 2.5 to 10 years (in the spectral estimates) indicates that shifts in demand tend to identify supply over these cycles. Since a major systematic shift determinant in demand is monetary and fiscal policy, this evidence is consistent with the hypothesis that business cycle links between unemployment and inflation are linked to systematic macroeconomic policy.

To test the sensitivity of these findings to alternative specifications, three extensions were considered. First, regression estimates of the Phillips curve often measure unemployment as the deviation from the natural rate. The natural unemployment rate series for the United States from 1953 to 1981 is very close to a pure trend. Since the unemployment variable was detrended prior to estimation, such a modification would not alter our estimates. Second, previous estimates of the Phillips curve usually include a measure of inflationary expectations, implying that with inflation as the dependent variable the coefficient on unemployment defines a short-run Phillips curve. Estimation of the cross-spectrum from unemployment to inflation conditional on inflationary expectations (from the

Table 2: Cross Spectral Estimates between U.S. Inflation and Unemployment (1953I - 1981IV)

Period in Years	Coherence Square	Gain	Phase (Units of the Circle)
∞	.011	.31	.88
10	.547*	2.05*	.35 ± .13
5	.718*	1.68*	.35 ± .08
3.33	.650*	1.39*	.35 ± .10
2.5	.422*	.91*	.36 ± .19
2	.220	.52	.37
1.67	.043	.34	.32
1.43	.124	.94	.33
1.25	.168	1.34	.38
1.11	.257	1.16	.66
1.00	.437*	1.28	.69 ± .18

* significant at the five percent level.

Note: Unemployment is the input and plus/minus value on phase is 95 percent confidence interval.

University of Michigan survey) reduces at intermediate cycles the magnitude of the bivariate gain from unemployment to inflation, as predicted. Finally, Levi and Makin (1980) include inflationary uncertainty as a shift determinant of supply. Estimation of the cross-spectrum from unemployment to inflation conditional on the variance of expected inflation yields gain estimates at intermediate cycles which are significant but slightly lower than corresponding bivariate ones.[12]

CONCLUSIONS

Direct, conclusive evidence regarding the long-run links between inflation and unemployment does not exist. This study explores the links for the United States from 1953 to 1981. Time series tests of the lead-lag relationship demonstrate directly the significance of both demand and supply links in the short-run, and indicate that supply is best captured by relating inflation to past rates of unemployment. Frequency domain tests of the steady-state relationship provide clear support for the monetary natural unemployment rate hypothesis. Finally, we find no substantive structural changes in these findings, even for the 1970's.

FOOTNOTES

[1]The theoretical exposition of this section draws heavily from Gordon (1976).

[2]The role of inflationary expectations can be traced back at least to Lerner (1949) and Von Mises (1953). A complicating factor is that demand-induced fluctuations in inflation may increase price uncertainty and unemployment.

[3]For an assessment of the research, see Laidler and Parkin (1975), Santomero and Seater (1978), Parkin (1980), and Gordon (1981).

[4]See Pippenger (1982) for the usefulness of spectral methods in analyzing the theoretical steady-state properties of a dynamic IS-LM model. Scully (1974) investigates empirically the Phillips curve using spectral analysis, but does not relate the analysis to the natural unemployment rate hypothesis and misinterprets the evidence, especially the phase statistic.

[5]For purposes of identification we assume that the structure of aggregate supply and demand is recursive--due to sluggish labor markets, the unemployment response to supply shocks occurs substantially after the price response. In this case there is no contemporaneous demand link between unemployment and inflation, although there may be a contemporaneous supply link.

[6]See Friedman (1968).

[7]Each series (obtained from the Organization for Economic Cooperation and Development) was prewhitened by regressing the series on a distributed lag of past values out eight quarters. The approach produced "whiter" residuals than estimating an ARIMA model.

[8]Chi-square tests of "causality" based on one side of the cross-correlation function are valid as a test of lagged single-equation influence because our two-equation model is identified by its recursive structure. (For a discussion of the identification problem in tests of causality, see Jacobs, Leamer and Ward (1979).) The cumulative sum of the lagged cross-correlation estimates is computed for a test of the sign of the link since the chi-square test is silent about the sign.

[9]For a development of spectral analysis, see Jenkins and Watts (1968).

[10]The coherence square between inflation and inflationary expectations increases with the length of cycle, and exceeds 0.9 for the longest cycles. The U.S. data on inflationary expectations (and the variance of the series used below) are from the University of Michigan Survey Research Center. For a discussion of the series, see Juster and Comment (1978).

[11]For the spectral tests, the same second-order autoregressive filter (with seasonal component) was applied to each series so that recoloring was unnecessary to capture the longer-run effects. Since the empirical coherence square based on these filtered series is very similar to the empirical choherence square based on the prewhitened data, gain and phase estimates in Table 2 appear robust (theoretical coherence square, unlike theoretical gain and phase, is invariant to filtering).

[12]Estimates conditional on expected inflation, or on the variance of expected inflation, could not be computed at the longest cycle due to insufficient degrees of freedom.

APPENDIX 1:

Derivation of Theoretical Cross Spectral Statistics

The model is supply eq. (2). For identification, we assume the variation in z(t) is small relative to the variation in the error term in the demand equation.

The cross spectrum from U(t) to I(t), $\Gamma_{UI}(iw)$, is the expectation of the product between the Fourier transform of I(t) and the complex conjugate of the Fourier transform of U(t):

$$\Gamma_{UI}(iw) = \frac{-(a+b) + b\cos w + i\, b\sin w}{(a+b)^2 + b^2 - 2(a+b)\, b\cos w}\, \Gamma_U(w)$$

where w is frequency in radians, Γ_u is the autospectrum of U(t), and i equals $\sqrt{-1}$.

The frequency response function from U(t) to I(t) is the cross spectrum from U(t) to I(t) divided by the autospectrum of U(t):

$$\frac{\Gamma_{UI}(iw)}{\Gamma_U(w)} = \frac{-(a+b) + b\cos w + i\, b\sin w}{(a+b)^2 + b^2 - 2(a+b)\, b\cos w}$$

The gain from U(t) to I(t), $G_{UI}(w)$, is

$$G_{UI}(w) = \frac{|\Gamma_{UI}|}{\Gamma_U(w)} = \frac{1}{\sqrt{(a+b)^2 + b^2 - 2(a+b)\, b\cos w}}$$

where $|\Gamma_{UI}|$ is the modulus of $\Gamma_{UI}(iw)$.

The phase from U(t) to I(t), $Ph_{UI}(w)$, is

$$Ph_{UI}(w) = \tan^{-1}\left(\frac{-Im\Gamma_{UI}}{Re\Gamma_{UI}}\right) = \tan^{-1}\left(\frac{b\sin w}{-(a+b) + b\cos w}\right)$$

where $Im\Gamma_{UI}$ is the imaginary part of $\Gamma_{UI}(iw)$, and $Re\Gamma_{UI}$ is the real part of $\Gamma_{UI}(iw)$.

The coherence square from U(t) to I(t), $K^2(w)$, is

$$K^2(w) = \frac{|\Gamma_{UI}|^2}{\Gamma_U(w)\Gamma_I(w)} = \frac{1}{1 + \dfrac{\Gamma_z(w)}{G^2_{UI}(w)\, \Gamma_U(w)}}$$

where $\Gamma_z(w)$ is the autospectrum of error term z(t). We assume that exogenous shocks in supply diminish in importance as the length of cycle increases, i.e., $\Gamma_z(w)/\Gamma_U(w)$ declines by cycle. If $\Gamma_z(w)/\Gamma_U(w)$ is assumed constant over cycles, the theoretical coherence square for the Keynesian model would be constant.

To obtain the theoretical predictions in Fig. 2 resulting from eq. (2), evaluate $G_{UI}(w)$, $Ph_{UI}(w)$, and $K^2_{UI}(w)$ by cycle. To obtain the predictions for the Keynesian (monetarist) model, evaluate the statistics after setting the parameter b (a) to zero. Since the gain and phase, unlike the coherence square, do not depend on $\Gamma_z(w)/\Gamma_U(w)$, only the gain and phase are used to discriminate among alternative models.

BIBLIOGRAPHY

Box, G.E.P. and G.M. Jenkins, Time Series Analysis: Forecasting and Control (Revised Version), Holden-Day, 1976.

Engle, R.F., "Hypothesis Testing in Spectral Regression; the Lagrange Multiplier Test as a Regression Diagnostic," in Evaluation of Econometric Models, J. Kmenta and J.B. Ramsey, eds, New York: Academic Press, 1980, 309-321.

Friedman, M., "Comment," in G.P. Shultz and R.Z. Aliber, eds., Guidelines, Informal Controls, and the Market Place, Chicago: Univ. of Chicago Press, 1966.

__________, "The Role of Monetary Policy," American Economic Review, 58 (1968) 1-17.

Gordon, R.J., "Recent Developments in the Theory of Inflation and Unemployment," Journal of Monetary Economics, 2 (1976) 185-219.

__________, "Output Fluctuations and Gradual Price Adjustment," Journal of Economic Literature, 19/2 (1981) 493-530.

Howrey, E.P. "The Role of Time Series Analysis in Econometric Model Evaluation," in Evaluation of Econometric Models, J. Kmenta and J.B. Ramsey, eds, New York: Academic Press, 1980, 275-308.

Jacobs, R.L., E.E. Leamer, and M.P. Ward, "Difficulties with Testing for Causation," Economic Inquiry (1979) 401-13.

Jenkins, G.M. and D.G. Watts, Spectral Analysis and its Applications, San Francisco: Holden-Day, 1968.

Juster, F.T. and R. Comment, "A Note on the Measurement of Price Expectations," unpublished paper, Institute for Social Research, Univ. of Michigan, 1978.

Laidler, D. and M. Parkin, "Inflation: A Survey," Economic Journal, 85 (1975) 741-809.

Lerner, A.P., "The Inflationary Process -- Some Theoretical Aspects," Review of Economics and Statistics, 31 (August 1949) 193-200.

Levi, M.D. and J.H. Makin, "Inflation, Uncertainty, and the Phillips Curve," American Economic Review, 70/5 (December 1980) 1022-27.

Organization for Economic Cooperation and Development, Main Economic Indicators, Various issues.

Parkin, M, "Unemployment and Inflation: Facts, Theories, Puzzles and Policies," in E. Malinvaud and J.P. Fitoussi, eds., Unemployment in Western Countries, New York: St. Martin's Press, 1980.

Phelps, E.S., "Phillips Curves, Expectations of Inflation and Optimal Unemployment over Time," Economica, 34 (August 1967) 254-81.

Pippenger, J. "Monetary Policy, Homeostasis, and the Transmission Mechanism," American Economic Review, 72/3 (June 1982) 545-554.

Santomero, A.M. and J.J. Seater, "The Inflation-Unemployment Trade-off: A Critique of the Literature," Journal of Economic Literature, 56:2 (June 1978) 499-544.

Scully, G.W., "Static vs. Dynamic Phillips Curves," Review of Economics and Statistics, 41/3 (August 1974) 387-389.

Stein, J.L., "Unemployment, Inflation, and Monetarism," American Economic Review, 64:6 (December 1974) 867-87.

Van Order, R., "Unemployment, Inflation and Monetarism: A Further Analysis," American Economic Review, 67:6 (December 1974) 867-87.

Von Mises, L., The Theory of Money and Credit, New Haven: Yale University Press, 1953.

TIME SERIES ANALYSIS: Theory and Practice 4
O.D. Anderson (editor)

ESTIMATION AND FORECASTING OF EQUATIONS WITH EXPECTATIONS VARIABLES USING MULTIPLE INPUT TRANSFER FUNCTIONS

Sergio G. Koreisha

Graduate School of Management
University of Oregon
Eugene, Oregon
U.S.A.

In the article we present an efficient way of dealing with adaptive expectations models, one which makes use of all the information which is available in the data. Our procedure is based on multiple input transfer functions (MITF): by calculating lead and lag cross correlations between innovations associated with the variables in the model, it is possible to determine which periods have the greatest effects on the dependent variable. If information about k periods ahead is required, then using fitted values for the expectations variables, k-period ahead forecasts can be generated. These in turn can be used in the estimation of the transfer function equation which not only contains the usual lagged variables, but also allows for incorporation of lead fitted values for the expectations variables. The MITF identification and estimation procedures that we utilize are based on the Corner Method. We also contrast our method with the Almon distributed lag approach, using a model relating stock market prices to interest rates and expected corporate profits.

1. INTRODUCTION

The usage of the adaptive expectations hypothesis to deal with models which contain variables representing current information about the future, not only forces autoregressive lag structures on the expectation variables, but also imposes specific lag distributions which may not be valid. It is well known that an arbitrary choice of length of a lag distribution can bias parameter estimates (Feige and Pearce, 1979), and that imposition of a lag distribution, e.g., Almon, may give the impression of there being a distributed lag when in effect there is not one (Schmidt and Waud, 1973). Clearly, imposing an unsubstantiated structure on the expectation variables may lead to spurious results.

In this article we suggest a more efficient way of dealing with adaptive expectations models, one which makes use of all the information which is available in the data. In Section 2 we describe our procedure for constructing models with expectations variables. The method requires using Multiple Input Transfer Functions (MITF). In Section 3 we show how to identify and estimate MITF models via the Corner Method. In Section 4 we contrast our method with the Almon distributed lag procedure using a model relating stock market prices to interest rates and expected corporate profits. Finally, in Section 5 we offer some concluding remarks.

2. PROCEDURE FOR ESTIMATING EQUATIONS CONTAINING EXPECTATIONS VARIABLES

The particular information about the future which is needed to explain present conditions can be obtained directly from the data. By calculating lead and lag cross correlations between the innovations associated with the variables in the model it is possible to determine which periods have the greatest effects on the dependent variable. If information about k periods ahead is required, then using fitted values of the expectations variables (obtained from the ARMA filters used to derive the innovations series associated with these variables),[1] k-period ahead forecasts can be generated. These in turn can be used in the estimation of the transfer function equation which not only contains the usual lagged variables (X), but also allows for incorporation of lead fitted values for the expectation variables (E^f), i.e.,

$$Y_t = v_1(B)X_t + v_2(B)E^f_{t+k} + n_t \tag{1}$$

where $v_i(B) = v_{i0} + v_{i1}B + \cdots + v_{im}B^m$; B is the backshift operator such that $B^b a_t = a_{t-b}$; and n_t is the noise corrupting the model, assumed to be generated by an ARMA process which is statistically independent of the input variables.

The advantage of this procedure over the classical ways of dealing with expectations variables are at least threefold:

(i) information contained in the data about past and future is not wasted,

(ii) it reduces the numbers of leads and lags which need to be estimated since no apriori distributed lag specification is imposed, i.e., the data set itself is used to determine an appropriate lead/lag structure, and

(iii) parameter estimates and predictions are not as susceptible to the consequences of residual autocorrelation since transfer function methods unlike distributed lag regression procedures deal explicitly with the error term (noise).

The actual steps used in the formulation of a model containing expectation variables can be summarized as follows:

a. Construct the ARIMA models associated with each variable in the model to derive the innovation series associated with them.

b. Calculate the cross correlations between innovations series associated with the dependent variable and those of the independent variables. Verify, via Haugh (1976) causal tests (see Appendix), that the specification based on theory can be supported empirically, i.e., for the expectation variables, only the lead cross correlations (estimated from the actual values of the variables) should collectively be significant; whereas, for other explanatory variables, only the lag cross correlations should collectively be significant.[2]

c. Construct fitted-value series for the expectation variables using the filters in Step a.

d. Apply the Corner Method identification procedure (to be described in the next section) to identify the lead/lag order of transfer function equation (1).

e. Estimate the multiple-input transfer function equation.

---

[1]The conventions and notations used in this article follow those in Box and Jenkins (1970).

[2]Significant lead and lag cross correlations would be an indication of feedback between variables.

3. IDENTIFICATION AND ESTIMATION OF THE TRANSFER FUNCTION EQUATION

The methodology used for the construction of bivariate transfer function models is well known and well documented (Box and Jenkins, 1970; Nelson, 1973; Montgomery and Johnson, 1976). It consists of four stages: identification, estimation, diagnostic checks, and forecasting.[3] In the *identification* phase potential models are singled out for further analysis by examining auto, partial, and cross correlations. Parameters for these potential models are then *estimated*, and afterwards subjected to various *diagnostic* checks to determine which model will be used for *forecasting* purposes.

The methodology used for the construction of multiple input transfer functions, on the other hand, is not as well-known nor as well documented (Koreisha, 1980; Liu and Hanssens, 1982; Koreisha, 1982). In this section we will focus on the corner method, primarily because of its relative simplicity, and also because of its applicability to situations which require using fitted values.[4] We will first describe how the method works for situations not involving expectation variables, and then will note the minor modifications that are necessary when dealing with such variables.

The origins of the Corner method for identifying possible multiple-input transfer function model structures stem from identification problems associated with univariate ARMA processes. Beguin, Gourieroux and Montfort (BGM) (1980)[5] have proven that:

> The process x_t has a minimal ARMA (p,q) representation, if and only if, the sequence of autocorrelations $[\rho_h, h \varepsilon Z]$ satisfies a linear difference equation of minimal rank q+1:

$$\rho_h - \phi_1\rho_{h-1} - \cdots - \phi_p\rho_{h-p} \begin{cases} = 0 & h>q+1 \text{ with } \phi_p \neq 0 \\ \neq 0 & h=1. \end{cases} \quad (2)$$

The term "minimal" refers to the smallest possible values for both p and q such that (2) will be satisfied. That this theorem is reasonable can be seen by comparing the theorem's conditions to the usual Box-Jenkins characterizations of pure AR and MA processes.

As one can readily see, this difference equation characterization for an ARMA (p,q) process is not of much practical use because the values of ϕ_i are not known in advance. However, BGM have shown that the minimal order and rank conditions, and, thus, the characterization of the x_t process is equivalent to the conditions:

(1) $\Delta(i,j) = 0 \quad \forall_i \geq q+1$ and $\forall_j \geq p+1$

(2) $\Delta(i,p) \neq 0 \quad \forall_i \geq q$

(3) $\Delta(q,j) \neq 0 \quad \forall_j \geq p$

---

[3]See also Anderson (1977) for dicussion of some extra stages.

[4]The residual cross correlation method (Koreisha, 1980) for example, requires the use of innovation series to iteratively construct the model structure, thus making it inappropriate for constructing models for which fitted values are an integral part of the equation.

[5]Gray, Kelly, and McIntire (1978) have also proposed similar criteria for model identification.

where $\Delta(i,j)$ is the determinant of

$$D(i,j) = \begin{bmatrix} \rho_i & \rho_{i-1} & \cdots & \rho_{i-j+1} \\ \rho_{i+1} & \rho_i & \cdots & \rho_{i-j+2} \\ \vdots & \vdots & & \vdots \\ \rho_{i+j-1} & \rho_{i+j-2} & \cdots & \rho_i \end{bmatrix} ; \; i \geq 0; \; j \geq 1.$$

A natural identification procedure, therefore, can be developed as follows:

(a) estimate the values for K autocorrelations (K should be greater than either p or q);

(b) build the array of determinants $\Delta(i,j)$, $i=1,\ldots,K$; $j=1,\ldots,K$ using the autocorrelation estimates;

(c) observe the entries in the array of determinants which have zero values, and deduce the order of the process.

The process x_t has a minimal ARMA (p,q) representation if, and only if, the array of determinants has the structure in Table 1.

TABLE 1

ARRAY OF DETERMINANTS FOR AN ARMA (p,q) PROCESS

i \ j	1	2	...	p	p+1	...	K
1	$\Delta(1,1)$	$\Delta(1,2)$		$\Delta(1,p)$	$\Delta(1,p+1)$		$\Delta(1,K)$
2	$\Delta(2,1)$			.	.		
.				.	.		
q	$\Delta(q,1)$			x	x		x
q+1	$\Delta(q+1,1)$			x	0	...	0
.				.	.		.
.				.	.		.
K	$\Delta(K,1)$			x	0	...	0

The x indicates that the corresponding term is different from zero.

Statistical tests of significance have been developed to check if the values of the determinants are different from zero (BGM, 1980). However, visual inspection is generally sufficient to select a small number of possibilities for the estimation stage. Finally, we note that the corner method procedure for both pure AR and MA processes is linked to the usual criteria of auto and partial autocorrelations, since $\Delta(h,1) = \rho_h$ and $(-1)^{h-1} \frac{\Delta(1,h)}{\Delta(0,h)} = \phi_{hh}$.

Extension of the BGM corner method for identification of multiple input transfer function models is relatively straightforward. In a manner analogous to the estimation of the K autocorrelation values for building the D(i,j) matrix, Liu and Hanssens (1982) suggest that estimates for the transfer function weights of rational form equations such as

$$y_t = C+(v_{10}+v_{11}B+\cdots+v_{1k_1}B^{k_1})x_{1t}+\cdots+(v_{m0}+v_{m1}B+\cdots+v_{mk_m}B^{k_m})x_{mt}+\varepsilon_t$$

be derived by least squares for arbitrarily large values of k_i (sufficiently large to be greater than r_i and s_i).[6] After standardizing the estimates of v_{ij}, by dividing them by $v_{i,max}$, where $v_{i,max}$ is the maximum value of $|v_{ij}|$ for $j=0,\ldots,k_i$, i.e., calculating $\eta_{ij} = v_{ij}/v_{i,max}$ (η_{ij} is analogous to ρ_i), m matrices $D_i(f,g)$ and their corresponding array of determinants $\Delta_i(f,g)$ are constructed where[7]

$$D_i(f,g) = \begin{bmatrix} \eta_{if} & \eta_{i,f-1} & \cdots & \eta_{i,f-g+1} \\ \eta_{i,f+1} & \eta_{i,f} & \cdots & \eta_{i,f-g+2} \\ \vdots & & & \vdots \\ \eta_{i,f+g-1} & \eta_{i,f+g-2} & \cdots & \eta_{i,f} \end{bmatrix} \qquad \begin{array}{l} f \geq 0, \quad g \geq 1 \\ \eta_{ij} = 0, \ \forall_j > 0 \\ i = \{1,\ldots,m\}. \end{array}$$

The arguments f and g in the arrays of determinants represent the orders of the polynomials $\delta_i(B)$ and $\omega_i(B)B^{b_i}$ respectively, just as i and j represent the orders of $\phi(B)$ and $\theta(B)$ respectively in the univariate case (Table 1). Note that the corner method only serves to identify the order of the polynomials of the transfer function; the error component in (1) may still need to be modelled if after fitting the transfer function component some systematic process can still be identified.

Thus, the characterization of the transfer function relating y_t and x_{it} could be inferred by observing the appropriate values of s_i, r_i, and b_i which satisfy the following conditions:

(1) $\Delta_i(f,g) = 0 \qquad \forall f < b_i$

(2) $\Delta_i(f,r) \neq 0 \qquad \forall f > s_i+b_i$

(3) $\Delta_i(s_i+b_i,g) \neq 0 \quad \forall g \geq r_i$

(4) $\Delta_i(f,g) = 0 \qquad \forall f \geq s_i+b_i+1$ and $\forall g \geq r_i+1$.

In other words, the transfer function weights v_{ij} have a representation $\omega_i(B)B^{b_i}/\delta_i(B)$ with orders r_i, s_i, and b_i, if, and only if, the array of determinants $D_i(f,g)$ have the structure in Table 2. The noise portion of the model, $n_t = y_t-\Sigma_i v(B)x_{it}$, if necessary, can be modelled using the usual Box-Jenkins (1970) univariate procedures.

Note that in the arrays of determinants, f is allowed to take values from 0 to N_i, while g is restricted to values only between 1 and N_i. This is because of normalization rules: the output value must appear with a coefficient of 1 at time t while the input variables may appear in the equation with any value for its starting lag.

---

[6]The values of k_i should also be sufficiently large to avoid truncation bias.

[7]Stipulation of $\eta_{ij} = 0$ for $j < 0$ implies that the model is causally adequate.

TABLE 2

ARRAY OF DETERMINANTS FOR A TRANSFER FUNCTION COMPONENT X_i

WITH POLYNOMIAL ORDERS b_i, s_i, and r_i

f \ g	1	2	...	r_i	r_i+1	...	N_i*
0	0	0	...	0	0	...	0
1	0	0	...	0	0	...	0
.	.	.		.	.		.
.	.	.		.	.		.
.	.	.		.	.		.
b_i-1	0	0	...	0	0	...	0
b_i	$\Delta(b_i,1)$	$\Delta(b_i,2)$	...	$\Delta(b_i,r_i)$	$\Delta(b_i,r_i+1)$		$\Delta(b_i,N_i)$
.	.	.		.	.		.
.	.	.		.	.		.
.	.	.		.	.		.
s_i+b_i	$\Delta(s_i+b_i,1)$	$\Delta(s_i+b_i,2)$		x	x	...	x
s_i+b_i+1	$\Delta(s_i+b_i+1,1)$	$\Delta(s_i+b_i+1,2)$	...	x	0	...	0
.	.	.		.	.		.
.	.	.		.	.		.
.	.	.		.	.		.
N_i	$\Delta(N_i,1)$	$(N_i,2)$	...	x	0	...	0

The x indicates that the corresponding term is different from zero.

*N_i is chosen to be greater than r_i and s_i. (Due to losses in degrees of freedom one has to be somewhat cautious about the magnitude of N_i.)

It should also be noted that obtaining estimates for v_{ij} through Ordinary Least Squares (OLS) estimation using the actual data may be problematical when one of the input series contains an AR factor with roots close to one, i.e., if the series is approaching nonstationarity, then the corresponding X'X matrix may be nearly singular. Liu and Hanssens (1982) recommend that prior to estimating the transfer function in rational form by OLS, one should construct ARIMA models for the input series to see if AR factors are found and if the roots of the factors are close to one. If there are processes with roots close to one, then they recommend choosing a common filter from the AR factors and using it to filter all the input and output series. The use of a common filter does not alter the transfer function weights if the series are stationary. Furthermore, if the noise series ε_t in the transfer function equation is not white noise, then Generalized Least Squares (GLS) should be used instead of OLS.

Identification of models for which the coefficients of some lagged right hand side (RHS) variables are zero within a specified range, would be very difficult with this method. The best that one can do is to identify the maximum lag variable, and then in the estimation of the transfer function equation observe that some of the coefficients of some lagged variables within the specified range are insignificant.

For equations containing expectation variables, instead of estimating equations with just current and lagged values for the RHS variables, we estimate regression equations which also contain lead fitted values for the expectations variables. The length of the lead structure (and for that matter the lag structure) can be inferred directly from the values of the cross correlations estimated in Step b of Section 2 in conjunction with the ARIMA filters (Step a) used in the construction of the innovation series. If the expectation variables are the only RHS variables containing AR factors, and if one or more of them is solely governed by an AR process, then fitted values should be constructed from a parsimonious MA approximation to the AR model. (See discussion of Psi-weights in Box and Jenkins, 1970.)

4. EXPECTATIONS AND STOCK MARKET PRICES

In order to illustrate our procedure with real data we will use some of Keran's (1971) assertions about expectations and the stock market. Very briefly, his model is based on the assumption that the average investor's evaluation of expected returns (expected dividends and expected changes in stock prices discounted to present value) will determine the price of the stock. Expected changes in capital gains in turn are assumed to be affected primarily by the rate at which retained earnings are plowed back into the firm. Noticing that dividends and earnings are related, and that during most normal periods, expectations about them are influenced by current and recent past experiences, Keran theorized the following relationship between stock prices (SP), interest rates (R), and current expectations about future corporate earnings (E^e),

$$SP_t = a_0 + \sum_{i=0}^{1} a_i R_{t-i} + a_2 E_t^e + n_t \tag{3}$$

where

$$E_t^e = \sum_{i=0}^{n} \omega_i E_{t-i}.$$

This adaptive expectations model has been criticized by others for many reasons (Pesando, 1974; Rogalski and Vinso, 1977), but of concern to us at this stage is that it *forces* a lag distribution which may neither be adequate nor reasonable. (Keran estimated an Almon lag polynomial of degree 6 and of order 19 for the actual values of corporate earnings and a lag polynomial of degree 2 for interest rates.)

Thus, it may be possible to obtain a more adequate specification and, consequently, more reliable forecasts for stock prices by constructing a model which contains a specific lead structure for fitted values of corporate profits. In this section we will compare our thus derived model with one which adheres to the adaptive expectations hypothesis, i.e. one which replicates Keran's Almon lag procedure with our data base.[8] The comparison will be made in terms of which model generates the "best" forecasts. Models will be fitted with only a portion of the data and then used to make predictions for periods for which data are available.

Using the following quarterly data covering the period from 1947.1 to 1977.1 (121 observations):[9]

SP_t = Standard & Poor's Stock Price — Composite (common stock) — Series: JS&PINDNS

---

[8]The Durbin-Watson statistic in Keran's equation was very low suggesting that his results might be spurious.

[9]Source of Data: Data Resources Incorporated. Accessed through Harvard Business School Computer Center. Ten more data points are available for forecasting comparisons.

R_t = Yield on Moody's AAA Corporate Bonds;
Series: RMMBCAAANS

E_t = Corporate Profits After Tax Excluding Inventory Valuation Adjustment in billions of current dollars;
Series: ZA;

we first estimated the appropriate ARIMA models governing each of the series and then calculated the cross correlations between each pair of innovations associated with the actual series. Tables 3 and 4 respectively contain a summary of the models and the estimates of the cross correlations. The Haugh (1976) tests for

TABLE 3

ARIMA MODELS FOR STOCK MARKET DATA

$$\nabla SP_t = (1 + \underset{(.081)}{.441}B + \underset{(.083)}{.175}B^8)\beta_t$$

$$\nabla R_t = (1 + \underset{(.076)}{.539}B)\alpha_{1t}$$

$$\nabla CP_t = (1 + \underset{(.084)}{.290}B + \underset{(.087)}{.311}B^6)\alpha_{2t}$$

TABLE 4

CROSS CORRELATIONS BETWEEN THE INNOVATION SERIES

(Stock Market Data)

CROSS CORRELATIONS:

Between Innovations β_t and α_{1t}

Zero Lag = -.27

Leads Future $\beta_t, \alpha_{1,t+k}$ Standard Error = .09

1-12	.06	.12	-.03	-.08	-.03	.18	-.04	.14	.02	.18	.00	.10
13-24	.07	.03	-.04	-.25	.06	.04	.07	-.11	.00	.02	.01	.04

Lags Past $\beta_t, \alpha_{1,t-k}$

1-12	-.43	.01	.14	-.05	-.11	.13	.10	.03	.08	.05	-.04	.08
13-24	-.05	.03	.09	-.17	-.14	-.01	.06	-.19	.04	-.04	-.02	.00

Between Innovations β_t and α_{2t}

Zero Lag = .01 Standard Error = .09

Leads Future $\beta_t, \alpha_{2,t+k}$

1-12	.31	.29	.01	-.04	.01	-.23	.02	-.11	-.14	-.05	-.11	-.12
13-24	.11	.14	-.10	-.07	-.04	.13	.18	-.05	.04	-.06	.01	-.14

Lags Past $\beta_t, \alpha_{2,t-k}$

1-12	-.19	.11	-.04	-.10	-.11	-.04	-.12	.05	-.03	.04	-.15	.01
13-24	.03	-.06	.06	.02	.10	-.03	.10	-.06	-.02	-.04	.00	.04

series indepencence and causal direction (Table 5) conducted from the estimates of the cross correlations corroborate the assertion that expectations of future corporate earnings indeed affect current stock prices. Thus, using the moving average filter of Table 3 on the series $\{E_t\}$ we obtained the fitted values for the corporate profit series $\{E_t^f\}$ to use in the estimation of OLS and GLS regressions of the form:[10]

$$\nabla SP_t = C + \sum_{k=0}^{m_1} a_k \nabla R_{t-k} + \sum_{k=0}^{m_2} b_k \nabla E^f_{t+k} + n_t.$$

TABLE 5

HAUGH DIRECTIONAL TESTS BETWEEN THE STOCK MARKET INNOVATION SERIES

Innovation Pair	——> $122\sum_{i=-8}^{-1}$	——> $122\sum_{i=-12}^{-1}$	<—— $122\sum_{i=1}^{8}$	<—— $122\sum_{i=1}^{12}$	<——> $122\sum_{i=-8}^{8}$	<——> $122\sum_{i=-12}^{12}$	r(0)/SE
$\beta_t - \alpha_{1t}$	9.7	14.9	30.1	32.2	48.8	56.0	3.0
$\beta_t - \alpha_{2t}$	30.2	36.1	11.0	14.1	41.2	50.2	.1
$\alpha_{1t} - \alpha_{2t}$	21.9	26.6	4.9	9.3	27.5	36.7	.9
	$\chi^2_{.95}=15.5$	$\chi^2_{.95}=21.0$	$\chi^2_{.95}=15.5$	$\chi^2_{.95}=21.0$	$\chi^2_{.95}=27.6$	$\chi^2_{.95}=37.6$	

Arrows indicate causal direction.

The results associated with the various regressions are found in Table 6. Those associated with the GLS estimates for the regression containing 10 leads and lags and an AR(1) error component 1 (column 6), were used to construct the arrays of determinants (Table 7) for the corner method, from which the following maximum order transfer function representation was tentatively identified:

$$\nabla SP_t = C + (\omega_{10}-\omega_{11}B-\cdots-\omega_{19}B^9)\nabla R_t + \frac{(\omega_{20}-\omega_{21}B-\cdots-\omega_{26}B^6)}{(1-\delta_{21}B-\delta_{22}B^2)} \nabla E^f_{t+10} + n_t$$

where n_t could be any ARMA process.

Since in general we are not used to thinking in terms of lead RHS variables, the partitioning of nonzero values in the array of determinants associated with ∇E_t^f requires some explanation. If a transfer function model contains denominator parameters as well as lead RHS numerator parameters, then it is possible to have nonzero v-weights for leads greater than the lead of the RHS numerator parameter, which will then give rise to patterns in the array of determinants similar to the one associated with ∇E_t^f. To see this, all one has to do is equate the polynomials of (1) with those of the alternative form:

$$Y_t = \frac{\omega_1(B)B^{b_1}X_t}{\delta_1(B)} + \frac{\omega_2(B)B^{b_2}E^f_{t+k}}{\delta_2(B)} + n_t$$

[10]The equations were estimated in terms of changes in the variables because in order to induce stationarity; all univariate models required first differencing.

to derive the set of simultaneous equations relating the v-weights to the numerator and denominator polynomials. For example, associated with the simple model

$$Y_t = \frac{(\omega_0 - \omega_1 B)}{(1-\delta B)} X_{t+b} + n_t,$$

with b > 0, are the polynomial equations

$$\begin{aligned} &\vdots \\ v_{-(b+1)} &- \delta v_{-(b+2)} = 0 \\ v_{-(b)} &- \delta v_{-(b+1)} = \omega_0 \\ v_{-(b-1)} &- \delta v_{-(b)} = -\omega_1 \\ v_{-(b-2)} &- \delta v_{-(b-1)} = 0 \\ &\vdots \end{aligned}$$

TABLE 6

ESTIMATES OF THE TRANSFER FUNCTION WEIGHTS

(Stock Market Data)

	OLS		OLS		GLS		GLS	
Transfer Function Weights for Variable	Parameter Estimate	Standard Error	Parameter Estimate	Standard Error	Parameter Estimate	Standard Error	Parameter Estimate	Standard Error
Intercept	1.247	0.469	1.452	0.482	1.252	0.567	1.456	0.568
VR(t)	-6.103	2.163	-6.090	2.242	-6.743	2.048	-6.175	2.198
VR(t-1)	-5.767	2.359	-5.664	2.389	-5.937	2.122	-5.698	2.251
VR(t-2)	0.030	2.374	0.836	2.472	-0.013	2.156	0.754	2.318
VR(t-3)	0.140	2.382	-0.413	2.503	0.434	2.156	-0.297	2.319
VR(t-4)	3.854	2.415	3.721	2.483	3.135	2.144	3.661	2.275
VR(t-5)	-3.900	2.460	-3.135	2.527	-3.341	2.166	-3.174	2.317
VR(t-6)	1.455	2.451	2.178	2.500	1.322	2.171	2.148	2.288
VR(t-7)	4.077	2.491	3.434	2.563	4.250	2.217	3.464	2.348
VR(t-8)	0.209	2.502	-0.128	2.573	-0.286	2.228	-0.124	2.351
VR(t-9)	4.060	2.593	3.900	2.692	4.751	2.317	3.970	2.468
VR(t-10)	0.269	2.338	0.908	2.678	0.424	2.227	0.820	2.459
VR(t-11)	—	—	-1.352	2.653	—	—	-1.230	2.448
VR(t-12)	—	—	0.921	2.392	—	—	0.615	2.356
$VE^f(t)$	0.085	0.130	0.076	0.134	0.101	0.119	0.076	0.131
$VE^f(t+1)$	0.199	0.130	0.239	0.134	0.105	0.118	0.232	0.136
$VE^f(t+2)$	0.136	0.118	0.211	0.123	0.171	0.116	0.207	0.128
$VE^f(t+3)$	-0.025	0.110	0.014	0.113	0.026	0.104	0.010	0.116
$VE^f(t+4)$	-0.047	0.104	-0.117	0.120	-0.079	0.089	-0.110	0.122
$VE^f(t+5)$	-0.049	0.097	-0.081	0.109	-0.103	0.089	-0.080	0.110
$VE^f(t+6)$	-0.136	0.093	-0.135	0.096	-0.107	0.087	-0.133	0.096
$VE^f(t+7)$	-0.250	0.095	-0.257	0.100	-0.284	0.090	-0.260	0.099
$VE^f(t+8)$	0.021	0.093	-0.023	0.096	0.032	0.089	-0.020	0.096
$VE^f(t+9)$	-0.176	0.096	-0.193	0.098	-0.210	0.090	-0.194	0.097
$VE^f(t+10)$	-0.197	0.093	-0.189	0.095	-0.140	0.087	-0.187	0.095
$VE^f(t+11)$	—	—	-0.076	0.092	—	—	-0.073	0.093
$VE^f(t+12)$	—	—	-0.140	0.090	—	—	-0.129	0.089
ϕ	—	—	—	—	-0.223	0.097	-0.186	0.100

TABLE 7

ARRAYS OF DETERMINANTS FOR TRANSFER FUNCTION COMPONENTS

(Stock Market Data)

Transfer Function Component ∇R_t

f \ g	1	2	3	4	5	6	7	8
t	-1.00	1.00	-1.00	1.00	-1.00	1.00	-1.00	1.00
t-1	-0.92	0.97	-1.06	1.73	-2.84	4.28	-5.27	6.90
t-2	0.12	-0.03	0.57	0.00	-0.63	3.37	1.75	0.41
t-3	-0.05	-0.07	-0.30	0.20	-0.14	2.91	-0.32	-2.20
t-4	0.59	0.33	0.19	0.34	0.92	2.50	3.71	5.17
t-5	-0.51	0.06	0.25	-0.27	0.06	0.98	-2.61	4.61
t-6	0.35	0.41	0.43	0.17	0.29	0.44	0.62	1.97
t-7	0.56	0.32	0.45	0.35	0.16	0.02	0.19	0.54
t-8	-0.02	-0.36	0.20	0.32	0.06	-0.07	0.04	0.15
t-9	0.64	0.42	0.35	0.26	0.16	0.10	0.06	0.04
t-10	0.13	0.15	0.08	0.03	0.02	0.01	0.01	0.00

Transfer Function Component ∇E_t^{f}*

f \ g	1	2	3	4	5	6	7	8
t+12	-0.49	0.24	-0.12	0.06	-0.03	0.01	-0.01	0.00
t+11	-0.28	-0.28	-0.00	0.14	-0.04	-0.03	0.02	0.02
t+10	-0.72	0.31	-0.33	0.33	-0.23	0.12	-0.11	0.12
t+9	-0.74	0.50	-0.64	0.26	-0.23	0.31	-0.18	0.34
t+8	-0.08	-0.74	-0.86	-0.23	0.13	0.43	0.66	0.81
t+7	-1.00	0.96	-0.89	0.62	-0.51	0.34	-0.47	1.50
t+6	-0.51	-0.04	-0.23	0.31	0.39	-0.29	-0.43	1.44
t+5	-0.31	-0.12	-0.08	0.31	-0.48	0.75	-1.28	2.07
t+4	-0.42	0.19	-0.19	0.19	-0.01	0.18	-0.21	0.73
t+3	0.05	0.34	0.01	0.11	0.07	0.04	0.07	0.21
t+2	0.79	0.59	0.20	0.06	-0.07	-0.02	0.02	0.05
t+1	0.89	0.56	0.30	0.15	0.07	0.02	0.01	0.00
t	0.29	0.09	0.03	0.01	0.00	0.00	0.00	0.00

*Dotted lines indicate another possible characterization of the array.

Thus, since $\delta \neq 0$ by construction, it follows that one of several v-weight patterns associated with the above model includes declining nonzero values for $v_{-(b+k)}$, where $k \geq 1$. (A negative subscript associated with a v-weight refers to a lead value.) Consequently, the necessary nonzero values in the positions used to denote denominator parameters in the array of determinants matrix, can be of the form shown below since the first column of the matrix is nothing more than the normalized v-weights themselves:

g \ f	1	2	3	4
t+k	x			
.	.		0	
.	.			
.	.			
t+a	x	x	x	x
t+a-1	x	x	x	x
.	.			
.	.		0	
.	.			
t	x			

A tentative third order polynomial denominator parameter associated with ∇E^f_{t+10} could also have been included. Our preliminary trials, however, indicated that this parameter was unnecessary.[11]

As expected, estimation of (3) with n_t modelled as an AR(1) process yielded an overparameterized equation. Subsequent refining steps yielded the following more parsimonious equation:

$$\nabla SP_t = \underset{(.42)}{1.55} + (\underset{(1.81)}{-6.52} - \underset{(1.83)}{7.32}B + \underset{(2.02)}{3.80}B^7 + \underset{(2.10)}{4.70}B^9)\nabla R_t$$

$$+ \frac{(\underset{(.06)}{-.092} - \underset{(.07)}{.148}B - \underset{(.10)}{.342}B^3 - \underset{(.08)}{.212}B^5 - \underset{(.07)}{.118}B^6)}{(1 - \underset{(.12)}{.469}B + \underset{(.11)}{.865}B^2)} \nabla E^f_{t+10} + \frac{a_t}{(1 - \underset{(.11)}{.245}B)} \quad (4)$$

where the numbers within parentheses refer to the standard errors of the parameters.

The negative signs associated with the coefficients of the fitted lead corporate profits may at first glance appear counterintuitive. However, if we consider the fact that the post 1973 era was a period of unprecedented high inflation in the US,[12] then the following argument might help explain the signs. High corporate earnings during this period were brought about by changes in the economy which also created high inflation. Inflation, as well as inflation expectations, lead to higher interest rates which tend to depress stock prices.[13] Thus, expected changes in corporate profits may have acted as proxies for expected inflation.

It should also be noted that very little is actually known about the mechanisms guiding the post 1973 economy. Currently, James, Koreisha, and Partch (1982) are in the process of studying the relational mechanisms between inflation and stock prices, through key variables such as industrial production, money supply, and interest rates.

The cross correlations between the residuals and the RHS variables of equation (4) are white noise, an indication of model adequacy.

In order to contrast our results with those using Keran's approach we first attempted to estimate the best Almon distributed lag model using our data. Based on actual data all our trials produced equations with extremely low Durbin Watson (DW) statistics (below .2), no matter the lag length or the polynomial order

---

[11]At present Taylor (1982) is studying the asymptotic distributional properties of the array of determinants in order to derive tests of significance for the entries. Until final details are worked out, judgment will be necessary. However, the number of tentative possibilities will in general be very small.

[12]Post World War II.

[13]It is possible that in the early stages of an inflation, when expectations of continued inflation are not yet strong, for stock prices to rise. However, when inflation continues long enough so that major decision-making units in the economy expect further inflation, then stock prices drop.

imposed on the corporate profits variable.[14] In order to improve the DW statistic we estimated an Almon distributed lag model in terms of first differences. This transformation markedly improved the DW statistic (from .15 to 1.6). In checking the residuals associated with first difference equations, however, we observed that these were still somewhat autocorrelated ($.15 < r_1 < .2$). As a result we applied the Cochrane-Orcutt procedure to estimate the representative first difference equation which is given below:

$$\nabla SP_t = \underset{(.40)}{1.57} + \sum_{i=0}^{1} - \underset{(2.12)}{15.82}\nabla R_{t-1} + \sum_{i=0}^{18} - \underset{(.05)}{.157}\nabla E_{t-i} + n_t \tag{5}$$

$N = 121$

$SSR = 1040.6$ $\qquad n_t = \underset{(.08)}{.15}n_{t-1} + a_t$

$SE = 3.05$

$DW = 1.82$

Degree of Polynomial Associated with $\nabla E_{t-1} = 6$[15]

Degree of Polynomial Associated with $\nabla R_{t-1} = 2$

$\nabla E_{t+1} \neq 0;\ \nabla E_{t-n} = 0$

$\nabla R_{t+1} \neq 0;\ \nabla R_{t-n} = 0$

---

[14] If the error term is autocorrelated, then the least squares estimates are no longer efficient, and the least square variance will also be biased. In addition, the prediction formula $\hat{Y} = \underline{X}\hat{\beta}$ will be incorrect if the residuals are autocorrelated. The correct prediction formula should be $\hat{Y} = \underline{X}\hat{\beta} + V'\psi^{-1}(Y-X\hat{\beta})$ where $\underline{X}$ is the matrix of future explanatory variables assumed known, $\sigma_e^2\psi$ is the covariance matrix of future disturbances and $\sigma_e^2 V$ is a matrix of covariances between past and future disturbances. If the residuals are known to follow an AR(1) process then the prediction formula simplifies to $\hat{Y}_{T+h} = \underline{X}'_{T+h}\hat{\beta} + \rho_h(Y_T - X_T'\hat{\beta})$ where $\hat{\beta}$ is the GLS estimate. (See Judge et al., 1980, p. 210.)

[15] The degrees of the polynomials and lag lengths were chosen to minimize the severity of the Almon constraints. Furthermore, since the purpose of this estimation is for contrasting forecasts, consideration was also given to polynomials and lag lengths which produced the smallest SSR values, best DW-statistics, and smallest regression standard errors. Almon (1965) established the convention that the maximum degree of the polynomial should be set at one more than the number of lags of the independent variables up to 5. Since then several other approaches have been suggested in the literature (Judge et al., 1980), none of which is truly reliable (nor for which much is known about the properties of the resulting estimates). Without prior information about polynomial degrees and lag lengths, these procedures generally generate biased results. (For a discussion of the consequences of applying incorrect restrictions see Trivedi and Pagan, 1976.) We did not consider estimating an Almon distributed lag model for both RHS variables in terms of changes because of the large number of permutations (associated with degree order and lag length) which would have to be considered, and also because in light of the above this is not an adequate way of dealing with the problem. Since our approach provides a systematic procedure for zeroing in on the appropriate lag structure, there is no need to embark on such a fruitless exercise.

In order to generate one step ahead forecasts from equations (4) and (5) we first derived the one step ahead forecasts for the RHS variables using the ARIMA models of Table 3,[16] and then systematically updated the parameters of the equations to generate forecasts for ∇SP. These forecasts are found in Table 8.[17]

TABLE 8

ONE STEP AHEAD FORECASTS

Equation (4) Forecasts

Period	Forecast	Actual	95% Confidence Limit Lower	95% Confidence Limit Upper	Percent Deviation from Actual
1977.2	102.86	99.03	96.64	109.09	+ 3.87
1977.3	98.83	98.05	92.54	105.11	+ .79
1977.4	97.67	93.95	91.42	103.91	+ 3.96
1978.1	92.21	89.35	85.96	98.46	+ 3.20
1978.2	85.93	95.93	79.70	92.15	-10.40*
1978.3	95.40	101.67	88.95	101.84	- 6.17
1978.4	99.75	97.13	93.61	105.89	+ 2.70
1979.1	93.64	99.35	87.15	100.13	- 5.75
1979.2	106.09	101.18	99.67	112.50	+ 4.85
1979.3	102.45	106.22	95.96	108.94	- 3.55

Equation (5) Forecasts

Period	Forecast	Actual	95% Confidence Limit Lower	95% Confidence Limit Upper	Percent Deviation from Actual
1977.2	101.58	99.03	94.33	108.82	+ 2.57
1977.3	96.91	98.05	89.72	104.10	- 1.17
1977.4	97.09	93.95	89.85	104.33	+ 3.35
1978.1	90.52	89.35	83.30	97.75	+ 1.31
1978.2	85.44	95.93	78.22	92.67	-10.93*
1978.3	95.36	101.67	87.90	102.81	- 6.20
1978.4	101.24	97.13	93.75	108.72	+ 4.23
1979.1	90.96	99.35	83.43	98.49	- 8.44*
1979.2	93.15	101.18	85.59	100.73	- 7.94*
1979.3	98.68	106.22	91.00	106.35	- 7.11

*Actual values are outside the 95% forecast confidence band.

As can be seen from Table 8, forecasts from equation (4) appear to be markedly better than those derived from equation (5). The one step ahead mean squared error of the forecasts were 25.4 and 37.75 for equations (4) and (5) respectively.

Furthermore, if confidence in the predictability of the model is to be established, then the percentage of forecasts falling within a set of confidence intervals must be the same as the percentage confidence of the intervals themselves. For the 10 forecast periods in Table 8 we see that only one

---

[16]Note that for equation (4) 11 forecast values were needed at each step for the fitted profits variable since that variable appears with 10 leads in the equation. These were derived from ARIMA models of Table 3.

[17]Forecasts from equation (5) reflect the necessary adjustments due to the autocorrelated error term.

prediction derived from equation (4) fell outside the 95% confidence band whereas three predictions from equation (5) fell outside the 95% confidence limits. Theoretically, of course, only .5 points (one forecast) should have fallen outside the 95% confidence limits.[18]

5. CONCLUDING REMARKS

In this article we have shown that using multiple input transfer functions, it is possible to more adequately estimate equations with expectations variables, and thus derive better forecast values than with using more conventional distributed lag models. Unlike other classical methods our procedure does not force autoregressive lag distributions on the expectations variables, nor does it impose specific lag distributions. It permits the data to determine the appropriate processes governing the stochastic behavior of the variables as well as the structure of the equation. Furthermore, as demonstrated, the method is not as susceptible to estimation inefficiencies as are other approaches.

APPENDIX

Haugh (1976) has shown that if n observations are available for two jointly covariance-stationary linear processes x_t and y_t which have ARMA univariate representations where α_t and β_t are corresponding white noise innovation series, and, moreover, if the two series are independent, then $\sqrt{n}P_{\alpha\beta}$ and $\sqrt{n}R_{\alpha\beta}$ have the same asymptotic distribution N(0,I), where $P_{\alpha\beta} = (\rho_{\alpha\beta}(k_1),\ldots,\rho_{\alpha\beta}(k_m))'$ and $k_1,\ldots,k_m$ are different integers, and $R_{\alpha\beta}$ are the estimates of the cross correlations of the innovations.

Therefore, since asymptotically the set of lagged cross correlations between α and β follow a normal distribution, chi-square tests can be used to check for series independence. In fact Haugh has proposed two test statistics for determining series independence:

$$S_{MN} = n \sum_{k=-M}^{N} r_{\alpha\beta}^2(k) \qquad \text{(a)}$$

and

$$S_{MN} = n^2 \sum_{k=-M}^{N} (n-|k|)^{-1} r_{\alpha\beta}^2(k). \qquad \text{(b)}$$

Both statistics would be compared with a critical value from a chi-square distribution with (M+N+1) degrees of freedom. Statistic (a) is appropriate for large samples whereas statistic (b) has been shown to be more appropriate for small samples (Haugh, 1976). The particular lag range [-M,N] which should be used in these statistics depends on the analyst's own knowledge about the phenomenon under study.

Tests for causality directions can, therefore, be made by limiting the range of k in either (a) or (b) to positive or negative integer values. Verification for possible instantaneous causality can be made by testing the significance of $r_{\alpha\beta}(0)$.

A nomenclature for categorizing causality events based on restrictions on the values of the estimated cross correlations of the whitened series, has been developed by Pierce and Haugh (1977).

---

18 Since for a minimum mean square error forecast, the one step ahead forecast errors are uncorrelated, the probability of having a set number of points outside the confidence interval is given by the binomial distribution.

REFERENCES

ALMON, S. (1965). The distributed lag between capital approximations and expenditures. Econometrica 33, 178-196.

ANDERSON, O.D. (1977). Time series analysis and forecasting: another look at the Box-Jenkins approach. The Statistician 26, 285-303.

BEGUIN, J.M., GOURIEROUX, C. and MONTFORT, A. (1980). Identification of a mixed autoregressive moving average process: the corner method. In Time Series (edited by O. D. Anderson). Amsterdam, North Holland, 423-435.

BOX, G. and JENKINS, G. (1970). Time Series Analysis: Forecasting and Control. San Francisco, Holden Day.

Data Resources Incorporated Computer Files; Tapes available through Harvard Business School Computer Center.

FEIGE, E. and PEARCE, D. (1979). The causal relationships between money and income. Journal of Economics and Statistics LXI, 521-533.

GRAY, H., KELLEY, G. and MCINTIRE, D. (1978). A new approach to ARMA modeling. Communications in Statistics B 7, 1-77.

HAUGH, L.D. (1976). Checking the independence of two covariance-stationary time series: a univariate residual cross correlation approach. Journal of the American Statistical Association 71, 378-385.

JAMES, C., KOREISHA, S. and PARTCH, M. (1982). Structural relations between inflation and stock prices. (working manuscript) University of Oregon.

JUDGE, G.G., GRIFFITHS, W.E., HILL, R.C. and LEE, T.C. (1980). The Theory and Practice of Econometrics. New York, Wiley & Sons.

KERAN, M.W. (1971). Expectations, money, and the stock market. Review-Federal Reserve Bank of St. Louis, 16-31.

KOREISHA, S.G. (1980). The integration of transfer functions with econometric modeling. (unpublished dissertation) Harvard University.

KOREISHA, S.G. (1982). The relationships between causality, transfer functions and econometric modeling. In Applied Time Series Analysis (edited by O. D. Anderson & M. R. Perryman). Amsterdam, North Holland, 139-166.

KOREISHA, S.G. (forthcoming). Causal implications: the linkage between time series and econometric modeling. Journal of Forecasting.

LIU, L.M. and HANSSENS, D. (1982). Identification of multiple-input transfer function models. Communications in Statistics A 11, 297-314.

MONTGOMERY, D. and JOHNSON, L. (1976). Forecasting and Time Series Analysis. New York, McGraw-Hill.

PESANDO, J.E. (1974). The supply of money and common stock prices: further observations on the econometric evidence. Journal of Finance, 909-922.

PIERCE, D.A. and HAUGH, L.D. (1977). Causality in temporal systems. Journal of Econometrics 5, 265-293.

ROGALSKI, R.J. and VINSO, J.D. (1977). Stock returns, money supply and the direction of causality. Journal of Finance, 1017-1030.

SCHMIDT, P. and WAUD, R.N. (1973). The Almon lag technique and the monetary versus fiscal policy debate. Journal of the American Statistical Association, 11-19.

TAYLOR, S. (1982). Topics in time series analysis. (unpublished dissertation) University of Oregon.

TIAO, G.C. and BOX, G.E.P. (1981). Modeling multiple time series with applications. Journal of the American Statistical Association 76, 802-816.

TRIVEDI, P.K. and PAGAN, A.R. (1976). Polynomial distributed lags: a unified treatment. (working paper) Australian National University.

VANDAELE, W. and KOREISHA, S. (1977). Formulation of Box-Jenkins models for U. S. residential construction. Harvard Business School Working Paper 77-50.

TIME SERIES ANALYSIS: Theory and Practice 4
O.D. Anderson (editor)

TIME SERIES MODELING: A COMPARISON OF THE MAXIMUM χ^2 AND BOX-JENKINS APPROACHES

James Tipton

Department of Finance
Baylor University
Waco, Texas
U.S.A. 76798

James T. McClave

Department of Economics
University of Florida
Gainesville, Florida
U.S.A. 32611

This paper presents four simple forecasting comparisons between the maximum χ^2 and Box-Jenkins approaches. Each time, the "max χ^2" model is shown to be either as good or better than the Box-Jenkins one. Our main explanation for the max χ^2 superiority is its ability to select an "optimal" autoregressive order, which allows longer AR models to be chosen (when appropriate) than does Box-Jenkins.

INTRODUCTION

The selection among alternative approaches for use in prediction has, until recently, been primarily an arbitrary choice of the forecaster. Box and Jenkins (1970) argue that the decisions concerning models and approaches need not be arbitrary. The decision, they say, should involve a series of logical steps. The model building process for each approach should be designed to transform the residuals of the model to white noise, uncorrelated with any other known input.

Box and Pierce (1981) recently restated the Box-Jenkins philosophy for models and modeling time series. They offer a paradigm of two stages for any statistical or scientific investigation:

1) to build a model for the data under study;
2) to use that model to supply answers to whatever it is that we want to know."

The purpose of this paper is to answer the question, "Which method: the max χ^2 or Box-Jenkins, yields the smallest relative root-mean-square-error, RMSE, in <u>ex-ante</u> forecasting?"

NOTATION AND BACKGROUND

The Box-Jenkins method of forecasting is certainly no stranger to the literature in several disciplines. Its four distinct phases: Identification, Estimation, Diagnostic Checking, and Forecasting, have remained intact in the commercial marketing of many computer software packages.

The more recent advent of the max χ^2 method occasions a much smaller dissemination. It is therefore useful to partially summarize the max χ^2 approach of McClave (1975).

For the stationary AR model, $X_t + \alpha_1 X_{t-1} + \alpha_2 X_{t-2} + \ldots + \alpha_m X_{t-m} = \varepsilon_t$, with ε_t an uncorrelated series with mean zero and variance σ^2, the possibility exists that some of the lag coefficients $(\alpha_1, \alpha_2, \ldots, \alpha_m)$ may be zero. If lag coefficients do equal zero, then the model becomes a subset AR model of pth order with a maximum lag m if exactly p of the lag coefficients are nonzero, $\alpha_m \neq 0$. We refer to the preceding model as an AR model of order p max m with lags $(j_1, j_2, \ldots, j_p)$, where $(j_1, j_2, \ldots, j_p)$ are the lags having nonzero coefficients.

Note that $j_p = m$.

The parameter set θ for the subset AR model of order p max m can be written as the union of two disjoint parameter sets: $\theta = \theta_1 \cup \theta_2$, where

$$\theta_1 = \{p; j_1, j_2, \ldots, j_p\} \text{ and } \theta_2 = \{\alpha_{j_1}, \alpha_{j_2}, \ldots, \alpha_{j_p}; \sigma^2\}.$$

Conveniently, θ_1 is the set of order parameters, and θ_2 is the set of model parameters.

Briefly, the identification of θ_1, the first step, is accomplished by sequential testing of higher order polynomials with likelihood ratio tests. The "best" AR model of each order $k \in [1,2,\ldots,K]$, where K is chosen so that one is confident that $m \leq K$(m = maximum lag of true AR model), is defined as that subset AR model (with $m \leq K$) with minimum residual variance, $\hat{\sigma}_k^2$.

The comparison of the AR model of order k max k with the AR model of order (k+r) max (k+r) is accomplished with a modification of the test statistic,

$$\hat{T}_{k,r} = \{N-(k+r)-s\} \frac{\hat{\sigma}_k^2 - \hat{\sigma}_{k+r}^2}{\hat{\sigma}_{k+r}^2} \xrightarrow[H_0]{D} T_{k,r} \quad [1]$$

where $T_{k,r}$ is a χ^2 random variable with r degrees of freedom (for any k). In [1], D/H_0 reads "converges in distribution when H_0 is true," and s is the number of functionally independent parameters estimated in detrending.

The order identification step in the max χ^2 approach follows the work of Quenouille (1947), Walker (1952), Whittle (1952), Bartlett and Diananda (1950), Hannan (1970), and Anderson (1971). Its sequential testing method is separated from point-estimation approaches, e.g., Akaike (1969, 1970) and Parzen (1974a, 1974b).

The second step involves the estimation of θ_2, the lag coefficients. Estimation is achieved by a simple modification of the Yule-Walker equations used for the standard p max p situation. This modification allows the constraint that some intermediate lags are zero and is detailed in McClave (1978).

The estimation of lag coefficients in the max χ^2 approach follows the work of Mann and Wald (1943), Huber (1967), and Pagano (1974). This estimator of θ_2 is fully efficient in the sense that the asymptotic covariance matrix is the inverse of the Fisherian information matrix. Both are convex functions of the true underlying distribution.

The final step, given order identification and parameter estimation of the stationary series, is a deterministic extrapolation. These extrapolations are genuine ex-ante forecasts in which the actual future path is predicted using projected future values. This is distinguished both from: (a) ex-post forecasting, and (b) "hypothetical" forecasting.

The forecasting is accomplished using ordinary-least-squares, OLS, techniques for which we offer neither justification nor refutation. Conceptually, forecasting utilizing the Yule-Walker equations may be preferred, Tipton (1982).

THE TEST

Comparison is made for monthly exponential rates of growth of four economic series: inflation, money stock, income, and interest rates.

The Consumer Price Index (US Department of Labor, Bureau of Labor Statistics) represents the inflation series, P. Cash and demand deposits, M1A (Federal Reserve Bulletins) represent the money stock series, M. The income series, Y, is the aggregate personal income series (Survey of Current Business). Finally, one month, end-of-the-month, secondary market yields on a bond equivalent basis for US Treasury bills (Wall Street Journal) represent the interest rate series, r.

The models were fitted from the sample period, January 1959 to December 1978. The order selection of the max χ^2 and ARIMA models is presented as Table 1 and Table 2, respectively. Ex-ante forecasts were for the period, January 1979 to August 1980.

Since some models are built primarily for forecasting, while others are built primarily for descriptive purposes and hypothesis testing, attempts to make performance comparisons across various models and techniques are difficult. It is not at all clear that any absolute measures of quality can be agreed upon. Even when a standardized basis can be agreed, the comparison attempt may be critized for its neglect of the "tender loving" care of the model builder.

No small part of the disagreement stems from three characteristics of ex-ante forecasting. First, genuine ex-ante forecasting attempts do not allow repeated sampling. Therefore, a model designed for forecasting purposes should have as small a standard error of forecast as possible, while other statistics (t,F,R^2,DW) are more important in a model designed to test a specific hypothesis or measure the estimation period. Other measures, such as the ability to predict turning points or respond to large changes in economic variables, in addition to the size of the forecast error, become more important as measures of forecasting performance.

Second, use of the same data in specification and estimation of the two approaches means the summary measures are not statistically independent. It is possible and highly probable in practice that the different models are merely different representations of the same structure.

Finally, the accuracy of ex-ante forecasting can only be described in an ex-post sense. This causes two additional measurement problems. First, the ex-ante error would most likely be larger than the ex-post error. Second, ranking ex-ante models based on the size of the standard error of forecast is very difficult and inaccurate. This is because the forecast errors are compounded by the feedback structure of the model.

Common substitutes for the forecast error comparison are performance comparisons of the mean absolute error, MAE, and the root-mean-square-error, RMSE. Table 3 gives the mean of the forecast series, the MAE, and the RMSE for each of the four series.

Both approaches did equally well in forecasting inflation. Neither, according to the RMSE criterion, dominates the other. Such is not the case, however, for the money stock, aggregate personal income, and interest rates. The max χ^2 approach dominates the Box-Jenkins approach for these three series.

The independence of the models, thus allowing a preference ranking, is established by testing for correlation between corresponding one-step ahead forecast errors. This test for method difference becomes a test for zero correlation, $\rho(u,v) = 0$, where

$$u = e^{(1)}_{n,1} + e^{(2)}_{n,1},\ v = e^{(1)}_{n,1} - e^{(2)}_{n,1} \text{ and } e^{(1)}_{n,1},\ e^{(2)}_{n,1}$$

Table 1. Order Selection for the Maximum χ^2 Approach

Subset Autoregressive Models: Optimal Model According to the Specified Criterion: Max Chi-Square level = 0.05.

Model 1: Consumer Price Index, P_t:

Lag order which yields minimum residual variance: 1, 2, 6.
Residual Variance = 0.85083E+00
Akaike's FPE = 0.87246E+00
Max CHISQ = 0.176E+02
Prob = 0.001

Model 2: Money Stock (M1A), M_t:

Lag order which yields minimum residual variance: 1, 3, 5, 17.
Residual Variance = 0.13260E+01
Akaike's FPE = 0.13711E+01
Max CHISQ = 0.129E+02
Prob = 0.007

Model 3: Aggregate Personal Income, Y_t:

Lag order which yields minimum residual variance: 3, 5, 8, 9.
Residual Variance = 0.13834E+01
Akaike's FPE = 0.14305E+01
Max CHISQ = 0.99E+01
Prob = 0.033

Model 4: US Treasury Bill Yields, r_t:

Lag order which yields minimum residual variance: 1, 13.
Residual Variance = 0.34535E+01
Akaike's FPE = 0.35117E+01
Max CHISQ = 0.103E+02
Prob = 0.031

Table 2. Order Selection for the Box-Jenkins Approach

Time Series Models: Selection According to the Autocorreclation and Partial Autocorrelations Functions.

Model 1: Consumer Price Index, P_t: ARIMA(1,2,0)
Model 2: Money Stock (M1A), M_t: ARIMA(3,1,0)
Model 3: Aggregate Personal Income, Y_t: ARIMA(0,1,0)
Model 4: US Treasury Bill Yields, r_t: ARIMA(1,1,0)

are the one-step forecast errors from the first and second model, respectively. Table 4 presents the probabilistic results derived from testing for correlation under the null hypothesis that rho is zero. These Pearson product-moment coefficients indicate significant correlation is present at the .10 level, which implies a significant difference in the SSE values from the two forecasts. The significance probability of the correlation is indicated in brackets.

The combined implication of Tables 3 and 4 is dominance of the max χ^2 approach over the Box-Jenkins approach in ex-ante forecasting. It could be argued that this dominance is due to the "screening" ability of the likelihood ratios on a set of orders which includes higher order polynomials. When the higher orders are not optimal, this "screening" selects lower order polynomials which translate to be low order ARIMA models. Note the higher order polynomials have some intermediate lags constrained to zero.

The order selection process for the max χ^2 approach reveals an "efficient" set of possible polynomials at the .05 level that could fit the data. Usually, lower orders of the Box-Jenkins type can be found within the "efficient" set. An argument is advanced concerning behavior of time series analysts using the Box-Jenkins approach: seldom do they look beyond those lower order polynomials in the "efficient" set, due to the complexity of the identification process. They typically underfit the series. When the Box-Jenkins approach does involve larger order polynomials, seldom are intermediate lags constrained, or omitted. Thus, they could be said to overfit the series in terms of the max χ^2 criterion.

Table 3. Results of Forecasting Models

		Max CHISQ	B-J
P_t	mean	.005	.005
	MAE	.006	.006
	RMSE	.006	.006
M_t	mean	.004	.004
	MAE	.007	.008
	RMSE	.010	.011
Y_t	mean	.007	.007
	MAE	.004	.004
	RMSE	.005	.006
r_t	mean	.008	.008
	MAE	.122	.126
	RMSE	.172	.181

Table 4. Correlation Results
(Correlation Coefficients/Prob > | R | Under H_0: R = 0 | N=20)

P_t	M_t	Y_t	r_t
.28318	.13666	-.1910	-.09688
(.2263)	(.5656)	(.4197)	(.6932)

SUMMARY AND CONCLUSIONS

A comparisons of two time series approaches, max χ^2 and Box-Jenkins, has been presented. Specifically, the differently selected models forecast monthly four series: inflation, money stock, aggregate personal income, and interest rates. The preliminary parameters and distributed lag patterns were estimated from the period, January 1959 to December 1978. The ex-post forecast period used in the test was January 1979 to August 1980. During this ex-post forecast period, the preliminary parameters and distributed lag patterns for the max χ^2 approach were not updated. The same models for the Box-Jenkins approach were kept, but the parameters were updated.

The main conclusion, based on minimization of the RMSE, is that the better method to use in ex-ante forecasting is the max χ^2 approach. The principal explanation offered for the superiority of the max χ^2 approach is its ability to select an "optimal" autoregressive order which allows longer AR models to be chosen (when appropriate) than does Box-Jenkins.

Two interesting observations for related questions are the implications that: 1) published seasonally adjusted data is incorrectly fitted (due to the spurious lag patterns); and 2) theories based on random, or low order AR models, e.g., efficient markets theory (economics and finance), may be due to the lack of proper order identification by the analysts.

REFERENCES

AKAIKE, H. (1969). Fitting autoregressive models for prediction. Annals of Institute of Statistical Mathematics, 21, 243-247.

AKAIKE, H. (1970). Statistical predictor identification. Annals of Institute of Statistical Mathematics, 22, 203-217.

ANDERSON, T.W. (1971). The Statistical Analysis of Time Series. Wiley, New York.

BARTLETT, M.S. and DIANANDA, P.H. (1950). Extensions of Quenouille's tests for autoregresive schemes. Journal of Royal Statistical Society B, 12, 108-115.

BOX, G.E.P. and JENKINS, G.M. (1970). Time Series Analysis: Forecasting and Control. Holden-Day, San Francico. Revised Edition (1976).

BOX, G.E.P. and PIERCE, D.A. (1981). Estimating current trend and growth rates in seasonal time series. Special Studies Paper 156, Division of Research and Statistics, Federal Reserve Board, Washington, D.C.

GRANGER, C.W.J. and NEWBOLD, P. (1977). Forecasting Economic Time Series. Academic Press, New York.

HANNAN, E.J. (1970). Multiple Time Series. Wiley, New York.

MADDALA, G.S. (1977). Econometrics. McGraw-Hill, New York.

MANN, H.B. and WALD, A. (1943). On the statistical treatment of linear stochastic difference equations. Econometrica, 11, 173-220.

McCLAVE, J. (1975). Subset autoregression. Technometrics, 17, 213-220.

McCLAVE, J. (1978). Estimating the order of autoregressive models: The max χ^2 method. Journal of American Statistical Association, 73, 122-128.

PAGANO, M. (1974). An algorithm for optimization under linear constraints. Technical Report 12, Statistical Science, SUNY at Buffalo.

PARZEN, E. (1974a). Some solutions to the time series modeling and prediction problem. Technical Report 5, Statistical Science, SUNY at Buffalo.

PARZEN, E. (1974b). Some recent advances in time series analysis. IEEE Transactions on Automatic Control, AC-19, 723-730.

QUENOUILLE, M.H. (1947). A large-sample test for the goodness of fit of autoregressive schemes. Journal of Royal Statistical Society A, 10, 123-129.

TIPTON, J. (1982). Forecasting short-term interest rates. 57th Annual Conference of Western Economic Association, Los Angeles, California.

WALKER, A.M. (1952). Some properties of the asymptotic power functions of goodness-of-fit tests for linear autoregressive schemes. Journal of Royal Statistical Society B, 14, 117-134.

WHITTLE, P. (1952). Test of fit in time series. Biometrika, 39, 309-318.

TIME SERIES ANALYSIS: Theory and Practice 4
O.D. Anderson (editor)

MULTIPLE TIME SERIES MODELING OF MACROECONOMIC SERIES

Jack Y. Narayan

Department of Mathematics
State University of New York at Oswego
Oswego, New York 13126
U.S.A.

In this investigation we use procedures from multiple time series analysis to investigate the relationships between the state of the economy as measured by GNP, and a number of economic indicators. Bivariate autoregressive moving average models for GNP and each of the indicator series are constructed. A comparison of ex ante forecasts from these models is made with the ex ante forecasts resulting from a comparable univariate model for GNP. We then discuss whether or not it is advantageous to use these multiple time series methods to predict future values of the GNP series.

INTRODUCTION

According to Granger and Newbold (1977) causality is present if the past of a series x causes the present or future of another series y, so that the forecasts of y can be improved by utilizing the information provided by series x. Feedback is present if there is causality from y to x and also from x to y. Causal relationships between economic variables have been investigated by many authors. These include Granger (1969), Sims (1972), Pierce and Haugh (1977), Ledolter (1977), Thury (1980) and Fey and Jain (1981). In some of these studies relationships have been identified and claims have been made that variations in a certain economic variable can be better explained if the information set is expanded to include not only the past history of the variable of interest but also the past history of other related economic variables. This type of reasoning serves as the motivation for the use of leading indicators in making forecasts for the state of the economy.

The purpose of this investigation is to determine whether models for certain macroeconomic series, based on these expanded data sets, can produce more accurate short term forecasts for periods outside of the data set than those models which depend only on the past history of a single series. In particular, we will determine whether forecasts for the Gross National Product modeled by a univariate ARIMA procedure can be improved by using a bivariate ARIMA procedure in which GNP is modeled simultaneously with each of thirteen so called leading indicators. Thus we will determine how successful the use of a leading indicator is, in obtaining improved forecasts for the GNP series.

We analysed fourteen macroeconomic series obtained from Business Conditions Digest. These were seasonally adjusted quarterly data for the period 1949-1 to 1979-4, and thus each series consisted of 120 data points. The series were: (1) Gross National Product in 1972 dollars, (2) Vendor Performance, (3) Money Supply (M1-B) in 1972 dollars, (4) Index of Stock Prices, (5) Layoff Rate manufacturing per 100 employees, (6) Rate of Capacity Utilization manufacturing, (7) Percent Change in total Liquid Assets, (8) Average Workweek of production workers manufacturing, (9) Index of Net Business Formation, (10) Percent Change in Sensitive Prices, (11) Building Permits private housing, (12) Value of

Manufacturer's New Orders for consumer goods and materials, (13) Change in inventories on hand and on order, 1972 dollars, (14) Contract and Orders for plant and equipment in 1972 dollars.

The procedure used is as follows: eight data points were withheld in order to compare the univariate and multivariate procedures on their ability to forecast GNP. A univariate ARIMA model was fitted to the 112 points of the GNP series and a forecast was made for period 113. The data set was then updated to include 113 points, the model was reestimated and a forecast for period 114 was obtained. The process of updating and forecasting was continued until eight such ex ante forecasts were obtained. The sum of squared errors for these ex ante forecasts was then calculated. Finally univariate ARIMA models were fitted to each of the fourteen series on the entire data set of 120 points for the following reasons. First, a comparison of the two procedures could be made on how well they fit the GNP data. Secondly, knowledge of the univariate structures of the other thirteen series would be helpful in the identification of the multivariate models.

Thirteen bivariate models, in each of which GNP was modeled simultaneously with one of the other series, were estimated using 112 time periods and a forecast was made for the next period. Eight such ex ante forecasts were obtained by the process of updating the data sets and reestimating the model. Finally the models were reestimated using all 120 time periods.

The remainder of the paper is arranged as follows: A summary of the univariate results is given, followed by a brief review of the multivariate methodology and an example to illustrate the procedure. The results of the multivariate analysis are then summarized. Finally a comparison of the methods is made and a brief conclusion given.

UNIVARIATE RESULTS

The basic univariate nonseasonal ARIMA (p,d,q) model has the following form

$$\phi_p(B)(1-B)^d x_t = \theta_0 + \theta_q(B)a_t \tag{1}$$

Here x_t denotes an observed time series and a_t is a sequence of identically independently and normally distributed random variables with mean zero and constant variance. The symbol B is the backshift operator defined as $B^m x_t = x_{t-m}$. The term θ_0 is a constant which can be used to represent a deterministic trend of degree d. The terms $\phi_p(B)$ and $\theta_q(B)$ are polynomials in the backshift operator B, given by

$$\phi_p(B) = 1 - \phi_1 B - \ldots \phi_p B^p \quad \text{and} \quad \theta_q(B) = 1 - \theta_1 B - \ldots \theta_q B^q .$$

Finally d represents the order of differencing that is needed to achieve stationarity.

The steps of model identification, model estimation and diagnostic checking outlined by Box and Jenkins (1976) and Thury (1981) were carried out in the modeling process. The logarithm and first difference of all the series with the exception of series 10 (Percent Change in Sensitive Prices) and series 13 (Change in inventories on hand and on order) in which only the first difference was taken. The model obtained for GNP on 120 data points was an ARIMA (1,1,4). A new model on 112 points was sought to produce the one step-ahead forecasts for GNP. This also turned out to be an ARIMA (1,1,4) model. However the coefficients were different. One step-ahead forecasts were then obtained for periods 113 to 120 by reestimating the model at each fresh data point.

TABLE 1. SUMMARY OF UNIVARIATE MODELS

SERIES	ARIMA MODEL	PARAMETERS
GROSS NATIONAL PRODUCT	(1,1,4)	$\phi_1, \theta_1, \theta_4$
VENDOR PERFORMANCE	(5,1,0)	ϕ_1, ϕ_2, ϕ_5
MONEY SUPPLY (M1-B)	(1,1,0)	ϕ_1
INDEX OF STOCK PRICES	(0,1,1)	θ_1
LAYOFF RATE MANUFACTURING	(0,1,5)	θ_4, θ_5
RATE OF CAPACITY UTILIZATION MANUFACTURING	(2,1,5)	$\phi_1, \phi_2, \theta_4, \theta_5$
PERCENT CHANGE IN TOTAL LIQUID ASSETS	(0,1,3)	θ_1, θ_3
AVERAGE WORKWEEK OF PRODUCTION WORKERS MANUFACTURING	(1,1,0)	ϕ_1
INDEX OF NET BUSINESS FORMATION	(1,1,0)	ϕ_1
PERCENT CHANGE IN SENSITIVE PRICES	(1,1,4)	ϕ_1, θ_4
BUILDING PERMITS PRIVATE HOUSING	(2,1,4)	$\phi_1, \phi_2, \theta_3, \theta_5$
VALUE OF MANUFACTURER'S NEW ORDERS FOR CONSUMER GOODS AND MATERIALS	(0,1,4)	θ_4
CHANGE IN INVENTORIES ON HAND AND ON ORDER	(4,1,5)	ϕ_4, θ_5
CONTRACT AND ORDERS FOR PLANT AND EQUIPMENT	(0,1,4)	$\theta_2, \theta_3, \theta_4$

MULTIVARIATE ARIMA MODELS

Let $\underline{Z}_t = (y1_t, y2_t, \ldots, ym_t)$ be an m-dimensional column vector whose components represent m individual time series. For simplicity we assume that non-linear transformations and differencing operations have already been performed on the original time series to, respectively, stabilize the variance and induce normality, and to achieve stationarity. Then the multiple ARMA (p,q) model is a multivariate generalization of the univariate time series model (see Quenouille 1957).

$$\underline{Z}_t = \underline{\Phi}_1 \underline{Z}_{t-1} + \underline{\Phi}_2 \underline{Z}_{t-2} + \cdots + \underline{\Phi}_p \underline{Z}_p + \underline{a}_t - \underline{\Theta}_1 \underline{a}_{t-1} - \underline{\Theta}_2 \underline{a}_{t-2} - \cdots - \underline{\Theta}_q \underline{a}_{t-q}$$

where the $\underline{\Phi}_k$ are $m \times m$ constant matrices whose elements Φ_{kij} are the autoregressive parameters. The $\underline{\Theta}_k$ are also $m \times m$ constant matrices whose elements Θ_{kij} are the moving average parameters. Finally $\{\underline{a}_t\}$ with $\underline{a}_t = (a_{1t}, \ldots, a_{mt})'$ is a sequence of random shock vectors identically independently and normally distributed with mean zero and time-invariant covariance matrix. For example, a multiple ARIMA (1,1) model has the form

$$\begin{bmatrix} Z1 \\ Z2 \end{bmatrix}_t = \begin{bmatrix} \phi_{111} & \phi_{112} \\ \phi_{121} & \phi_{122} \end{bmatrix} \begin{bmatrix} Z1 \\ Z2 \end{bmatrix}_{t-1} + \begin{bmatrix} a1 \\ a2 \end{bmatrix}_t - \begin{bmatrix} \theta_{111} & \theta_{112} \\ \theta_{121} & \theta_{122} \end{bmatrix} \begin{bmatrix} a1 \\ a2 \end{bmatrix}_{t-1}$$

If the parameters ϕ_{121} and θ_{121} are both zero we have causality from Z2 to Z1 . Feedback is present if at least one of these parameters is significantly different from zero. Thus in multivariate time series modeling there is no need to discriminate between causality and feedback before hand.

As with univariate models, certain restrictions must be imposed on the autoregressive and moving average parameters to guarantee stationarity and invertibility (see Tiao and Box 1981). The Wisconsin multiple time series package (Tiao et al 1979) was used to carry out the steps of model identification, parameter estimation and verification of model adequacy, and to obtain forecasts.

The sample cross correlation matrices for the original data are helpful in the identification process. For example, an early cut off in the cross correlations is indicative of a low order moving average process. A generalized partial cross correlation matrix function, which is estimated by fitting successive autoregressive models of increasing order to the data, is similarly used for specifying the order of a purely autoregressive model. A test statistic, based on the Chi-square distribution, indicates whether the coefficient matrix of the autoregressive process at a particular lag is significant. Once the order of the model has been tentatively selected, efficient estimates of the associated parameter matrices are obtained by maximizing the likelihood function. Standard errors for the estimates indicate whether the individual parameters are significant.

Diagnostic tests based on the cross correlations of the residual series are then used to criticize the model. These tests are motivated from the consideration that, if the residual series were white noise, then for large n the cross correlations would be normally distributed with mean 0 and variance 1/n .

AN EXAMPLE

A bivariate model, in which GNP is the first component and the Index of Net Business Formation is the second, is developed to illustrate the modeling procedure. The log transformation and first differencing have been performed on each component to stabilize the variance and induce normality, and to achieve stationarity. The cross correlations are recorded in Table 2.

TABLE 2. CROSS CORRELATIONS OF GNP AND INDEX OF NET BUSINESS FORMATION.

LAGS	-13	-12	-11	-10	-9	-8	-7	-6	-5	-4	-3	-2	-1
CORR	-.03	.02	-.05	-.07	-.24*	.21*	-.13	-.07	-.05	.02	.08	.25*	.46*

LAGS	0	1	2	3	4	5	6	7	8	9	10	11	12
CORR	.48*	.19	-.08	-.15	-.28*	-.13	-.06	-.09	-.12	-.07	-.06	-.09	-.01

The asterisks indicate those cross correlations which are significant. The correlations seem to die out quickly except for small but significant correlations at lags -8 and -9. Possible models are a low order autoregressive or a mixed model with a low order autoregressive operator.

The results of the generalized partial cross correlation analysis are listed in Table 3. As was remarked previously autoregressive proceses of successively higher order were fitted. The Chi-square statistic indicates whether the coefficient matrix is significant and the + and - indicate the significace of the

TABLE 3. GENERALIZED PARTIAL CROSS CORRELATIONS

LAG	STD. PARTIAL	AR. COEFF.	SIGNIF.	RESIDUAL VARIANCES	CHI-SQ.
1	1.77	4.05	• +	.740E-04	40.32*
	.02	4.44	• +	.301E-03	
2	-.05	-.12	• •	.740E-04	7.89
	-2.10	-.56	- •	.282E-03	
3	-.06	.22	• •	.739E-04	.32
	-.36	-.08	• •	.281E-03	
4	-2.4	1.03	- •	.702E-04	11.19*
	-3.42	1.70	- •	.253E-03	

individual elements. The residual variances give an idea of the improvement in fit as each new term is added. The generalized partial cross correlations indicate significant terms at lags 1 and 4. Several models were specified and estimated. These included an ARMA (1,4) with $\underline{\Theta}_1 = \underline{\Theta}_2 = \underline{\Theta}_3 = \underline{0}$, an ARMA (1,4) with $\underline{\Theta}_2 = \underline{\Theta}_3 = \underline{0}$ and an ARMA (4,0) with $\underline{\Phi}_2 = \underline{\Phi}_3 = \underline{0}$. An analysis of the residual series from each model was done in order to specify the final model. The cross correlations and the generalized partial correlations of the residual series from an ARMA (1,4) model $\underline{\Theta}_2 = \underline{\Theta}_3 = \underline{0}$ were all insignificant. This model was therefore selected and an exact likelihood estimation procedure produced the following equation

$$\begin{bmatrix} \nabla y \\ \nabla x \end{bmatrix}_t = \begin{bmatrix} .995 & -.178 \\ 0 & .44 \end{bmatrix} \begin{bmatrix} \nabla y \\ \nabla x \end{bmatrix}_{t-1} + \begin{bmatrix} a1 \\ a2 \end{bmatrix}_t - \begin{bmatrix} .779 & -.365 \\ 0 & 0 \end{bmatrix} \begin{bmatrix} a1 \\ a2 \end{bmatrix}_{t-1} - \begin{bmatrix} .220 & 0 \\ 0 & 0 \end{bmatrix} \begin{bmatrix} a1 \\ a2 \end{bmatrix}_{t-4}$$

Here ∇y and ∇x represent the first differences of the transformed series. Since the coefficient matrices are upper triangular there is one-way causality from the Index of Net Business Formation to GNP.

The error variance for the GNP component was .7734E-04 which represents a 11.5% reduction from that for the univariate model. There was virtually no change in error variance for the second component. This agrees with the one-way causality exhibited by the model.

Note that the first component included the parameters of the univariate model for GNP as well as the parameters reflecting the causality. The second component has exactly the same form as its univariate counterpart. It was found that the inclusion of the univariate structure in the bivariate model was useful in the identification process.

RESULTS FROM BIVARIATE ANALYSIS

In a manner similar to that of our example, the other twelve bivariate models of GNP with each of the other indicator series, were obtained. A summary of the results is recorded in Table 5 where the form of model and significant parameters are given. The causal relationships identified by the modeling procedure are listed in Table 4. These are the types of relations upon which forecasters hope to capitalize in order to get better forecasts. It should be noticed that no causality was found to exist from money supply to GNP. This is in disagreement with the findings of Sims (1972) and Fey and Jain (1981) who used not only different data but also different modeling procedures.

In all but three cases of the data fitting, the multivariate models resulted in smaller error variances. The question naturally arises as to whether these reductions are significant. Granger and Newbold (1977) suggest a test to ascertain whether there is any significant difference between the forecasts of two competing models. The test is based on one-step forecast errors under certain reasonable assumptions.

Suppose (e_t, a_t) $t = 1,2, \ldots, N$, constitute a random sample from a bivariate normal distribution with mean zero, standard deviations σ_1 and σ_2 and correlation coefficient ρ. In particular, then, it is assumed that the individual forecasts are unbiased and the forecast errors are not

TABLE 4. CLASSIFICATION OF CAUSALITY ON FITTED DATA

ONE-WAY CAUSALITY FROM GROSS NATIONAL PRODUCT TO

(1) LAYOFF RATE MANUFACTURING
(2) AVERAGE WORKWEEK OF PRODUCTION WORKERS MANUFACTURING

ONE-WAY CAUSALITY TO GROSS NATIONAL PRODUCT FROM

(1) INDEX OF STOCK PRICES
(2) RATE OF CAPACITY UTILIZATION MANUFACTURING
(3) INDEX OF NET BUSINESS FORMATION
(4) PERCENT CHANGE IN SENSITIVE PRICES

TWO-WAY CAUSALITY OF GROSS NATIONAL PRODUCT WITH

(1) PERCENT CHANGE IN TOTAL LIQUID ASSETS
(2) VALUE OF MANUFACTURER'S NEW ORDERS FOR CONSUMERS GOODS AND MATERIALS

NO CAUSALITY OF GROSS NATIONAL PRODUCT WITH

(1) VENDOR PERFORMANCE
(2) CONTRACT AND ORDERS FOR PLANT AND EQUIPMENT
(3) MONEY SUPPLY (M1-B)
(4) CHANGE IN INVENTORIES ON HAND AND ON ORDER

autocorrelated, properties that hold in the present analysis. Their test is motivated by observing that the expected value of the product of the two random

TABLE 5. SUMMARY OF BIVARIATE MODELS

SERIES	ARMA	SIGNIFICANT PARAMETERS
GNP WITH....		
VENDOR PERFORMANCE	(5,4)	$\phi_{111}, \phi_{122}, \phi_{222}, \phi_{522}, \theta_{111}, \theta_{411}$
MONEY SUPPLY (M1-B)	(1,8)	$\phi_{111}, \phi_{122}, \theta_{111}, \theta_{411}, \theta_{822}$
INDEX OF STOCK PRICES	(2,4)	$\phi_{111}, \phi_{122}, \phi_{212}, \phi_{222}, \theta_{111}, \theta_{211}, \theta_{312}, \theta_{411}$
LAYOFF RATE MANUFACTURING	(1,4)	$\phi_{111}, \theta_{111}, \theta_{121}, \theta_{411}, \theta_{422}$
RATE OF CAPACITY UTILIZATION MANUFACTURING	(2,4)	$\phi_{111}, \phi_{122}, \phi_{212}, \phi_{222}, \theta_{111}, \theta_{112}, \theta_{322}, \theta_{422}$
PERCENT CHANGE IN TOTAL LIQUID ASSETS	(8,4)	$\phi_{111}, \phi_{121}, \phi_{122}, \phi_{221}, \phi_{812}, \theta_{111}, \theta_{322}, \theta_{411}, \theta_{412}$
AVERAGE WORKWEEK OF PRODUCTION WORKERS MANUFACTURING	(1,4)	$\phi_{111}, \phi_{122}, \theta_{111}, \theta_{121}, \theta_{411}$
INDEX OF NET BUSINESS FORMATION	(1,4)	$\phi_{111}, \phi_{112}, \phi_{122}, \theta_{111}, \theta_{112}, \theta_{411}$
PERCENT CHANGE IN SENSITIVE PRICES	(1,4)	$\phi_{111}, \phi_{122}, \theta_{111}, \theta_{112}, \theta_{411}$
BUILDING PERMITS PRIVATE HOUSING	(4,4)	$\phi_{111}, \phi_{112}, \phi_{122}, \phi_{222}, \phi_{312}, \phi_{422}, \theta_{111}, \theta_{322}, \theta_{411}$
VALUE OF MANUFACTURER'S NEW ORDERS FOR CONSUMERS GOODS AND MATERIALS	(1,4)	$\phi_{111}, \phi_{121}, \theta_{111}, \theta_{412}, \theta_{411}, \theta_{422}$
CHANGE IN INVENTORIES ON HAND AND ON ORDER	(4,5)	$\phi_{111}, \phi_{422}, \theta_{111}, \theta_{411}, \theta_{522}$
CONTRACT AND ORDERS FOR PLANT AND EQUIPMENT	(4,4)	$\phi_{111}, \phi_{121}, \phi_{112}, \phi_{422}, \theta_{111}, \theta_{411}$

TABLE 6. BIVARIATE MODELLING: PERCENTAGE REDUCTION IN ERROR VARIANCE

SERIES	DATA FITTING	EX ANTE FORECASTING
GNP WITH....		
VENDOR PERFORMANCE	.3	2.3
MONEY SUPPLY (M1-B)	0	-2.4
INDEX OF STOCK PRICES	7.9	-7
LAYOFF RATE MANUFACTURING	0	3.9
RATE OF CAPACITY UTILIZATION MANUFACTURING	6.9	-1.4
PERCENT CHANGE IN TOTAL LIQUID ASSETS	4.5	-2.6
AVERAGE WORKWEEK OF PRODUCTION WORKERS MANUFACTURING	0	3.3
INDEX OF NET BUSINESS FORMATION	11.9	.7
PERCENT CHANGE IN SENSITIVE PRICES	6.3	-11.1
BUILDING PERMITS PRIVATE HOUSING	12.6	13
VALUE OF MANUFACTURER'S NEW ORDERS FOR CONSUMERS GOODS AND MATERIALS	2.9	15
CHANGE IN INVENTORIES ON HAND AND ON ORDER	3.7	-1.31
CONTRACT AND ORDERS FOR PLANT AND EQUIPMENT	3.7	-1.32

variables $(e_t + a_t)$ and $(e_t - a_t)$ is $\sigma_1^2 - \sigma_2^2$. Thus the two error variances, and hence, given the assumption of unbiasedness, the two expected squared errors, will be equal if and only if this pair of random variables is uncorrelated. The usual test for zero correlation based on the sample correlation coefficient between the two variables is then applicable. This test was used on the residuals of the univariate and multivariate models. It was found that, with respect to data fitting, there was no significant difference between the two types of models for GNP.

Finally, a comparison of the forecasting abilities of the univariate and multivariate models was made. In only six cases did the multivariate model prove better. Moreover, the improvement was very small, the largest reduction in the sum of squared errors being 15%. In seven cases the multivariate forecasts were slightly worse. Table 6 contains the percentage reduction in error sum of squares achieved by the bivariate models in both data fitting and forecasting.

CONCLUSION

The results of our data fitting indicate that some relations exist between GNP and certain indicator series. However, ex ante forecasts for GNP from the bivariate ARIMA models which incorporated these relations were not significantly better than the ex ante forecasts from corresponding univariate models. This small improvement in the accuracy of forecasts for macroeconomic series achieved by analysing a larger information set was noted by Pierce (1977) and Ledolter (1977). The latter offered some explanation in an attempt to reconcile the time series results with the fact that certain causes and effects are known to exist among macroeconomic variables. Based on our experience here, we conclude that leading indicators can be of little value in short term forecasting by a multiple time series procedure.

REFERENCES

BOX, G.E.P. and JENKINS, G.M. (1976). Time Series Analysis Forecasting and Control (Revised Edition). Holden Day, San Francisco.

FEY, R.A. and JAIN, N.C. (1982). Identification and Testing of Optimal Lag Structures and Causality in Economic Forecasting. In Applied Time Series Analysis (Proceedings of the International Conference held at Houston, Texas, August 1981). Eds: O.D. Anderson and M.R. Perryman, North-Holland, Amsterdam and New York, 65-73.

GRANGER, C.W.J. (1969). Investigating Causal Relations by Econometric Models and Cross Spectral Methods. Econometrica 37, 424-438.

GRANGER, C.W.J. and NEWBOLD, P. (1977). Forecasting Economic Time Series. Academic Press, New York.

LEDOLTER, J. (1977). A Multivariate Time Series Approach to Modeling Macroeconomic Sequences. Empirical Economics, Vol. 2, Issue 4, 225-243.

PIERCE, D.A. (1977). Relationships - and the Lack thereof - Between Economic Time Series, with Special Reference to Money and Interest Rates, (with discussion). Journal of the American Statistical Association 72, 11-16.

PIERCE, D.A. and HAUGH, L.D. (1977). Causality in Temporal Systems: Characterization and a Survey. Journal of Econometrics 5, 265-293.

QUENOUILLE, M.H. (1957). The Analysis of Multiple Time Series. Griffin, London.

SIMS, C. (1972). Money Income, and Causality. American Economic Review, 62, 540-552.

THURY, G. (1980a). The Use of Transfer Function Models for the Purposes of Macroeconomic Forecasting: Some Empirical Evidence for Austria. In Analysing Time Series (Proceedings of the International Conference held on Guernsey, Channel Islands, October 1979). Ed: O.D. Anderson, North-Holland, Amsterdam and New York, 349-364

THURY, G. (1980b). Time Series Analysis of the Relationship Between Private Consumer Expenditure and Disposable Personal Income. Empirica 2, 169-198.

THURY, G. (1981). Discrimination Between Alternative ARIMA Model Specifications. Empirica 1, 3-24.

TIAO, G.C., BOX, G.E.P., GRUPE, M.R., HUDAK, G.B., BELL, W.R., and CHANG, I. (1979). The Wisconsin Multiple Time Series Analysis Program, University of Wisconsin, Madison.

TIAO, G.C. and BOX, G.E.P. (1981). An Introduction to Applied Multiple Time Series Analysis. Journal of the American Statistical Association. Vol. 76, No. 376, 802-816.

TIME SERIES ANALYSIS: Theory and Practice 4
O.D. Anderson (editor)

OUTPUT FLUCTUATIONS AND RELATIVE PRICE ADJUSTMENT: A VECTOR MODEL APPROACH

Houston H. Stokes

University of Illinois at Chicago
Department of Economics, Box 4348
Chicago, Illinois 60680
USA

A vector autoregressive model is estimated for a reduced form model and a structural equations model to measure the effect of unexpected and expected changes in price and the variance of prices on output. The structural equations model consists of a generalized Cobb Douglas production function with the change in price and variance of price variables as shift parameters. Differences were found in the effects of expected versus unexpected variables.

INTRODUCTION

Economic theory suggests that a fully anticipated change in all prices, such as the widely anticipated change in the units of the French franc in the late 1950's, will have no effect on the real magnitudes such as output. When price changes are unanticipated or where there are shifts in relative prices, real magnitudes can be affected in the short run. The findings reported in this study extend work done by Blejer and Leiderman (1980), Fisher (1981), Gorden (1981), Parks (1978), and others and are an attempt to study the effects of anticipated and unanticipated changes in prices and the variance of prices on output. In addition to a reduced-form model, a structural-equation, production-function model is developed and estimated using the vector autoregressive procedure.

Changes in prices and the variance of prices are viewed as shift parameters of a generalized Cobb Douglas production function. Feedback is allowed from the inputs, labor and capital, to the shift parameters and from output to the inputs, and to the shift parameters. In the spirit of Sims (1980), no prior restrictions are placed on the direction of causality. This study sets the stage for further work with more disaggregate data. After a brief review of the theory underlying the model, the statistical techniques used, the functional form estimated, and the results are discussed in detail. Finally, implications for future research are mentioned.

BRIEF SURVEY OF THE LITERATURE

Following the pioneer work of Muth (1961), Lucus (1973) and others began to study the effects on endogenous variables of expected and unexpected changes in exogenous and endogenous variables. An important survey of some of this work is contained in Gorden (1980), who, in addition, estimated a model that related the percentage change in the GNP deflator to a proxy variable that was constructed to be the difference between the rate of change of nominal GNP and the rate of change of natural real GNP. This proxy variable was an attempt to measure unanticipated changes in GNP. A weakness of the approach lies in the need to construct the proxy variable for natural real GNP. One objective of this research was to test the sharp contrasts between the implications of Phillips curve-type models, which assert that changes in nominal variables could influence real magnitudes, and those of models such as the quantity theory in monetary economics that asserted that in the long run monetary changes are neutral with respect to real quantities.

In related research, Parks (1978) studied the relationship between the variance of price changes and the change in aggregate prices. Parks' major finding was that there was a linear relationship between the variance of relative price changes and a squared measure of surprise, or unanticipated, inflation. Recent work by Fischer (1981) using a reduced-form, vector-autoregressive model in the period 1956-1980, found that inflation and relative price variability were correlated (due to energy and food supply disturbances), relative price variability did not influence real GNP growth, and the effect of monetary shocks, although significantly related to price variability, was of secondary importance. It is unknown how much this latter finding was a function of the period of the data selected.

Vining and Elwertowski (1976), in work predating Parks (1978), found a positive relationship between the variance of relative price changes and the amount of price change instability. In a related area, Stokes and Neuburger (1979) used Box and Jenkins transfer function models to measure the effect of monetary changes on interest rates, via income effects, while Stokes (1982) studied the effect of monetary variables on relative producer prices.

Blejer and Leiderman (1980) used the Parks (1978) data to test the effect of price changes and the variance of price change on real output. Following Parks they define from the n components of the consumer price index (CPI)

$$DP_t = \sum_{i=1}^{n} w_{it}^{*} Dp_{it} \tag{1}$$

and

$$VP_t = \sum_{i=1}^{n} w_{it}^{*} (Dp_{it} - DP_t)^2, \tag{2}$$

where w_{it}^{*} = the average expenditure share on the ith commodity in the years t-1 and t and $Dp_{it} = \ln P_{it} - \ln P_{it-1}$. Since $(Dp_{it} - Dp_t)$ is the rate of change in the ith relative price, VP_t becomes the variance of relative price changes. Unlike Parks (1978), who asserted that DP_t was a random walk (implying that the expectation of DP_t $[EDP_t]$ was equal to DP_{t-1}), Blejer and Leiderman (1980, footnote 9) formed their expectations of DP_t from the equation

$$DP_t = .020 + .688\, DP_{t-1} - .339 DP_{t-2}. \tag{3}$$

Using OLS estimation methods, Blejer and Leiderman (1980) related output to lagged output and $(DP_t - EDP_t)$ and various combinations of the variables VP_t, EDP_t, SVP1, and SVP2, where $SVP1 = VP_t - VP_{t-1}$ and $SVP2 = SVP1 + VP_{t-2}$. They argue that because Parks (1978) found that unanticipated inflation $(DP_t - EDP_t)$ is positively correlated with relative price variability (VP_t), if VP_t is omitted from the equation, the coefficient of the unanticipated inflation variable will be biased toward zero. Blejer and Leiderman argue that unanticipated inflation should have a positive effect on output, while price variability has a negative effect on output.

Their findings are provocative and are further refined and tested in this paper. They will be approached in a number of ways. Since one major limitation of their study was that they used a reduced-form model for output, in this study a structural equation model for output will be estimated. In alternative models, the expected values and values of VP_t and DP_t are entered into a Cobb Douglas production function as shift parameters. The coefficients of the inputs and the shift parameters are assumed to be up to fourth order unconstrained polynomials in the lag operator B. Since a vector autoregressive model is estimated, a test can be made for feedback from output to the shift parameters and to the inputs.

Inspection of these terms allows the dynamic relationship between inputs and the shift parameters to be determined.

In the Blejer and Leiderman study, where expected values of DP_t were used, VP_t was used in level form. In the proposed model, both the expected and unexpected form of VP_t and DP_t are modelled and the results are contrasted.

THE APPROPRIATE STATISTICAL MODEL

In a previous study, Sinai and Stokes (1972) used the Christensen and Jorgenson (1970) data on output (Table 3, pp. 30-31, col. 1), labor (pp. 28-34), and capital (Table 5, col. 4, p. 36) to estimate a production function containing labor capital and real money balances. In this paper, the Christensen and Jorgenson data will be used to estimate a production function for the period 1930-1967 for quality-adjusted inputs of capital and labor and shift parameters of, in one case, expected values of DP, and VP, and, in another case, the unexpected values of DP and VP. Before discussing the models estimated, a few comments are needed on the interpretation of these variables.

If we assume, following Parks (1978), that DP follows a random walk and extend this assumption to VP, and if the present values of DP_t and VP_t do not enter a simple Cobb Douglas production of the form

$$Q = Ae^{B_3 t + B_4 DP_t + B_5 VP_t} L^{B_1} K^{B_2}, \tag{4}$$

we can interpret logged values of DP_t and VP_t as expected values in a more general production function that contains polynomials in the lag operator B $(B^i X_t = X_{t-i})$ in place of the OLS coefficients B_5 and B_4 in equation 4.

An alternative model, after that of Lucus (1973), would continue to assume that $EDP_t = DP_{t-1} + e_1$ and $EVP_t = VP_{t-1} + e_2$, where e_1 and e_2 are normally distributed error terms having mean zero, but would replace the values of DP_t and VP_t with their first differences. The terms $(DP_t - DP_{t-1})$ and $(VP_t - VP_{t-1})$ can be interpreted as the unexpected components of DP_t and VP_t, respectively. If these unexpected component proxy variables are entered into a generalized reduced form model and a generalized form structural equation model containing polynomials in the lag operator B in place of the OLS coefficients, the pattern and significance of the terms within the polynomials will indicate the dynamic relationship between the variables. Before discussing the specific models estimated, a few comments on vector autoregressive models are in order.

Quenouille (1957) argued that a dynamic system of simultaneous equations could be written and estimated in the form

$$H(B)X_t = F(B)u_t, \tag{5}$$

Where X'_t is a suitably differenced (to achieve stationarity) row vector of the t th observation of k series. H(B) and F(B) are, respectively, the k by k autoregressive matrix and the k by k moving average matrix. Each element in H(B) and F(B) is a polynomial in the lag operator B. Equation 5 assumes that $|H(B)|$ and $|F(B)|$ are on or outside the unit circle (the invertability condition), and the expected value of the error vector u_t is a null vector $(Eu_t = 0)$. Following Granger and Newbold (1977), the typical elements of H(B) and F(B) are

$$H_{ij}(B) = \sum_{k=0}^{MAR} h_{ij,k}B^k \quad (6)$$

and
$$F_{ij}(B) = \sum_{k=0}^{MMA} f_{ij,k}B^k, \quad (7)$$

Where MAR = the maximum order on the AR matrix and MMA = the maximum order on the MA matrix. Typically, MAR ≠ MMA. Assuming the above-mentioned invertability condition, the vector autoregressive moving average (VARMA) form of the model (equation [5]) can be written as a vector autoregressive model (VAR)

$$A(B)X_t = u_t \quad (8)$$

or a vector moving average model (VMA)

$$X_t = M(B)\ u_t, \quad (9)$$

where
$$M(B) = (H(B))^{-1}F(B) \quad (10)$$

and
$$A(B) = (F(B))^{-1}H(B). \quad (11)$$

If the model is estimated in the VAR form (equation [8]), with a finite order on the matrix coefficients, the random shock or VMA form of the model equation will, in general, have an infinite number of terms since $(M(B) = (A(B))^{-1}$. Sims (1980) employed this procedure and proceeded to interpret the MA coefficients, although he was forced to admit that he was unable to obtain significance tests on the individual terms in the VMA model directly because they were obtained from the inverse of A(B). While it is possible to estimate a model in the VARMA form (equation [5]), since many coefficients will be constrained to be equal to zero, there is always the possibility that the interpretation will be misleading, single alternative H(B) and F(B) matrices can give rise to the same general A(B) or M(B) matrix. When both H(B) and F(A) matrices are in the model, economic interpretation is difficult at best. If the VARMA form is translated to the VAR form (equation [8]) or VMA form (eq [9]) for interpretation, the problems Sims (1980) faced reappear; there are no significance tests on the coefficients in M(B) or A(B), even though these matrices probably were calculated from matrices H(B) and F(B), which all contained significant terms. In summary, the model estimated in this paper will be of the unconstrained VAR form, owing to a desire to both obtain significance tests for all coefficients and to avoid making judgments on which terms within the elements of A(B) to set to zero.

Estimation will be performed with the WMTS-1 program, which was developed by Tiao, et al. (1980) and which was modified by Stokes (1981). Two basic VAR models will be estimated, a reduced form model and a structural equations model, and for each type of model both the expected form of the variables DP_t and VP_t and the unexpected form of the variables DP_t and VP_t will be tried. The reduced form of the model will be discussed first. Expanding equation (8) we write

$$\begin{bmatrix} A_{11}(B) & A_{12}(B) & A_{13}(B) \\ A_{21}(B) & A_{22}(B) & A_{23}(B) \\ A_{31}(B) & A_{32}(B) & A_{33}(B) \end{bmatrix} (1-B) \begin{bmatrix} \ln Q \\ DP_t \\ VP_t \end{bmatrix} = \begin{bmatrix} u_{1t} \\ u_{2t} \\ u_{3t} \end{bmatrix} \tag{12}$$

Following what Granger and Newbold (1977, p. 223) have called model A, we restrict the zero order terms in $A_{ij}(B)$ to one for cases where $i=j$ and zero for cases $i \neq j$. Such a setup implies that instantaneous causality is seen in the off-diagonal elements in the covariance matrix, which is not assumed to be a diagonal matrix. This form of equation (12) has advantages that are well known for estimation purposes but that sometimes are seen as peculiar by economists who are interested in the simultaneous equations form of equation (8), which does not restrict the zero order elements in $A_{ij}(B)$ to be zero for the case $i \neq j$ but which does restrict the covariance matrix of the error terms to be a diagonal matrix. Using this later form of the model, instantaneous causality is seen as significant zero order $A_{ij}(B)$ terms, where $i \neq j$.

The terms $A_{ii}(B)$ in equation (12) control for the effects of past values of the variable. If, for example, the expected value of DP_t is significantly positively related to the natural log of real output, terms at some yet unspecified lag in the MAR order polynomial $A_{12}(B)$ will be significantly positive. In the estimation carried out in this paper MAR is set to 4 because yearly data were used, and because the cross correlations of the residuals were, in general, clear after a 4-order polynomial was fit. In the case of equation (12), this involved estimating 36 parameters jointly.

If equation (8) is to be written in a form to test the effect of unexpected changes in DP_t and VP_t, the only change needed is to impose one more level of differencing on DP_t and VP_t than on ln Q.

In the Blejer and Leiderman (1980) study, discussed earlier, the effect of lagged output on output was constrained to be a first order process. In terms of equation (8) this would be equivalent to asserting, without testing, that $A_{11}(B) = (1 - 0B)$ where 0 is the estimated coefficient for the lagged output term.

The structural form of the model to be estimated is a generalized form of the Cobb Douglas production function given in equation (4). This can be written in vector notation as

$$\begin{bmatrix} A_{11}(B) & A_{12}(B) & A_{13}(B) & A_{14}(B) & A_{15}(B) \\ A_{21}(B) & A_{22}(B) & A_{23}(B) & A_{24}(B) & A_{25}(B) \\ A_{31}(B) & A_{32}(B) & A_{33}(B) & A_{34}(B) & A_{35}(B) \\ A_{41}(B) & A_{42}(B) & A_{43}(B) & A_{44}(B) & A_{45}(B) \\ A_{51}(B) & A_{52}(B) & A_{53}(B) & A_{54}(B) & A_{55}(B) \end{bmatrix} (1-B) \begin{bmatrix} \ln Q \\ DP_t \\ VP_t \\ \ln L \\ \ln K \end{bmatrix} = \begin{bmatrix} u_1 \\ u_2 \\ u_3 \\ u_4 \\ u_5 \end{bmatrix}. \tag{13}$$

Instead of assuming all none zero-order coefficients equal zero, as was done in equation (4), in equation (13) the effect of an input, say labor, on output can be assumed to be a distributed lag and is measured by $A_{14}(B)$. Another advantage of equation (13) is that it allows modelling feedback of output on the inputs (terms $A_{41}(B)$ and $A_{51}(b)$) and feedback of output on the shift parameters DP_t and VP_t (terms $A_{21}(B)$ and $A_{31}(B)$). Since Blejer and Leiderman (1980) only estimated a single reduced form equation, they assumed feedback to be zero. As was the case with equation (12), equation (13) can be estimated as written to measure the effect of expected changes in DP_t and VP_t, or, if extra differencing is applied to DP_t and VP_t, in a form to measure the effect of unexpected changes in DP_t and VP_t on output. Before discussing these results, we first must estimate equation (4), using OLS (and GLS) procedures to test whether current levels of DP_t and VP_t significantly influence output.

THE RESULTS

The results of estimating a production function of the form of equation (4), containing labor and capital as inputs, and a trend term (T), a rate of inflation term (DP_t), and a variance of relative price term (VP_t) as shift parameters, are given in Table 1. The findings indicate that with just labor, capital, and time in the model (equation [4]), there is substantial autocorrelation (DW= .7890). After a second-order GLS correction (equation [15]), the serial correlation is reduced as measured by the Durbin Watson statistic (1.5636). In equation (15) labor, capital, and the time trend are significant. Equation (16) adds DP_t and VP_t as shifters to equation (14). In comparison with equation (14), there is no reduction in the serial correlation of the residual of the augmented equation in either its OLS (equation 16)) or second-order GLS from (equation [17]). In neither equation was DP_t or VP_t found to be significant.

This finding suggests that contemporaneous actual values of DP_t and VP_t do not affect the efficiency in which labor and capital are combined to produce output. In later tables showing vector autoregressive models, the effects of lagged DP_t or VP_t will be investigated. If lagged DP_t or VP_t are found to be significantly related to output; the finding can be viewed in two frameworks. The most straightforward interpretation is that if, for example, VP_{t-j} is significantly related to output, in equation (13), the efficiency by which labor and capital are combined to form output is changed by a change in the expected value of VP in period t-j+1. Another way to view the finding would be that the efficiency by which labor and capital are combined is changed by the level of VP in period t-j. Using the specification of equation (16) -(17), there is no way to tell how to interpret the coefficient B_4 or B_5, if they were significant. Since B_4 and B_5 were found not to be significant in equation (16) and (17), the level interpretation is not appropriate and possibly significant coefficients in the terms $A_{12}(B)$ and $A_{13}(B)$ in equation (12) and (13) can be interpreted as measuring the effect of expected values of DP_t and VP_t, respectively, since the current level (future expected value) was not significant.

Table 2 reports the results of estimating equation (12) for expected values of DP and VP. Standard errors are listed under the coefficients of the 4th order AR matrix A(B) and the significance of the terms in each element of the matrix have been summarized by +, -, . for positive significance, negative significance, and no significance. Significance is taken to mean that the t score is greater than 2 in absolute value (95% confidence value). The major findings of Table 2 are that the expected value for VP in period t-1 is positively related to output (see + significant term at lag 2 in $A_{13}(B)$), that changes in expected DP in period t-3 are negatively related to VP (see - significant term at lag 4 in $A_{32}[B]$), and that there is positive feedback from log output lagged 3 periods to VP (see + significant term at lag 3 in $A_{31}[B]$).

The first finding is counter to what was expected by Blejer and Leiderman (1980), while the relationship between expected DP and VP was what was expected. The feedback found between log output was not found by the former investigators because they used a single-equation, reduced-form model.

Table 3 adds the log of labor and capital to the model estimated in Table 2. The importance of this table is that the positive relationship between expected VP in period t-1 holds up and, in addition, the expected value of VP in period t is positively significant (see + significant terms at lag 1 and 2 in term $A_{13}[B]$). Table 3 also indicates a number of relationships between the inputs and output and between the inputs themselves. The two negative significant terms in $A_{14}(B)$ at lag 1 and 3 indicate that increases in labor 1 and 3 periods ago are associated with decreases in output today. A similar finding for capital, but with a different lag, indicates that an increase in capital 4 periods ago is associated with output decreasing today (see significant negative term in $A_{15}(B)$ at lag 4). These surprising results may be due to either some kind of business cycle or due to the fact that an increase in capital 3 periods ago is associated with a reduction in labor today (see significant negative term at lag 3 in $A_{45}(B)$. The only other observed findings are that increases in labor lagged three periods are negatively associated with changes in DP (see negative significant term in $A_{24}(B)$ at lag 3) and that, on balance, there is positive feedback from output to labor (see positive significant term at lag 2 and negative and smaller significant term at lag 4 in $A_{41}(B)$).

Results using a reduced form model and unexpected values of DP_t and VP_t are given in Table 4. The two positive significant terms in $A_{13}(B)$ at lags 2 and 3 indicate that unexpected movements in VP are positively associated with changes in output. The sign of this effect is consistent with what was found for the expected values of VP in Tables 2 and 3. Unexpected changes in DP lagged one period, were positively related to unexpected changes in VP, while lagged 4 period changes were negatively related. On balance the effect was positive (see significantly positive coefficient at lag 1 and negative coefficient at lag 4 in $A_{32}(B)$).

Results for the structural equations model for unexpected changes in DP_t and VP_t are given in Table 5. A major finding is that unexpected changes in DP_t are positively related to output (see significant positive coefficients at lag 1-4 in $A_{12}(B)$), and unexpected changes in DP are positively related to labor (see significant positive coefficients at lag 1-4 in $A_{42}(B)$). In contrast to the findings for the expected variables in Table 3, using unexpected values for DP_t and VP_t reveals many more relationships between real and nominal variables. A few of these relationships will be discussed. Changes in output at lags 1 and 3 are positively related to unexpected changes in prices (see significant positive coefficients at lag 1 and 3 in $A_{21}(B)$). On balance changes in labor and capital are negatively related to unexpected changes in prices (see negative term at lag 1 and positive and smaller term at lag 4 in $A_{24}(B)$ and negative terms at lags 3 and 4 in $A_{25}(B)$). In summary we can see that the unexpected changes in nominal variables have more effect on real variables than was found for expected changes in DP and VP.

Future work will fruitfully focus on the sensitivity of the results to different expectations models and different data periods such as monthly or quarterly data. A further question relates to the stability of the model over time.

ACKNOWLEDGEMENT

Computer time for this study was provided by the University of Illinois at Chicago. Diana Stokes provided editorial assistance.

Table 1

OLS and GLS Estimates of the Parameters of the Cobb-Douglas Production Function with and without price change and price variability shifters 1930-1967

$$LnQ = Ln\ A + B_1 Ln\ L + B_2 Ln\ K + B_3\ T + B_4\ (DP) + B_5\ (VP) + e$$

Eq	14	15	16	17
Ln A	-2.654 (.5286)	-1.5493 (.6666)	-2.6754 (.5640)	-1.4962 (.8799)
B_1	1.3008 (.09584)	1.14488 (.09914)	1.2937 (.12032)	1.1478 (.1478)
B_2	.25491 (.06600)	.1761 (.09763)	.2699 (.0837)	.1604 (.1041)
B_3	.00739 (.00278)	.01306 (.00413)	.00681 (.00297)	.01348 (.00486)
B_4			.038860 (.1750)	-.005161 (.1896)
B_5			-1.6055 (2.258)	-.7849 (.9809)
P1		.9668		.9630
P2		-.5771		-.5531
SEE	.03243	.03435	.03317	.03621
R^2	.995057	.994023	.994829	.993374
DW	.7890	1.5636	.7771	1.51744

Ln Q = natural log gross private domestic product, quantity index.
Ln L = natural log private domestic labor input, quantity index.
Ln K = natural log private domestic capital input, quantity index.
T = time trend 1929=0. Data source for Q, L and K Christensen and Jorgenson (1970)
DP = rate of inflation, source Parks (1978 p. 85)
VP = Variance of relative price change, source Parks (1978 p. 85)
Standard errors of the regression coefficients in parentheses.
DW = Durbin-Watson test statistic. R^2 reported has been adjusted.
If $p_1 = p_2 = 0$, the equation was estimated by OLS.

Table 2

Coefficients of Vector Autoregressive Model $A(B)\ X_t = e_t$ where $x_1 = \ln Q$ $x_2 = DP$ and $x_3 = VP$ 1930-1967

Term	lag 1	2	3	4	pattern of sig.
A_{11}	.239	.145	-.293	-.178	. . . .
	.214	.198	.156	.164	
A_{12}	-.520	-.286	-.121	-.207	. . . .
	.284	.299	.245	.223	
A_{13}	4.02	7.04	2.42	-1.38	. + . .
	2.47	3.14	3.46	2.46	
A_{21}	.118	.121	-.019	-.037	. . . .
	.137	.127	.100	.105	
A_{22}	-.158	-.528	.009	-.206	. - . .
	.182	.191	.157	.143	
A_{23}	1.000	-.337	-2.60	-1.46	. . . .
	1.58	2.00	2.21	1.57	
A_{31}	.015	.020	.020	-.011	. . + .
	.013	.012	.010	.010	
A_{32}	.023	-.033	-.014	-.047	. . . -
	.017	.018	.015	.014	
A_{33}	-.866	-.710	-.460	.051	- - - .
	.151	.192	.212	.150	

Residual Covariance Matrix

.00105		
.0004380	.00042800	
.0000104	.00000715	.00000393

Residual Correlation Matrix

1.00		
.65	1.00	
.16	.17	1.00

M(4) = 9.06 DF = 9 Probability = .5682

For data sources see text. The pattern of the significance takes the value +, -, . depending on whether the coefficient divided by the SE is greater than or equal to 2, less than or equal to -2, or otherwise. Diagnostic checking of the cross correlations of the residuals up to 12 periods back indicates 106 insignificant correlations and 2 significant negative correlations at lag 8 for p_{22} and at lag 11 for p_{23}. These have been ignored because they are so far back. SE is listed under the coefficient. The data vector X_t has been differenced to induce stationarity.

Table 3

Coefficients of the Vector Autoregressive Model $A(B)X_t = e_t$ where $x_1 = \ln Q$ $x_2 = DP$, $x_3 = VP$ $x_4 = \ln L$ and $x_5 = \ln K$ 1930-1967

Term	lag	1	2	3	4	pattern of sig.
A_{11}		.392	.739	.533	-.351	. + . .
		.386	.358	.348	.294	
A_{12}		-.001	-.047	.187	-.153	. . . .
		.358	.433	.283	.237	
A_{13}		5.71	9.54	7.48	-.643	+ + . .
		2.84	4.34	4.47	3.06	
A_{14}		-1.35	-.859	-.823	.647	− . − .
		.389	.490	.408	.429	
A_{15}		.491	-.398	-.416	-.668	. . . −
		.398	.335	.318	.264	
A_{21}		.295	.292	.478	-.158	. . . .
		.268	.248	.242	.204	
A_{22}		-.335	-.534	.076	-.175	. . . .
		.248	.300	.197	.165	
A_{23}		3.19	2.68	1.12	.233	. . . .
		1.97	3.01	3.11	2.13	
A_{24}		-.338	-.317	-.750	.106	. . − .
		.270	.340	.284	.298	
A_{25}		-.016	-.139	-.156	-.039	. . . .
		.276	.233	.221	.183	
A_{31}		-.001	.033	-.003	.013	. . . .
		.029	.027	.027	.022	
A_{32}		.034	-.039	-.017	-.034	. . . .
		.027	.033	.022	.018	
A_{33}		-.857	-.722	-.494	.078	− − . .
		.217	.331	.341	.234	
A_{34}		-.004	-.015	.039	-.049	. . . .
		.030	.037	.031	.033	
A_{35}		.021	-.003	-.003	.000	. . . .
		.030	.026	.024	.020	
A_{41}		.050	.640	.503	-.527	. + . −
		.345	.319	.311	.262	
A_{42}		-.021	-.180	.158	-.368	. . . .
		.319	.386	.253	.212	

Table 3 (continued)

Term	lag 1	2	3	4	pattern of sig.
A43	3.58 2.54	4.83 3.87	3.64 4.00	2.10 2.74	. . . .
A44	-.467 .348	-.655 .438	-.431 .365	.609 .383	. . . .
A45	.260 .355	-.427 .299	-.629 .284	.030 .236	. . - .
A51	-.105 .291	.199 .270	.293 .263	-.138 .222	. . . .
A52	-.209 .270	-.223 .326	-.136 .214	-.145 .179	. . . .
A53	3.58 2.14	3.44 3.27	1.60 3.38	-3.40 2.31	. . . .
A54	-.246 .294	-.497 .370	-.606 .308	.305 .324	. . . .
A55	.501 .300	.019 .253	-.044 .240	-.519 .199	. . . -

Residual Covariance Matrix

.000513				
.000251	.000248			
.000023	.000017	.000003		
.000421	.000260	.000020	.000410	
.000354	.000177	.000019	.000316	.000292

Residual Correlation Matrix

1.00				
.70	1.00			
.60	.62	1.00		
.92	.81	.57	1.00	
.91	.66	.65	.91	1.00

M(4) = 32. DF = 25 Probability = .8420

For data sources and for further explanation of the table see Table 1. Of 300 cross correlations of the residual calculated up to 12 periods back 297 were found to not to be significant. At lag 3, p_{13}, p_{43} and p_{53} were found to be negatively significant. The data vector X_t has been differenced to induce stationarity.

Table 4

Coefficients of Vector Autoregressive Model $A(B)X_t = e_t$ where $x_1 = \ln Q$ $x_2 = (1-B)DP$ and $x_3 = (1-B)VP$ 1930-1967. Note: x_2 and x_3 can be interpreted as expected values.

Term	lag 1	2	3	4	pattern of sig.
A_{11}	.193	.137	-.293	-.139	. . . .
	.221	.201	.199	.162	
A_{12}	-.358	-.434	-.190	-.216	. . . .
	.285	.294	.240	.207	
A_{13}	3.77	8.54	7.83	3.04	. + + .
	2.00	2.99	3.03	2.04	
A_{21}	.010	-.064	-.180	.053	. . . .
	.158	.144	.142	.116	
A_{22}	-.528	-.654	-.161	-.229	- - . .
	.204	.211	.172	.148	
A_{23}	.375	-.900	-3.50	-2.69	. . . .
	1.44	2.14	2.18	1.46	
A_{31}	-.006	.003	-.001	-.025	. . . -
	.013	.012	.012	.010	
A_{32}	.065	.019	.024	-.028	+ . . -
	.020	.018	.014	.012	
A_{33}	-1.41	-1.46	-1.10	-.280	- - - -
	.119	.178	.181	.121	

Residual Covariance Matrix

.00112		
.000681	.000578	
.000017	.000001	.000004

Residual Correlation Matrix

1.00		
.84	1.00	
.25	.03	1.00

M(4) = 32.75 DF = 9 Probability .9999

For Data Sources and a discussion of the theory behind the model see text. Diagnostic checking of the cross correlations of the residuals up to 12 periods back indicates 107 insignificant correlations and 1 negatively significant correlation at lag 3 for p_{21}. SE is listed under the coefficient. The data vector X_t has been differenced to induce stationarity.

Table 5

Coefficients of the Vector Autoregressive Model $A(B)X = e_t$ where $x_1 = \ln Q$ $x_2 = (1-B)DP$, $x_3 = (1-B)VP$, $x_4 = \ln L$ and $x_5 = \ln K$ 1930-1967. Note: x_2 and x_3 can be interpreted as expected values.

Term	lag 1	2	3	4	pattern of sig.
A_{11}	1.06	.126	.590	-.742	+ · · -
	.424	.372	.341	.299	
A_{12}	.735	.988	.975	.546	+ + + +
	.308	.342	.273	.203	
A_{13}	1.80	5.12	6.03	1.11	· · · ·
	1.90	3.40	3.29	2.21	
A_{14}	-1.86	-.011	-.725	1.38	- · · +
	.384	.442	.367	.395	
A_{15}	-.250	-.220	-.371	-.518	· · · -
	.503	.283	.313	.245	
A_{21}	.486	-.317	.991	-.280	+ · + ·
	.206	.181	.166	.145	
A_{22}	-.173	.038	.450	.291	· · + +
	.150	.166	.132	.099	
A_{23}	-1.97	-5.91	-8.10	-6.54	- - - -
	.925	1.65	1.60	1.08	
A_{24}	-1.32	-.017	-1.31	1.00	- · · +
	.187	.215	.178	.192	
A_{25}	.364	.079	-.360	-.466	· · - -
	.245	.137	.152	.119	
A_{31}	.013	.014	.013	.020	· · · ·
	.034	.030	.027	.024	
A_{32}	.081	.030	.027	-.013	+ · · ·
	.025	.027	.022	.016	
A_{33}	-1.53	-1.66	-1.37	-.529	- - - -
	.153	.272	.263	.178	
A_{34}	-.059	-.007	-.005	-.042	· · · ·
	.031	.036	.029	.032	
A_{35}	.023	-.009	-.017	-.039	· · · ·
	.040	.023	.025	.020	
A_{41}	.765	-.068	.533	-.767	+ · · -
	.367	.322	.295	.259	

Table 5 (continued)

Term	lag	1	2	3	4	pattern of sig.
A42		.769	.871	.997	.361	+ + + +
		.266	.296	.236	.176	
A43		.518	.585	-.671	-.951	. . . .
		1.65	2.94	2.84	1.92	
A44		-1.09	.207	-.562	1.26	- . . +
		.333	.383	.318	.342	
A45		-.417	-.212	-.527	-.002	. . . .
		.436	.245	.271	.212	
A51		.260	-.217	.393	-.320	. . . .
		.329	.289	.264	.232	
A52		.436	.477	.422	.179	. . . .
		.239	.265	.211	.157	
A53		2.31	3.55	3.29	-.663	. . . .
		1.48	2.63	2.55	1.72	
A54		-.735	-.137	-.765	.761	- . - +
		.298	.343	.285	.306	
A55		.162	.184	-.058	-.599	. . . -
		.391	.219	.242	.190	

Residual Covariance Matrix

.000420				
.000181	.000099			
.000017	.000004	.000003		
.000340	.000151	.000010	.000315	
.000288	.000121	.000015	.000246	.000253

Residual Correlation Matrix

1.00				
.89	1.00			
.51	.27	1.00		
.93	.85	.34	1.00	
.88	.76	.59	.87	1.00

M(4) = 60.23 DF = 25 Probability = .9999

For Data Sources and a discussion of the theory behind the model see text. Diagnostic checking of the cross correlations of the residuals up to 12 periods back indicates 293 insignificant correlations. Positive significant correlations were found at lag 2 for p_{31} and p_{32}. Negative significant correlations were found at lag 1 for p_{23} and at lag 3 for p_{13} p_{22}, p_{43} and p53. SE is listed under the coefficient. The data vector X_t has been differenced to induce stationarity.

REFERENCES

BLEJER, M.I. and LEIDERMAN, L. (1980). The Real Effects of Inflation and Relative Price Variability: Some Empirical Evidence. Review of Economics and Statistics 62, 539-544.

BOX, G.E.P. and JENKINS, G.M. (1976). Time Series Analysis, Forecasting and Control (Revised Edition). Holden-Day, San Francisco.

CHRISTENSEN, L.R. and JORGENSON, D.W. (1970). US Real Product and Real Factor Input 1929-1967, Review of Income and Wealth 16, 19-50.

FISCHER, S. (1981). Relative Shocks, Relative Price Variability and Inflation. in Brookings Papers on Economic Activity 2. Eds: W.C. Brainard and G.L. Perry, Brookings Institution, Washington, 381-431.

GORDEN, R.J. (1981). Output Fluctuations and Gradual Price Adjustment. Journal of Economic Literature 19, 493-530.

GRANGER, C.W.J. and NEWBOLD P. (1977). Forecasting Economic Time Series. Academic Press, New York.

LUCAS, R.E. (1973). Some International Evidence of Output-Inflation Tradeoffs. American Economic Review 63, 721-754.

MUTH, J.F. (1961). Rational Expectations and the Theory of Price Movements. Econometrics 29 No. 3, 315-35.

PARKS, R.W. (1978). Inflation and Relative Price Variability. Journal of Political Economy 86, 79-96.

QUENOUILLE, M.H. (1957). The Analysis of Multiple Time Series. Griffin, London.

SIMS, C. (1980). Macroeconomics and Reality. Econometrica 48, #1 1,48.

SINAI, A. and STOKES H.H. (1972). Real Money Balances: An Omitted Variable From the Production Function. Review of Economics and Statistics 54, 290-296.

STOKES, H.H. and NEUBURGER H. (1979). The Effects of Monetary Changes on Interest Rates: A Box-Jenkins Approach. Review of Economics and Statistics 51, 534-548.

STOKES, H.H. (1981). The B34S Data Analysis Program: A Short Writeup. Report FY 77-1 (Revised 1981) College of Business Administration Working Paper Series, University of Illinois at Chicago.

STOKES, H.H. (1982). The Effect of Monetary Variables on Relative Producer Prices. In Applied Time Series Analysis (Proceedings of the International Conference held at Houston, Texas, August 1981). Eds: O.D. Anderson and M.R. Perryman, North-Holland, Amsterdam & New York, 383-395.

TIAO, G., BOX, G.E.P., GRUPE, M.R., HUDAK, G.B., BELL, W.R., and CHANG, I. (1980). The Wisconsin Multiple Time Series (WMTS-1) Program: A Preliminary Guide. University of Wisconsin, Department of Statistics.

VINING, D.R. and ELWERTOWSKI (1976). The Relationship between Relative Prices and the General Price Level. American Economic Review 66, 699-708.

TIME SERIES ANALYSIS: Theory and Practice 4
O.D. Anderson (editor)

THE 1979 OIL PRICE SHOCK AND INFLATION IN FIVE INDUSTRIAL COUNTRIES: AN INTERVENTION ANALYSIS[1]

Ali T. Akarca

Department of Economics
University of Illinois at Chicago
Box 4348, Chicago, IL 60680, USA

Thomas V. Long, II

Technology 2000
P.O. Drawer 390
Chapel Hill, NC 27514, USA

The price of imported crude oil nearly doubled during 1979. Whether this has altered the behavior of the inflation series in France, Germany, Japan, UK and USA is investigated using the procedure of Box and Tiao (1976). For each country, forecasts made from a univariate ARMA model built on pre-1979 data are compared with actual observations. A significant divergence between these is taken as evidence of a change in the univariate model and the vector time series model underlying it. Continuing model adequacy is rejected only for Germany and UK. However, for both of these countries this is found to be attributable to domestic interventions.

1. INTRODUCTION

After having tripled during the Arab oil embargo of 1973-1974, the price of imported crude oil rose slowly until the end of 1978. However, following the disruption of supplies from Iran, it nearly doubled during 1979.[2] In this paper we examine the behavior of the inflation rates in five major industrial countries, Federal Republic of Germany (FRG), France, Japan, United Kingdom (UK) and United States of America (USA), to see whether they have changed significantly during 1979 and, if so, whether this can be attributed to the oil price shock created by the Iranian crisis.[3]

2. DESCRIPTION OF METHOD

The procedure utilized is that of Box and Tiao (1976) and is closely related to those of Box and Tiao (1975) and Feige and Pearce (1976). First we model the behavior of the inflation rate series in each country during the June 1974-

---

[1]Computer time for this study was provided by the University of Illinois at Chicago.

[2]For example, the refiner acquisition cost of imported crude oil in USA increased 14.40% between June 1974 and December 1978, from $13.06 to $14.94 per barrel. During 1979 however, it increased 93.51% to a level of $28.91 per barrel. A rise in the world petroleum demand was as much responsible for the increase in oil price as was the reduction in supply. However, this too must be attributed to the political turmoil in Iran since the shift in demand resulted mainly from attempts by the oil importing countries to raise their inventory levels in response to a greater uncertainty concerning the future supply of oil.

[3]Since imported oil is a factor in the aggregate production function, theoretically speaking, increases in its price will decrease the aggregate supply and thus create inflationary and recessionary pressures. The significance of these pressures, however, needs to be determined empirically.

December 1978 (pre-intervention) period.[4] To accomplish this we build for each country, using the methodology of Box and Jenkins (1970), an autoregressive moving average (ARMA) model of the following form:

$$\phi_1(B)\ \phi_{12}(B^{12})\ (z_t - \mu) = \theta_1(B)\ \theta_{12}(B^{12})\ a_t \tag{2.1}$$

where z_t is the inflation rate; $\{a_t\}$ is a sequence of independently and normally distributed random shocks with mean zero and variance σ_a^2; $\phi_1(B)$, $\theta_1(B)$, $\phi_{12}(B^{12})$ and $\theta_{12}(B^{12})$ are polynomials of finite orders in the lag operators B and B^{12} ($B^k z_t = z_{t-k}$) with all of their roots lying outside the unit circle; μ is a constant representing the mean or the level of the inflation rate series. We then use this model to generate one-step-ahead minimum-mean-square-error forecasts for the January 1979-December 1979 (post-intervention) period. The differences between these predictions and actual realizations (the one-step-ahead forecast errors) can be viewed as the estimates of innovations or shocks in the inflation series during the post-intervention period.[5] If the inflation process was not altered in 1979 we would expect these to have the same probability structure as the pre-intervention shocks. Hence, we take any departure from this structure as evidence of a change in our original model and the multivariate ARMA model which underlies it.[6] Following Box and Tiao (1976) we base our overall test of the continuing appropriateness of the pre-intervention model during the post-intervention period on the statistic:

$$\sum_{j=1}^{m} [z_j - \hat{z}_{j-1}(1)]^2 / m\ \hat{\sigma}_a^2 \tag{2.2}$$

in which $j=1,2,\ldots,m$ refer to the post-intervention period; $\hat{z}_{j-1}(1)$ stands for the forecast at time $j-1$ of z_j; $\hat{\sigma}_a^2$ is the estimate of σ_a^2. If the model remains unchanged, the prediction errors $\{z_j - \hat{z}_{j-1}(1)\}$ would be white noise with mean zero and variance σ_a^2. Then, the statistic (2.2) will be distributed approximately as F with m and n-p degrees of freedom, where p is the number of parameters in the model and n is the number of observations used in fitting it. If the model has changed in some way, we would expect the above statistic to be inflated. Thus, we test the continuing appropriateness of our model by referring the value attained by the statistic to an F table with 12 and 55-p degrees of freedom. When a change is indicated we also compare the forecast errors individually to $2\hat{\sigma}_a$ to determine more precisely the month(s) in which this change has occurred and to see whether it may be related to other known interventions of the period.[7]

---

[4]The data prior to June 1974 is excluded to avoid discontinuities caused by such major events as the Arab oil embargo, worldwide crop failures, devaluation of the dollar, wage-price control and decontrol in some of the countries.

[5]However, the same can not be said about the errors from multiple-step-ahead forecasts made from the same origin, December 1978. Although such forecasts would have the advantage of being uncontaminated by the events in the post-intervention period, they assume that all random shocks to the system in the forecast period take on their expected value of zero and their errors would in general be highly correlated giving a false impression of the consistency of discrepancies.

[6]For a discussion of the relationship between univariate and vector ARMA models see Zellner and Palm(1974).

[7]One caveat of the procedure we utilize is that the interventions with effects small relative to $\hat{\sigma}_a$ will not be detected.

3. EMPIRICAL RESULTS[8]

As a measure of inflation we take the rate of change in the consumer price index (CPI). More precisely, we let $z_t = (1-B)\ln CPI_t$. Following the Box-Jenkins (1970) methodology we fit the ARMA models given in Table 1 to represent the pre-intervention behavior of inflation rate series in the five countries being considered. The reported Ljung-Box (1978) chi-square statistics are all insignificant even at the 40% level indicating the adequacy of the fits obtained.[9] Twelve one-step-ahead forecasts generated from these models are displayed in Tables 2 through 6, together with actual realizations and forecast errors. Using these we obtain for the statistic (2.2) a value of 2.85 for FRG, 1.02 for France, 1.02 for Japan, 5.42 for UK and 1.56 for USA.[10] Thus, the null hypothesis of continuing model appropriateness in 1979 can not be rejected for France, Japan and USA, even at the 10% level. For these countries the

TABLE 1. ESTIMATED ARMA MODELS[11]

COUNTRY	MODEL
F.R.G.	$(1 - 0.66082\ B^{12})(z_t - 0.00111) = a_t$ $\hat{\sigma}_a = 0.19168 \times 10^{-2}$ $\quad Q = 15.241$ (d.f. = 22)
FRANCE	$(z_t - 0.00814) = (1 + 0.40227\ B)(1 + 0.35594\ B^{12})\ a_t$ $\hat{\sigma}_a = 0.21491 \times 10^{-2}$ $\quad Q = 18.470$ (d.f = 21)
JAPAN	$(1 - 0.67640\ B^{12})(z_t - 0.00041) = a_t$ $\hat{\sigma}_a = 0.63970 \times 10^{-2}$ $\quad Q = 22.723$ (d.f. = 22)
U.K.	$(1 - 0.27324\ B - 0.31017\ B^2)(1 - 0.36349\ B^{12})(z_t - 0.00701) = a_t$ $\hat{\sigma}_a = 0.47362 \times 10^{-2}$ $\quad Q = 13.681$ (d.f. = 20)
U.S.A.	$(1 - 0.57955\ B)(z_t - 0.00585) = a_t$ $\hat{\sigma}_a = 0.20047 \times 10^{-2}$ $\quad Q = 16.668$ (d.f. = 22)

---

[8]The data is obtained from the National Bureau of Economic Research (NBER) data bank. For a detailed description of the computer program utilized see Stokes (1977).

[9]Ljung-Box (1978) statistic is the small sample version of the better known Box-Pierce (1970) statistic.

[10]These figures should be compared to critical values from F distributions with (12, 53), (12, 52), (12, 53), (12, 51) and (12, 53) degrees of freedom, respectively.

[11]All parameters, except the means for FRG and Japan, are significant at the 5% level. Q stands for the Ljung-Box (1978) chi-square statistic calculated from the first twenty-four autocorrelations of the residuals.

TABLE 2. COMPARISON OF ONE-STEP-AHEAD FORECASTS WITH ACTUAL INFLATION RATES: FEDERAL REPUBLIC OF GERMANY

MONTH		ACTUAL x 10^2	FORECAST x 10^2	ERROR x 10^2
JANUARY	1979	0.9251	0.4146	0.5105
FEBRUARY	1979	0.4288	0.3709	0.0579
MARCH	1979	0.4878	0.2037	0.2841
APRIL	1979	0.5460	0.2446	0.3013
MAY	1979	0.2417	0.1616	0.0801
JUNE	1979	0.5417	0.2436	0.2981
JULY	1979	0.6581	-0.0856	0.7437
AUGUST	1979	0.0596	-0.0446	0.1042
SEPTEMBER	1979	0.2975	-0.0035	0.3010
OCTOBER	1979	0.2374	0.0790	0.1584
NOVEMBER	1979	0.3550	0.2436	0.1114
DECEMBER	1979	0.4714	0.2839	0.1875

TABLE 3. COMPARISON OF ONE-STEP-AHEAD FORECASTS WITH ACTUAL INFLATION RATES: FRANCE

MONTH		ACTUAL x 10^2	FORECAST x 10^2	ERROR x 10^2
JANUARY	1979	0.9413	0.7104	0.2309
FEBRUARY	1979	0.6496	0.8835	-0.2339
MARCH	1979	0.9265	0.7399	0.1866
APRIL	1979	0.9577	0.9318	0.0259
MAY	1979	1.0666	0.8505	0.2161
JUNE	1979	0.8218	0.8670	-0.0452
JULY	1979	1.3163	0.9462	0.3701
AUGUST	1979	1.0331	0.9142	0.1189
SEPTEMBER	1979	0.8340	0.8032	0.0307
OCTOBER	1979	1.2008	0.8749	0.3260
NOVEMBER	1979	0.6321	0.9100	-0.2780
DECEMBER	1979	0.8121	0.6239	0.1882

TABLE 4. COMPARISON OF ONE-STEP-AHEAD FORECASTS WITH ACTUAL INFLATION RATES: JAPAN

MONTH		ACTUAL x 10^2	FORECAST x 10^2	ERROR x 10^2
JANUARY	1979	0.0788	0.2887	-0.2099
FEBRUARY	1979	-0.3156	0.2876	-0.6031
MARCH	1979	0.7871	0.6399	0.1471
APRIL	1979	1.3627	0.7416	0.6211
MAY	1979	1.0387	0.3879	0.6508
JUNE	1979	0.0765	-0.3613	0.4378
JULY	1979	0.8757	0.2810	0.5947
AUGUST	1979	-1.0288	0.0667	-1.0955
SEPTEMBER	1979	1.2559	0.8360	0.4199
OCTOBER	1979	1.2404	0.1187	1.1217
NOVEMBER	1979	-0.3743	-0.7017	0.3274
DECEMBER	1979	0.5982	-0.0933	0.6915

TABLE 5. COMPARISON OF ONE-STEP-AHEAD FORECASTS WITH ACTUAL INFLATION RATES: UNITED KINGDOM

MONTH		ACTUAL x 10^2	FORECAST x 10^2	ERROR x 10^2
JANUARY	1979	1.4828	0.7325	0.7503
FEBRUARY	1979	0.8078	0.9430	-0.1352
MARCH	1979	0.8013	0.9660	-0.1647
APRIL	1979	1.6999	1.0545	0.6454
MAY	1979	0.7816	0.8938	-0.1122
JUNE	1979	1.7153	0.9826	0.7327
JULY	1979	4.2181	0.9170	3.3012
AUGUST	1979	0.7849	1.9800	-1.1951
SEPTEMBER	1979	0.9927	1.7409	-0.7482
OCTOBER	1979	1.0357	0.7553	0.2804
NOVEMBER	1979	0.8681	0.9315	-0.0634
DECEMBER	1979	0.7307	0.9249	-0.1942

TABLE 6. COMPARISON OF ONE-STEP-AHEAD FORECASTS WITH ACTUAL INFLATION RATES: UNITED STATES OF AMERICA

MONTH		ACTUAL x 10^2	FORECAST x 10^2	ERROR x 10^2
JANUARY	1979	0.8832	0.5036	0.3796
FEBRUARY	1979	1.1656	0.7578	0.4078
MARCH	1979	0.9611	0.9215	0.0396
APRIL	1979	1.1412	0.8030	0.3383
MAY	1979	1.2218	0.9074	0.3145
JUNE	1979	1.1609	0.9541	0.2069
JULY	1979	1.0563	0.9188	0.1375
AUGUST	1979	1.0000	0.8581	0.1419
SEPTEMBER	1979	1.0349	0.8255	0.2094
OCTOBER	1979	0.8913	0.8457	0.0456
NOVEMBER	1979	0.9274	0.7625	0.1649
DECEMBER	1979	1.0494	0.7834	0.2660

individual prediction errors, with the exception of one (which is only marginally significant), are also all smaller than the appropriate $2\hat{\sigma}_a$'s in absolute value. In contrast, the null hypothesis is rejected for FRG and UK at the 1% level, implying that structural changes have taken place there. Indeed, in both countries the actual inflation rate differs significantly from the predicted rate on two occasions: January and July 1979 in FRG and July and August 1979 in UK. However, these coincide with or are immediately preceded by major domestic policy interventions. In FRG income taxes were cut considerably in January 1979. In July 1979, parliament agreed to increase the value added tax (VAT) rate by 1% from 12% to 13% and an increase of 0.5% approved in November 1978 came into effect. In UK a new government took office in May 1979 which a month later lowered the income tax rates substantially and introduced a single VAT rate of 15% which replaced the rates of 8% and 12.5%. Many excise duties were also increased. Thus, all extraordinary shocks observed in the inflation rate series during 1979 can be attributed to domestic policy shocks, leaving none needing to be explained by the oil price shock.

If oil price increase indeed has contributed to the inflation in Western countries, it must have been small enough not to be detected by our method. The explanation for this may lie in the fact that imported crude oil is only a minor factor in the cost of consumer goods, and thus even a large change in its price can have only a small direct effect on the CPI. Substitution of other inputs (such as labor, capital and other forms of energy) for oil by producers, and substitution of less oil-intensive goods for more intensive ones by consumers may have reduced the impact of increased oil prices from what it would have been otherwise, thereby easing the pressure on consumer prices. It is also possible that following the Arab oil embargo, stabilization policies were designed systematically to counter the inflationary impact of the oil price shocks. Tightening of monetary policies during 1979 in all five countries is an indication of that. However, whether such feedback mechanisms exist and, if so, to what extent they have prevented oil price shocks from translating into inflationary shocks, can only be determined within the framework of a vector time series model which includes as variables the rate of inflation, the price of oil and the money supply.

4. CONCLUSIONS

Using a time series procedure we have investigated the inflationary impact of the 1979 oil price shock in five major industrial countries. We found that the observed monthly rates of inflation during 1979 in France, Japan and USA were not significantly different from what would have been expected given the past behavior of consumer prices. Thus, for these countries there is no indication that the inflationary shocks after December 1978 differed significantly from those that prevailed in or before that month. Although we have observed in the cases of FRG and UK some significant divergence between the actual and the expected rates of inflation, we have also pointed out that this is attributable to domestic interventions. Contrary to common belief, we have found no evidence to indicate that the existing patterns of inflation were altered by the rise in crude petroleum prices during 1979.

REFERENCES

BOX, G.E.P. and JENKINS, G. M. (1970). Time Series Analysis, Forecasting and Control. Holden-Day, San Francisco.

BOX, G.E.P. and PIERCE, D. A. (1970). Distribution of residual autocorrelations in autoregressive integrated moving average models. Journal of American Statistical Association 65, 1509-1526.

BOX, G.E.P. and TIAO, G.C. (1975). Intervention analysis with applications to economic and environmental problems. Journal of American Statistical Association 70, 70-79.

BOX, G.E.P. and TIAO, G.C. (1976). Comparison of forecast and actuality. Applied Statistics 25, 195-200.

FEIGE, E.L. and PEARCE, D.K. (1976). Inflation and incomes policy: an application of time series models. In The Economics of Price and Wage Controls. Eds: K. Brunner and A.H. Meltzer, North-Holland, Amsterdam, 273-302.

LJUNG, G.M. and BOX, G.E.P. (1978). On a measure of lack of fit in time series models. Biometrika 65, 297-303.

STOKES, H.H. (1977). The B34S data analysis program: a short write-up. Report number FY77-1, College of Business Administration, University of Illinios at Chicago.

ZELLNER, A. and PALM, F. (1974). Time series analysis and simultaneous equation econometric models. Journal of Econometrics 2, 17-54.

TIME SERIES ANALYSIS: Theory and Practice 4
O.D. Anderson (editor)

THE USE OF VECTOR ARMA MODELS IN MACROECONOMIC FORECASTING

Kenneth J. Jones
Daniel F. X. O'Reilly
Baldwin S. Hui
Katherine Sheehan

Florence Heller School
Brandeis University
Waltham, Massachusetts
U.S.A.

A twenty-four variable vector ARMA model is described as developed and used at Data Resources Incorporated. Its properties and track record, as observed over the past year, are discussed relative to a large simultaneous equation model.

INTRODUCTION

A number of studies Christ (1975), Prothero and Wallis (1976), Spivey and Wrobleski (1976), and Data Resources (1982), have compared the major macroeconomic models of the U.S. economy to ARMA models and found the latter quite inferior. Fewer studies Naylor (1972) and Nelson (1973) have found the contrary. These efforts are not as productive as they might seem since they have all compared univariate ARMA models to multivariate simultaneous equation models. As Zellner (1979) has pointed out, the latter would have to be poor indeed not to outperform the former.

A more interesting question to be addressed is whether one may build vector ARMA models which equal or excel the large macroeconomic systems in current use. Such a quest is, of course, fraught with potential hazards. Among these is how to compare the models themselves in terms of merit.

In July of 1981, a team of researchers set out to develop a vector ARMA model to forecast a subset of key variables included in the Data Resources, Inc. (DRI) model (Eckstein, Green and Sinai, 1974) of the U.S. economy. This model contains more than eight-hundred equations and forecasts more than one-thousand variables. As with almost all of these models, the DRI model has its roots in Keynesian theory (Keynes 1936); and is in the tradition of the early efforts by Tinbergen (1939) and Klein (1950). The DRI model is computed for quarterly data and updated on a monthly basis. Its forecasts are available each month in the DRI Review. The model's parameters are recomputed periodically. In addition, the forecasts produced are modified by a system of "add factors." These are qualitatively derived from subjective expectations of the future course of various policy events; however, daily and weekly quantitative data are input through this device to keep forecasts current.

THE MODEL

From the large macro model, a set of twenty-three variables was selected to be modeled by vector ARMA methods. The criteria for selection were that there be few variables, (less than thirty), and that they be of economic importance. The variables were selected by economists of the DRI National Forecasting Group (NFG), who are separate from the modeling team. In order to accommodate software then available, it was necessary to create subsets of the variables which were of ten variables or smaller. To do this we computed Sims' (1972) test of causality between pairs of variables. Using these test results, we partitioned the variables into nine overlapping clusters containing (as it turned out) five or six variables each. These clusters are listed in Table 1. The above procedure is more extensively described in (O'Reilly, Hui, Jones and Sheehan 1981). Clustering in this manner, of course, means that important predictors may be excluded from a cluster because their bivariate relationship may have been suppressed by uncontrolled variables. Some advantages of the procedure are that the modeling task may be split among several persons and the models themselves will contain a reasonable number of parameters in relation to the data set. The data were quarterly readings from 1960:1 through 1981:2.

Table 1. The Clusters

Cluster 1

C72 = Personal Consumption Expenditures
JAHEADJEA = Index of Hourly Earnings of Production Workers
JS&PNS = Standard & Poor's Composite Index
M72 = Imports of Goods & Services
RMFEDFUNDNS = Effective Rate on Federal Funds
WPI = Wholesale Price Index

Cluster 2

C72 = Personal Consumption Expenditures
EX72 = Exports of Goods & Services
IFIXNR72 = Gross Fixed Private Nonresidential Investment
M72 = Imports of Goods & Services
RMFEDFUNDNS = Effective Rate on Federal Funds
WPI = Wholesale Price Index

Cluster 3

C72 = Personal Consumption Expenditures
INV72CH = Change in Business Inventories
JATTC = Index of Consumer Sentiment
UCAPFRBM = Capacity Utilization - Manufacturing - Tool
YD72 = Disposable Personal Income

Cluster 4

C72 = Personal Consumption Expenditures
GF72 = Federal Govt. Purchases - All Goods & Services
GSL72 = State & Local Govt. Purchases - Goods & Services
M72 = Imports of Goods & Services
RMFEDFUNDNS = Effective Rate on Federal Funds

Cluster 5

C72 = Personal Consumption Expenditures
INV72CH = Change in Business Inventories
JAHEADJEA = Index of Hourly Earnings of Production Workers
RMFEDFUNDNS = Effective Rate on Federal Funds
RU = Unemployment Rate - All Civilian Workers
WPI = Wholesale Price Index

Cluster 6

C72 = Personal Consumption Expenditures
JQIND = Industrial Production Index
TP = Personal Tax & Nontax Payments
UCAPFRBM = Capacity Utilization - Manufacturing - Total
WPI = Wholesale Price Index
ZA72 = Corporate Profits after Taxes

Cluster 7

C72 = Personal Consumption Expenditures
EX72 = Exports of Goods & Services
JAHEADJEA = Index of Hourly Earnings of Production Workers
PGNP = Implicit Price Deflator - GNP
RMGBS3NS = Average Market Yield on U.S. Govt. 3-Mth Bills
WPI = Wholesale Price Index

Cluster 8

CPIU = Consumer Price Index
HUSTS = Housing Starts, Private including Farm - Total
IFIXR72 = Gross Investment in Residential Structures
IFIXNR72 = Gross Fixed Private Nonresidential Investment
RMFEDFUNDNS = Effective Rate on Federal Funds
SQTRCARS = Retail Unit Sales of New Passenger Cars

Cluster 9

C72 = Personal Consumption Expenditures
INV72CH = Change in Business Inventories
JAHEADJEA = Index of Hourly Earnings of Production Workers
JQIND = Industrial Production Index
WPI = Wholesale Price Index

The data were variously differenced to stationarity, and nine distinct vector ARMA models were developed. Forecasts were then made for all variables for 1981:3. Where a variable was in more than one cluster, its forecast was designated to be taken from the cluster where its standard error of forecast was smallest. The nine models were then given to the National Forecasting Group at DRI to be updated each quarter. The forecasts of the vector model are published each month in the DRI Review. The forecasts reported here are *ex ante*. The model has been held constant over the past year though the parameters are reestimated each month.

RESULTS

Table 2 compares three forecasts with the actual values for twenty-four variables. We include an additional variable, GNP, because it is of interest and is deterministic from the remaining set; although it was not specifically modeled. The values presented are root mean square errors over four forecasts from the third quarter of 1981 through the second quarter of 1982. The forecasts compared are (1) vector ARMA prior quarter data; (2) DRI prior quarter data plus add factors for approximately one month of the forecast quarter (DRI/1); and (3) DRI data for penultimate quarter only, but three months of add factors. The latter, (made on the last day of the quarter), does not have final data so may be considered a two-step ahead forecast (DRI/2). From the above, one deduces certain differences which make the forecasts not strictly comparable. DRI/1 has considerable advantages over vector ARMA due to the add factor information not available to the vector procedure. Vector ARMA has an advantage over DRI/2 in that it has one more quarter of data, albeit preliminary and subject to revision.

Table 2. Root Mean Square Errors for the Variables for Three Forecasts

Variable Code	Vector ARMA	DRI 1	DRI 2
1 GNP72	10.8	6.11	6.80
2 C72	8.80	6.00	7.30
3 IFIXNR72	4.80**	6.00	5.00
4 IFIXR72	0.67**	1.80	2.70
5 INV72CH	10.60**	10.70	9.10
6 EX72	0.74**	1.90	2.70
7 M72	4.50	3.20	3.50
8 GF72	3.70**	4.50	5.80
9 GSL72	1.40**	1.80	3.30
10 PGNP	0.038	.006	.008
11 CPIU	0.061	.018	.019
12 WPI	0.036*	.025	.040
13 JAHEADJEA	0.020	.005	.010
14 JQIND	0.065**	.142	.045
15 UCAPFRBM	0.021*	.012	.025
16 HUSTS	0.098*	.054	.185
17 SQTRCARS	0.634*	.618	.638
18 JATTC	0.062**	.080	.041
19 RU	0.428	.353	.403
20 RMFEDFUNDNS	2.53	1.30	2.21
21 RMGBS3N3	1.55*	0.89	2.00
22 JS PS	9.25	3.01	9.00
23 YD72	4.20*	2.50	6.00
24 ZA72	4.60**	4.90	3.50

**Vector ARMA superior to DRI 1.
*Vector ARMA superior to DRI only.

Table 3. Theil Coefficients for the Three Forecasts

Variable Code	Vector ARMA	DRI 1	DRI 2
1 GNP72	.95	.53	.59
2 C72	1.40	0.97	1.90
3 IFIXNR72	2.00	2.50	2.10
4 IFIXR72	0.34	0.61	0.87
5 INV72CH	0.80	0.79	0.68
6 EX72	0.23	0.60	0.86
7 M72	1.30	0.94	1.00
8 GF72	0.99	0.93	1.20
9 GSL72	1.20	1.60	2.80
10 PGNP	0.99	0.17	0.30
11 CPIU	1.20	0.36	0.37
12 WPI	2.10	1.50	2.30
13 JAHEADJEA	0.40	0.11	0.20
14 JOIND	1.50	3.30	1.10
15 UCAPFRBM	0.76	0.44	0.85
16 HUSTS	0.81	0.45	1.50
17 SQTRCARS	0.50	0.49	0.51
18 JATTC	1.40	1.70	0.90
19 RU	0.66	0.55	0.63
20 RMFEDRUNDNS	1.20	0.64	1.10
21 RMGBS2NS	0.89	0.51	1.10
22 JS & PNS	1.60	0.53	1.60
23 YD72	0.79	0.48	1.20
24 ZA72	0.63	0.68	0.49

Of the twenty-three modeled variables, the vector ARMA model produced forecasts for nine, with smaller mean square errors than the DRI one-step ahead forecasts. Fifteen out of the twenty-three modeled variables are superior to either one of the DRI two-step ahead forecasts. These results give neither procedure a clear stamp of objective superiority. However, the authors feel that when certain facts are considered, the vector ARMA procedure clearly becomes an attractive alternative to the usual simultaneous equation approach. These considerations are the following:

1. The use of add factors gives DRI/1 a clear edge. This is confirmed by its vast superiority in variables where daily data are available (e.g. T-bill rates). Here the twenty-eight days are a significant plus for the forecast. This is also true where the data, (which are available at a monthly interval for knowledge of trends in the later months of a quarter), are of value in modifying estimates of the next quarter; hence become an advantage even for the two-step (DRI/2) forecasts.

2. The vector model is by no means optimal since it is composed of a synthesis of nine smaller models. Using larger clusters which form on selected endogenous variables would most surely produce improved forecasts.

Table 3 presents the Theil coefficients (McNees 1975) which determine the forecast performance relative to a random walk prediction of no change. Values greater than one are due to net contrary predictions. Values less than one indicate that the root mean square one-step ahead forecast error is less than what would have been obtained from a no-change forecast. One can see that the vector ARMA has thirteen out of twenty-three series forecasts with values less than one while DRI/1 has seventeen and DRI/2 has eleven. We again do not include GNP since it was not modeled.

This record is somewhat sobering since several of the coefficients (though less than one), are not impressive.

There is also a suggestion that where DRI/1 does particularly well is where daily or weekly data are available such as T-bill rates, federal funds, stock market averages and unemployment.

As others have observed (McNees 1975), the add factors may be responsible for appreciable proportions of large macro model forecast accuracy. The Theil coefficients presented in Table 3 are somewhat poorer than are noted in McNees (1976) for the period 1970:3 to 1975:4. This is almost surely reflective of the particularly volatile nature of economic series in recent time. In addition, our interval is considerably shorter than was McNees'.

CONCLUSIONS

On the surface, one might conclude that the macro model procedure as exemplified at DRI is superior in forecast accuracy to the vector ARMA procedure. However, when one factors in all of the inequalities of the match, one could also conclude the contrary. It is suprising the vector ARMA procedure did so well given that (1) it was composed of a patchwork of smaller models; (2) the variables were not judiciously chosen so as to select a targeted set of endogenous series; (3) the parameters were recomputed monthly using preliminary data; and (4) no method was available to build in recent high frequency data or other "add factor" information. Clearly the investment of more effort is justified in developing the technology to compute larger and more specific vector ARMA models.

REFERENCES

CHRIST, C.F. (1975). Judging the performance of econometric models of the U.S. economy. International Economic Review 16 54-75.

DATA RESOURCES, INC. (July 1981). DRI Monthly Economic Review.

ECKSTEIN, O., GREEN, E.W. and SINAI, A. (1974). The Data Resources model: structure and analysis of the U.S. economy. International Economic Review 15 595-615.

KEYNES, J.M. (1936). The General Theory of Income and Unemployment, MacMillan, New York.

KLEIN, L.R. (1950). Economic Fluctuations in the United States: 1921-1941, Wiley, New York.

MCNEES, S.K. (1975). An evaluation of economic forecasts. New England Economic Review Nov/Dec 3-39.

MCNEES, S.K. (1976). An evaluation of economic forecasts: extension and update. New England Economic Review Sept/Oct 30-44.

MCNEES, S.K. (1979). Forecasting record for the 1970's. New England Economic Review Sept/Oct 35-52.

NAYLOR, T.H. and SEAKES, T.G. (1972). Box Jenkins methods: an alternative to econometric models. International Statistical Review 40 123-137.

NELSON, C.R. (1973). Applied Time Series Analysis for Managerial Forecasting. Holden-Day, San Francisco.

O'REILLY, D.F.X., HUI, B.S., JONES, K. and SHEEHAN, K. (1981). Macroeconomic forecasting with multivariate ARIMA. Data Resources, Inc., Lexington MA 02173.

PROTHERO, D.L. and WALLIS, K.F. (1976). Modelling macroeconomic time series. J. Royal Stat. Soc. A 139 468-494.

SPIVEY, W.A. and WROBLESKI, W.J. (1976). An analysis of forecast performance. (Proceedings of the Business and Economic Statistics Section, Joint Statistical Meetings A.S.A.).

TINBERGEN, J. (1939). Business Cycles in the United States of America: 1919-1932. Statistical Testing of Business - Cycle Theories 2 Geneva, League of Nations.

ZELLNER, A. (1979). Statistical analysis of economic models. JASA 74 628-651.

TIME SERIES ANALYSIS: Theory and Practice 4
O.D. Anderson (editor)

A MULTIPLE TIME SERIES ANALYSIS OF THE RELATIONSHIP BETWEEN ECONOMIC ACTIVITY AND WOMEN'S SKIRT GEOMETRY

Lucille M. Terry

Department of Home Economics
Bowling Green State University
Bowling Green, Ohio, U.S.A.

W. Robert Terry

Department of Industrial Engineering
University of Toledo
Toledo, Ohio, U.S.A.

Many persons in the fashion industry believe that changes in the width and fullness of women's skirts are associated with changes in the level of economic activity. This paper utilizes the state space approach to modeling multiple time series to examine how changes in these skirt geometry variables are associated with changes in the level of unemployment. An analysis of the fitted state space model provides no basis for believing that the unemployment rate can be useful in predicting either the length or the width of women's skirts.

1. INTRODUCTION

Many persons in the fashion industry believe that clothing reflects the economic conditions. Some theories advocated include the following: Horn and Gurel (1981, p. 128) "in times of depression or recession dress is somber and simple. When the economy prospers, clothes are likely to be lavish and elaborate." Greenwood and Murphy (1978)

> "when the economy is depressed, the price of fabric will decline. The trade is then in need of additional consumption to give it an economic boost. One way to solve this problem is to make skirts fuller and longer; this new style will create demand and money will be poured into a depressed industry. As the new lengths are accepted, the price of the fabric goes up. To compensate for the price increase and to keep cost figures within a reasonable range, skirts will then become shorter (p. 60-61)."

Further, many books on fashion include an illustration which attempts to graphically show a relationship between the rise and fall of skirt lengths and national income, stating that signs of prosperity are "rising skirts and loose waists" while a decline is reflected in "falling skirts and tight waist" (Troxell, 1971, p. 21). However, Horn and Gurel (1981) have noted that, although the relationship between economic activity and women's skirts has been widely discussed, it "has not been subjected to serious investigation or scientific research" (p. 128).

The purpose of this paper will be to utilize multiple time series analysis to analyze the relationship between economic activity and women's skirts. In particular it will utilize the state space approach to multiple time series analysis to examine how the length and fullness of women's skirts were related to the unemployment rate of the US during the period 1890 to 1936.

2. STATE SPACE ANALYSIS

The state space approach to building multivariate time series models differs from the approaches advocated by Box and Tiao (1978) and Jenkins and Alavi (1981) which restrict attention to only the input and output variables of a system. In contrast the state space approach introduces some intermediate variables, called state variables, which help to describe the internal state of a system.

The general form of the state space model is given by

$$v_{n+1} = Av_n + Bz_{n+1}$$

$$y_n = Cv_n$$

where

v_n = value of state vector at time n

y_n = value of observation vector at time n

z_n = value of white noise input vector at time n

A = transition matrix

B = input matrix

C = observation matrix.

The theoretical basis for identifying the most appropriate structure of the state space model was established by Akaike (1974). This paper suggested that a canonical correlation analysis between the set of the present and past observations and the set of the present and future observations could be utilized to obtain a reasonable estimate of the structure for the state space vector. However, in order to implement this approach it is necessary to have a procedure for determining the maximum number of past and future observations to include in the canonical correlation analysis. Akaike (1974) recommended that the quantity

$$\text{AIC} = -2 \log_e (\text{maximum likelihood}) + 2 (\text{number of independently adjusted parameters})$$

be used as a criterion for evaluating the fit of a statistical model.

The approach for developing a state space model, which is described in Akaike (1976), consists of the following steps. (1) A sequence of AR models of increasing order are fit and the order of the model with minimum AIC is used as the number of lags into the past to include in the canonical correlation analysis. (2) Canonical correlations of the past with an increasing number of steps into the future are calculated and the model with minimum AIC is used to determine the number of steps into the future to include in the state vector. (3) Preliminary estimates for the parameters of the transition and input matrices are obtained from the canonical correlation analysis. (4) The preliminary estimates of the matrix parameters are then used to obtain an infinite AR representation which in turn is used to obtain a sample estimate for the residual covariance matrix. (5) This sample estimate is then used to replace its theoretical counterpart in a log likelihood expression for the parameters of the transition and input matrices. (6) The non-linear equations which result from maximizing this expression are solved by the Newton-Raphson method.

3. MODELING RESULTS

3.1. Data Source

The data for the length and width of women's skirts came from a study by Richardson and Kroeber (1940). This study quantitatively measured six characteristics of women's dress for a three century period from 1605 to 1936. The six characteristics included: 1) length of skirt or dress, 2) length of waist, 3) length of decolletage, 4) width of skirt, 5) width of waist, and 6) width of decolletage. In order to measure these characteristics, full face or nearly full faced contemporary portraits and pictures of women, which could actually be dated within a specific year, and monthly fashion journals (ie: Vogue, Harper's Bazaar, Costume Royal) were used as the source of the dress. To reduce the variety of sizes to a common standard, the total length of the figure was measured and all measurements of each characteristic were calculated in terms of the ratio to the total length or height of the figure. These calculations were then converted into percentages and a year-by-year mean was calculated and is presented in their article. For the purposes of the present study only the two characteristics of length and width of skirt were utilized since they are the ones which are attributed as rising and falling with the economic conditions.

Several variables (national income, stock prices, and percentage unemployment) were considered as possible measures of aggregate economic activity. Unemployment was selcted on the basis of two considerations. First, the annual unemployment time series was longer than series of either national income or stock prices. Second, the unemployment series appeared to be stationary during the time period considered while the series for national income and stock prices were clearly nonstationary.

The scope of this study was restricted to those years for which both unemployment and skirt length and width data were available, 1890-1936. Since the US Bureau of Labor Statistics started reporting unemployment on an annual basis in 1890 and since the Richardson and Kroeber data spanned the period 1605 to 1936 this provided 47 data points for the multiple time series analysis. Figures (1) through (3) show plots of the data for unemployment, skirt length, and skirt width for the period 1890 to 1936.

3.2. Structure of State Vector

The analysis to determine the structure of the state vector resulted in a 4 dimensional vector with the following components:

Component	Symbol	Definition
1	W_t	width of skirt at hemline at time t
2	L_t	length of skirt at hemline at time t
3	U_t	percent unemployment at time t
4	$U_{t+1\|t}$	conditional expectation of percent unemployment at time t + 1 given that at time t

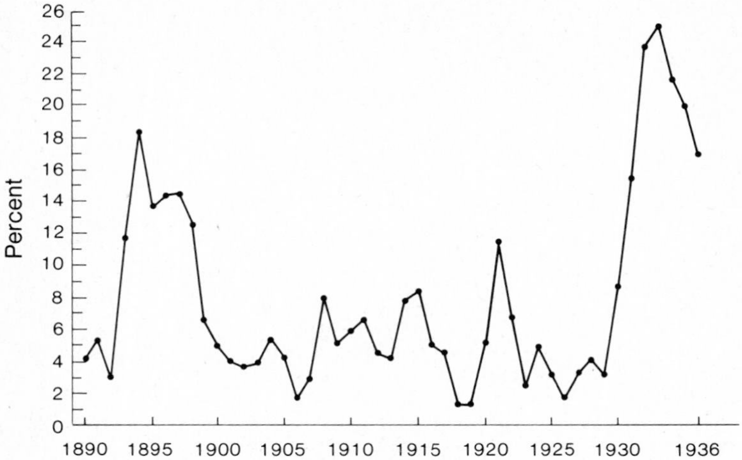

Figure 1. Unemployment

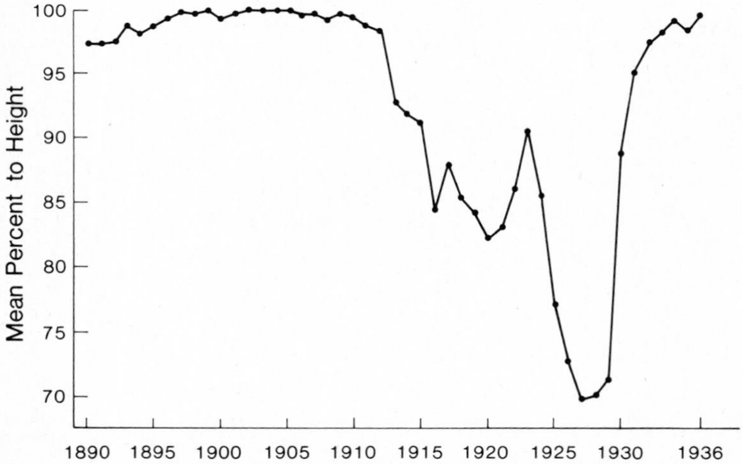

Figure 2. Skirt Length

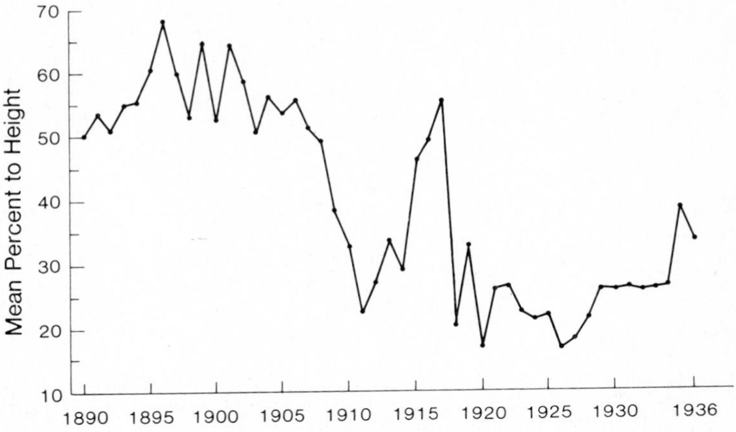

Figure 3. Skirt Width

3.3. Parameter Estimates

The estimates of the parameters of the transition and input matrices and the standard errors of these estimates are shown in Tables 1 and 2. In these matrices the odd numbered rows contain the parameter estimates, while the numbers in parentheses in the even numbered rows contain the standard errors of these estimates.

Table 1. Transition Matrix - A

	W_t	L_t	U_t	$U_{t+1\|t}$
W_t	0.765342	0.196519	0.107607	0
	(0.199093)	(0.348238)	(0.425672)	(0)
L_t	0.0781424	0.791332	0.125331	0
	(0.061577)	(0.109737)	(0.130862)	(0)
U_t	0	0	0	1
	(0)	(0)	(0)	(0)
$U_{t+1\|t}$	-0.00919	-0.051977	-0.456759	1.38023
	(0.0350635)	(0.0596512)	(0.11784)	(0.0624945)

Table 2. Input Matrix - B

	a_{Wt}	a_{Lt}	a_{Ut}
W_t	1 (0)	0 (0)	0 (0)
L_t	0 (0)	1 (0)	0 (0)
U_t	0 (0)	0 (0)	1 (0)
$U_{t+1\|t}$	0.0723258 (0.0269994)	0.296876 (0.0436999)	0.476436 (0.066719)

3.4. Diagnostic Checks

The adequacy of the fitted state space model was checked by examining plots of the autocorrelation and partial autocorrelation functions (acf and pacf) for residuals of the L_t, W_t and U_t time series. See Figures 4 to 6 for the acf's. Since none of the acf or pacf terms exceed two standard errors there is no reason to assume inadequacy in the fitted state space model.

```
  LAG    CORRELATION      -1 9 8 7 6 5 4 3 2 1 0 1 2 3 4 5 6 7 8 9 1
    1        0.32468                     .      ******
    2        0.12381                     .      **   .
    3       -0.04925                    .     *        .
    4       -0.19343                    .  ****        .
    5       -0.26639                    . *****        .
    6       -0.14565                    .   ***        .
    7        0.03689                    .       *      .
    8        0.11093                    .       **     .
    9        0.09613                    .       **     .
   10       -0.02922                    .     *        .
   11        0.04379                    .       *      .
   12       -0.01273                    .              .
   13        0.11039                    .       **     .
   14       -0.12667                    .   ***        .
   15       -0.07416                    .     *        .
   16       -0.06142                    .     *        .
   17       -0.18116                    .  ****        .
   18       -0.09178                   .     **         .
   19       -0.02264                   .                .
   20       -0.04635                   .      *         .
   21       -0.01277                   .                .
   22       -0.00125                   .                .
   23       -0.08298                   .     **         .
   24       -0.01140                   .                .
'.' marks two standard errors
```

Figure 4. Autocorrelation Function for Residuals of Length

```
LAG     CORRELATION       -1 9 8 7 6 5 4 3 2 1 0 1 2 3 4 5 6 7 8 9 1
  1        -0.23419                      .*****      .
  2         0.32615                      .      *******
  3        -0.06792                     .     *       .
  4         0.06477                     .       *     .
  5         0.02167                     .             .
  6         0.11794                     .       **    .
  7         0.03843                     .       *     .
  8         0.07249                     .       *     .
  9         0.16783                     .       ***   .
 10        -0.07099                     .     *       .
 11         0.13379                     .       ***   .
 12        -0.05813                     .     *       .
 13         0.03369                     .       *     .
 14         0.04146                     .       *     .
 15        -0.04800                     .     *       .
 16         0.10746                     .       **    .
 17        -0.23115                     . *****       .
 18         0.14162                     .       ***   .
 19        -0.25838                    .  *****        .
 20         0.17252                    .        ***    .
 21        -0.09381                    .     **        .
 22        -0.02431                    .               .
 23        -0.00729                    .               .
 24        -0.10380                    .     **        .
'.' marks two standard errors
```

Figure 5. Autocorrelation Function for Residuals of Width

```
LAG     CORRELATION       -1 9 8 7 6 5 4 3 2 1 0 1 2 3 4 5 6 7 8 9 1
  1         0.22097                      .      ****  .
  2        -0.07988                      .   **      .
  3         0.09561                      .      **   .
  4        -0.02287                      .           .
  5        -0.19257                      . ****      .
  6         0.00544                      .           .
  7         0.06099                      .      *    .
  8        -0.25729                      .*****      .
  9        -0.10369                     .    **       .
 10         0.09459                     .       **    .
 11         0.11400                     .       **    .
 12        -0.02278                     .             .
 13         0.06782                     .       *     .
 14         0.02888                     .       *     .
 15        -0.14183                     .   ***       .
 16        -0.05494                     .     *       .
 17         0.07537                     .       **    .
 18        -0.10069                     .    **       .
 19        -0.11072                     .    **       .
 20        -0.01433                     .             .
 21        -0.02275                     .             .
 22        -0.09705                     .    **       .
 23         0.04236                     .       *     .
 24         0.06210                     .       *     .
'.' marks two standard errors
```

Figure 6. Autocorrelation Function for Residuals of Unemployment

4. IMPLICATIONS

In the analysis all parameters with an absolute value of less than two standard errors are put at zero. After eliminating the nonsignificant terms, the following system equations were obtained:

$$W_{t+1} = 0.765\, W_t + a_{Wt} \tag{1}$$

$$L_{t+1} = 0.791\, L_t + a_{Lt} \tag{2}$$

$$U_{t+1} = U_{t+1|t} + a_{Ut} \tag{3}$$

$$U_{t+2|t+1} = -\,0.457\, U_t + 1.380\, U_{t+1|t} + 0.072\, a_{Wt} + 0.297\, a_{Lt} + 0.476\, a_{Ut} \tag{4}$$

The first two equations indicate that both the width and length of skirts approximate to first order autoregressive processes. Substituting (3) into (4) and simplifying yields

$$U_{t+2} - 1.38\, U_{t+1} - 0.457\, U_t = -\,0.904\, a_{U_t} + a_{U_{t+1}} + 0.072\, a_{Wt} + 0.297\, a_{Lt}. \tag{5}$$

Equation (5) indicates that unemployment is close to an ARMA (2,1) model with two additional terms which represent the noise components in the width and length time series. This system of equations provides no basis for believing that the unemployment rate can be useful in predicting either the length or the width of women's skirts.

REFERENCES

AKAIKE, H. (1974). A new look at the statistical model identification, IEEE Transactions on Automatic Control AC-19, 716-722.

AKAIKE, H. (1976). Canonical correlation of time series and the use of an information criterion. In System Identification: Advances and Case Studies, Eds: R.K. Mehra and D.G. Lainiotis, Academic Press, New York, 27-96.

CAMERON, A.V. (1981). State space models and forecasts for GNP and money supply. In Time Series Analysis, Eds: O.D. Anderson and M.R. Perryman, North Holland, Amsterdam, 43-52.

BOX, G.E.P. and TIAO, G.C. (1978). Multiple Time Series Model Building and Forecasting. University Associates, Princeton, New Jersey.

GREENWOOD, K.M. and MURPHY, M.F. (1978). Fashions Innovation and Marketing. Macmillan, New York.

HORN, M.J. and GUREL, L.M. (1981). The Second Skin. Houghton Miflin, Boston.

JENKINS, G.M. and ALAVI, A.S. (1981). Some aspects of modelling and forecasting multivariate time series. Journal of Time Series Analysis 2, 1-47.

MEHRA, R.K. and CAMERON, A.V. (1976). A multidimensional identification and forecasting technique using state space models, TIMS/ORSA Miami Conference.

RICHARDSON, J. and KROEBER, A.L. (1940). Three centuries of women's dress fashions, Anthropological Records 5, 111-153.

TROXELL, M.D. (1976). Fashion Merchandising. 2nd Edn., McGraw-Hill, New York.

U. S. Bureau of Labor Statistics. Employment and Earnings.

TIME SERIES ANALYSIS: Theory and Practice 4
O.D. Anderson (editor)

A VECTOR TIME SERIES ANALYSIS OF COTTON-POLYESTER PRICE COMPETITION

W. Robert Terry

Industrial Engineering, University of Toledo
Toledo, Ohio, USA

Shiv G. Kapoor

Mechanical and Industrial Engineering, University of Illinois
Urbana, Illinois, USA

This paper uses the methods of Pandit and Wu, and Kapoor, Madhok and Wu to develop a vector time series model of the market share for cotton and the cotton-polyester price ratio. It also discusses how such a model might be used to develop a system for determining how the price of cotton, relative to that of polyester, should be varied as a function of time so as to maximize profits for cotton producers.

I. INTRODUCTION

Both cotton and polyester have properties which are desired by consumers, but they also have properties which are undesirable. Garments made from 100% cotton fabrics tend to be relatively comfortable in both hot and cold weather, but they also wrinkle easily, require ironing and can shrink out of shape. On the other hand garments made from 100% polyester tend to be wrinkle resistant and virtually shrink free, but they also tend to be relatively uncomfortable in both hot and cold weather.

By blending cotton and polyester fibers it is possible to obtain fabrics which satisfy the consumers' desires for both comfort and ease of care. Typically it will be possible to satisfy these comfort and ease of care requirements with a large number of different blends. For example shirts with as little as 40% polyester and 60% cotton can be washed and then worn without ironing, while shirts with as little as 35% cotton and 65% polyester provide enough comfort to satisfy most consumers.

Comfort and ease of care properties, are not the only factors which are important to consumers. Price is also important. An increase in the price of cotton relative to that of polyester will tend to cause consumers to switch to blends which contain more polyester, while an increase in the price of polyester relative to that of cotton will tend to cause a switch in the opposite direction. State space analysis methods have been used by Terry and Terry (1981) for analyzing cotton-polyester market competition. In that paper four variables (cotton price, polyester price, cotton consumption, and polyester consumption) were used to characterize the cotton-polyester market competition.

The purpose of this paper will be to develop a model which will describe how the relative consumption of cotton and polyester is influenced by the present and past values of their relative prices. An autoregressive moving average vector (ARMAV) model will be utilized to model the relationship between the percent consumption of cotton and the cotton polyester price ratio. Section II will describe methods advanced by Pandit and Wu (1977) for modeling stationary ARMAV processes and by Kapoor, Madhok and Wu (1981) for modeling nonstationary ARMAV processes. Sections III and IV will describe the results of the nonstationary

ARMAV modeling process. Section V will briefly describe how a USAMV model such as the above might be utilized in the development of an optimal pricing policy.

II. 1. PANDIT AND WU METHOD

The Pandit and Wu method is based on the premise that the data should speak for itself. This method utilizes a highly flexible family of models and a systematic procedure for selecting an appropriate member of this family.

The theoretical foundations for the Pandit and Wu method is a theorem which states that any stationary discrete time stochastic vector process can be represented as closely as desired by a stochastic difference equation of the form

$$\mathbf{x}_t - \boldsymbol{\phi}_1\mathbf{x}_{t-1} - \boldsymbol{\phi}_2\mathbf{x}_{t-2} - \dots - \boldsymbol{\phi}_n\mathbf{x}_{t-n}$$
$$= \mathbf{a}_t - \boldsymbol{\theta}_1\mathbf{a}_{t-1} - \boldsymbol{\theta}_2\mathbf{a}_{t-2} - \dots - \boldsymbol{\theta}_{n-1}\mathbf{a}_{t-n+1} \tag{1}$$

where $\mathbf{a}_t$ is a p x 1 vector discrete time white noise process with $E[\mathbf{a}_t] = \mathbf{0}$ and $E[\mathbf{a}_t\ \mathbf{a}_{t-k}] = \delta_k\boldsymbol{\sigma}_a^2$; $\mathbf{x}_t$ is a p x 1 vector which represents the system response to the discrete time white noise input; $\boldsymbol{\phi}_1$, $\boldsymbol{\phi}_2$, ... , $\boldsymbol{\phi}_n$ are discrete autoregressive parameter matrices; $\boldsymbol{\theta}_1$, $\boldsymbol{\theta}_2$, ... , $\boldsymbol{\theta}_n$ are discrete moving average parameter matrices and δ_k is the Kronecker delta function. This model will be referred to as an autoregressive moving average vector model of order n, n-1 and denoted by ARMAV(n, n-1).

The Pandit and Wu method involves fitting successively higher order ARMAV(n,n-1) models to the data and testing whether the increase in the order fails to produce a statistically significant improvement in fit. A nonlinear optimization algorithm based on the work of Davidon (1959) and Fletcher and Powell (1963) is used to determine the values of the unknown parameters in the corresponding ARMAV (n,n-1) model which minimize the determinant of the conditional sum of squares and products matrix of residuals.

The F-test is used for checking the statistical significance of the reduction in the sum of squares of residuals which results when the order of the model is increased.

$$F = \frac{(1-\Lambda^{1/2})/ps}{\Lambda^{1/2}/(km-2\lambda)} \tag{2}$$

with $$k = N - r - \frac{p-s+1}{2}$$

$$m = \left(\frac{p^2s^2-4}{p^2+s^2-5}\right)^{1/2}$$

$$\lambda = \frac{ps-2}{4}$$

$$\Lambda = \frac{|A_0|}{|A_1|}$$

where A_0 = sum of squares and cross products matrix (CSP) of model with more parameters

A_1 = sum of squares and cross products matrix (CSP) of model with less parameters

p = number of series

N = number of (vector) observations

r = p x number of matrix parameters in model with more parameters

s = p x difference in the numbers of matrix parameters.

F, as defined above, has an F-distribution with ps and (km-2λ) degrees of freedom.

The other adequacy check involves a graphical check on the randomness of the multiple time series residuals by plotting their auto and cross-correlation functions at different lags. If their estimated values of the auto and cross-correlation functions lie between $\pm 2/N^{1/2}$, the time series are assumed to be independent, provided no systematic patterns show up in these functions

II. 2. KAPOOR, MADHOK AND WU METHOD

The Pandit and Wu method assumes that the vector time series are stationary. However, Kapoor, Madhok and Wu (1981) have developed a method which can be used to deal with nonstationarity. This method involves three stages. In the first stage an appropriate deterministic model is fit to the data. In the second stage the Pandit and Wu method is used to fit an ARMAV model to the residuals of the deterministic model developed in the first stage. In the third stage, a nonlinear estimation program is used to simultaneously estimate the values of the parameters of both the deterministic and ARMAV models.

III. DETERMINISTIC MODEL

Monthly data for the prices and consumption of cotton and polyester from January 1970 to December 1980 provided the basis for this modeling effort. This data was used to calculate the consumption of cotton as a percent of the total consumption of both cotton and polyester, X_{1t}, and the ratio of the price of cotton to the price of polyester, X_{2t}.

Plots of X_{1t} and X_{2t} are shown in Figures 1 and 2 respectively. These plots indicate that both X_{1t} and X_{2t} are nonstationary. In such cases the Kapoor, Madhok and Wu method calls for decomposing the time series into deterministic and stochastic components.

The general form of the deterministic components for both X_{1t} and X_{2t} were identified. First there were factors which could produce persistent long term trends. Second there were transient situations which could produce temporary behavioral aberrations.

The analyses of the factors which could induce deterministic behavior revealed that the behavior of both X_{1t} and X_{2t} could have been influenced by both persistent long term trends and temporary behavioral aberrations. Two factors were found which could have induced a persistent long term increasing trend in the price ratio X_{2t}. One was the discontinuation of the US Department of

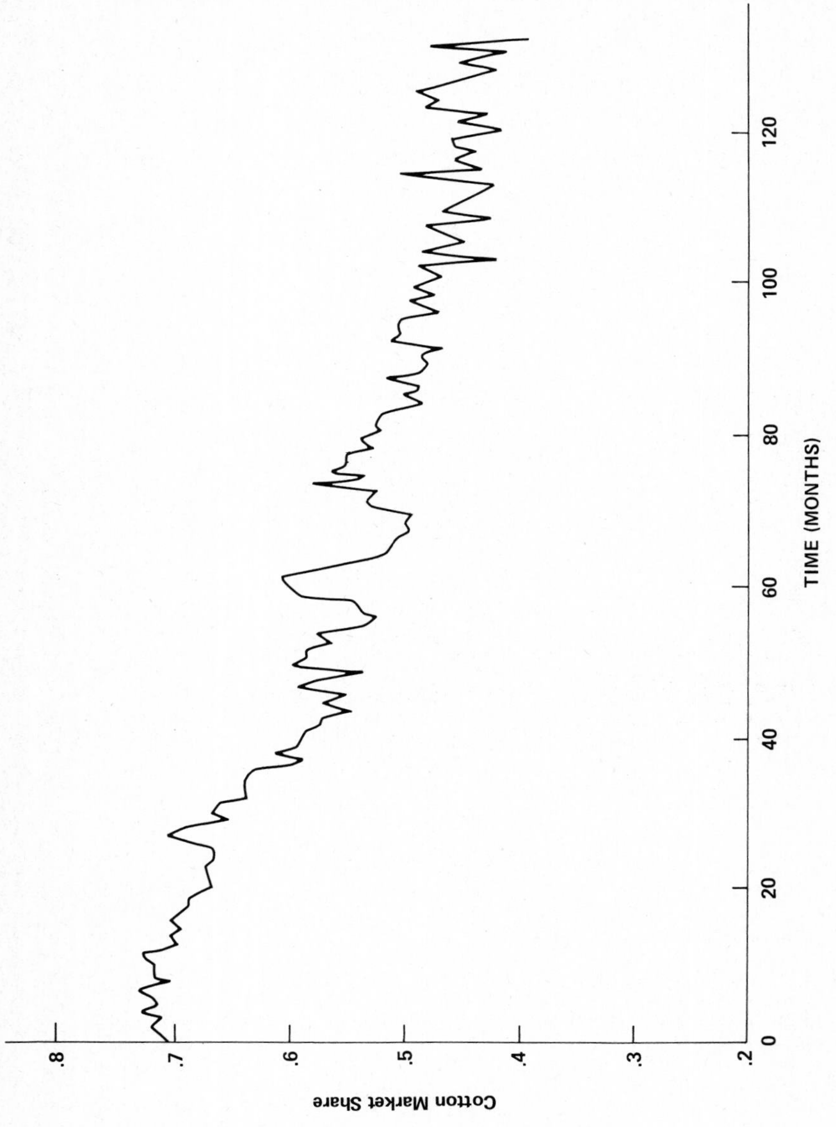

Figure 1 – Plot of Cotton Market Share Data

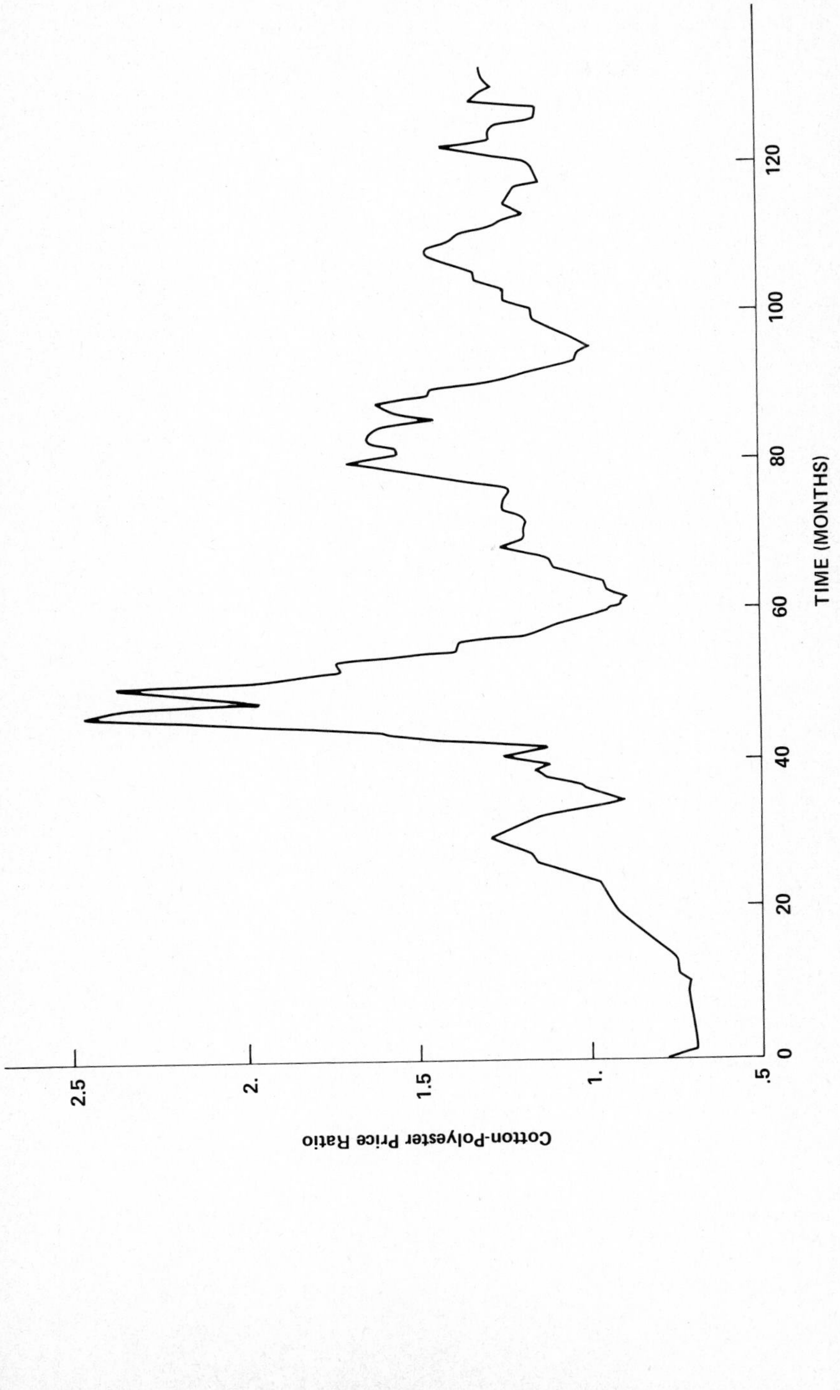

Figure 2 — Plot of Cotton-Polyester Price Ratio Data

Agriculture cotton stockpile. The major effect of the stockpile was thought to have artifically suppressed the price of cotton. The second factor was the creation of an organization in 1972 which has, among other things, made vigorous efforts to help cotton producers get a better price for their product. It was expected a priori that both of these factors would induce a persistent long term uptrend in X_{2t}. This expectation indicated the need for a model of the form

$$X_{2t} = a_2 + b_2 t \ .$$

In addition it was discovered that a world wide cotton shortage, which began in August 1973 and ended in April 1974, caused the price of cotton to increase temporarily from a level of 38 cents per pound to almost a dollar per pound. This shortage was caused by the following sequence of events. The Humbolt Current unexpectedly shifted in 1972 causing a drastic decrease in the Peruvian anchovie catch. Anchovies are used as a primary source of protein in cattle food. The shortage of anchovies which resulted caused a dramatic increase in the price of soybeans, an alternative protein source for cattle food. Since cotton and soybeans compete for land, the sharp increase in soybean prices caused agriculture businesses to shift land, which under normal circumstances would have been devoted to cotton, into soybeans. This shift of land from cotton to soybeans created a severe shortage of cotton, which became apparent when the US Department of Agriculture released its final crop estimate in August 1973. This shortage caused the price of cotton to increase by over two-fold. These high prices persisted until the sharp increase in energy cost which began in December 1973 caused a sharp decrease in the demand for textile products in early 1974. The sharp increase in energy cost also lead to higher prices of polyester, since polyester is derived from petroleum. These situations, when combined with those responsible for the long term trend, suggested that a model of the form

$$X_{2t} = a_2 + b_2 t + c_2 Z_t , \qquad (3)$$

where

$$Z_t = \begin{cases} 1 & \text{if } t = 44, 45, \cdots , 52 \\ 0 & \text{otherwise} \end{cases}$$

should be used to represent the structure of the deterministic component of X_{2t}.

Similar analyses were performed for X_{1t}. The analysis of factors revealed two factors which could produce a long term persistent downtrend in X_{1t}. The first is that current lifestyles have tended to make consumers value the ease of care properties of polyester more than the comfort of cotton. In particular women employed outside the home do not have the time to care for cotton garments. In addition, climate conditioned homes, schools, and workplaces have decreased the importance of comfort as a purchase criteria. The second factor is that the availability of polyester products at the retail level appears to have been limited by the rate at which polyester production capacity could be increased.

It was expected that the long term effect of these factors would be to induce a long term downtrend in the percent consumption of cotton. This expectation indicated the need for a model of the form

$$X_{1t} = a_1 - b_1 t \ .$$

In addition the world wide cotton shortage mentioned above could have induced a temporary reduction in the consumption of cotton, leading to a final model of the form

$$X_{1t} = a_1 - b_1 t - c_1 Z_t ,$$

where

$$Z_t = \begin{cases} 1 & \text{if } t = 44, 45, \cdots, 52 \\ 0 & \text{otherwise .} \end{cases} \quad (4)$$

The method of ordinary least squares was utilized to estimate the values of the unknown parameters in the deterministic models specified by equations (3) and (4). The estimates, all of which are significant at the 0.05 level, are shown in the following fits:

$$X_{1t} = 0.7078 - 0.002247t + 0.0255Z_t \quad (5)$$

$$X_{2t} = 1.022 + 0.0029033t - 0.9843Z_t \ . \quad (6)$$

The residuals of the above models denoted by r_{1t} and r_{2t} respectively are shown in Figures 3 and 4.

IV. ARMAV

The method of Pandit and Wu was utilized to fit an ARMAV model to the residuals of the deterministic models specified by equations (5) and (6). Table I below summarizes the models that were fitted in the process of developing an adequate ARMAV model.

TABLE I. ARMAV Modeling Summary

Model	Determinant of CSP Matrix	F_{cal}	$F(\nu_1, \nu_2)$
ARV(1)	.1569		
		5.3	2.93**
ARMAV(2,1)	.1136		
		0.099	2.93**
ARMAV(3,2)	.1128		
		5.71*	5.63***
ARV(2)	.1353		

* Compared to ARMAV(2,1)
** $F_{.95}(8,242)$
*** $F_{.95}(4,250)$

Results of the F tests shown in Table I indicate that: (1) ARMAV(2,1) represents a significant improvement over ARV(1); (2) the ARMAV(3,2) does not represent a significant improvement over ARMAV(2,1); and (3) ARMAV(2,1) represents a significant improvement over the ARV(2). This suggests that the ARMAV(2,1) is the most appropriate model.

The estimated values of the parameters, all of which are significant at the 0.05 level, for the ARMAV(2,1) model are shown below.

$$r_{1t} = -0.182\, r_{1t-1} - 0.642\, r_{2t-1} + 0.247\, r_{1t-2} + 0.459\, r_{2t-2} + a_{1t} + 0.735\, a_{1t-1} + 0.63\, a_{2t-1} \quad (7)$$

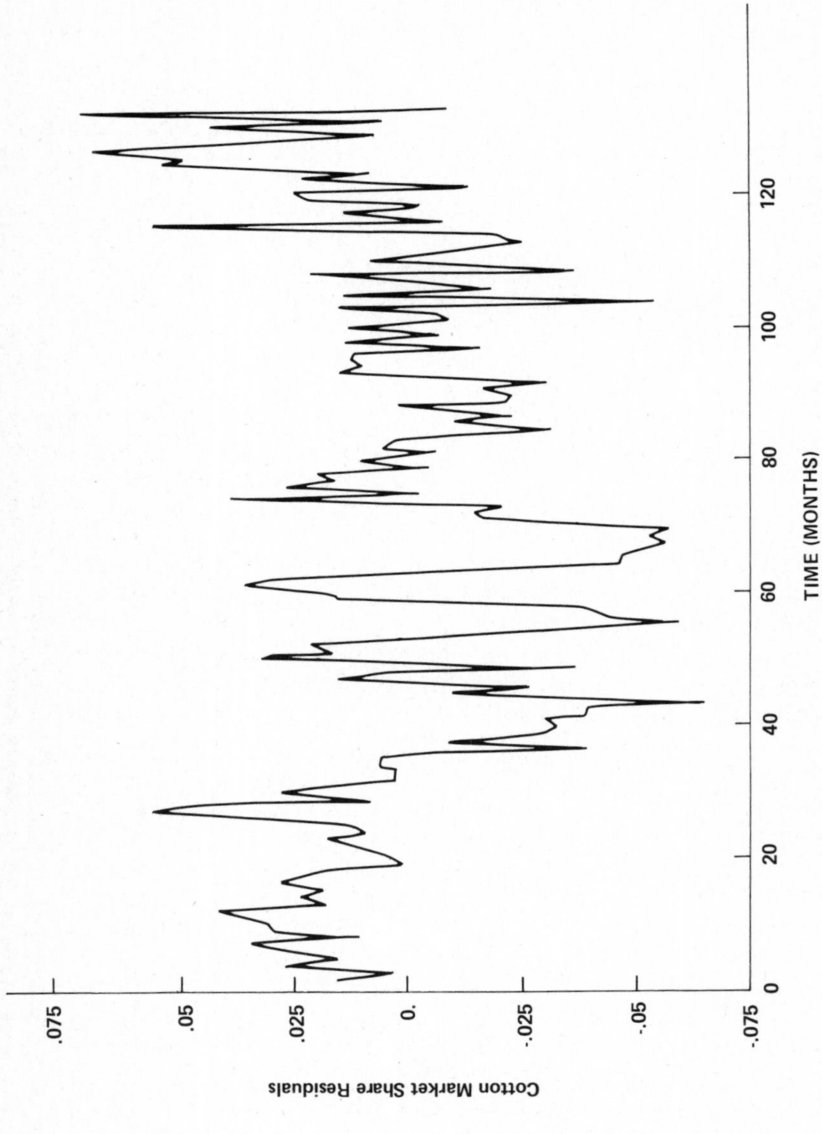

Figure 3 — Plot of Residuals for the Deterministic Model for Market Share

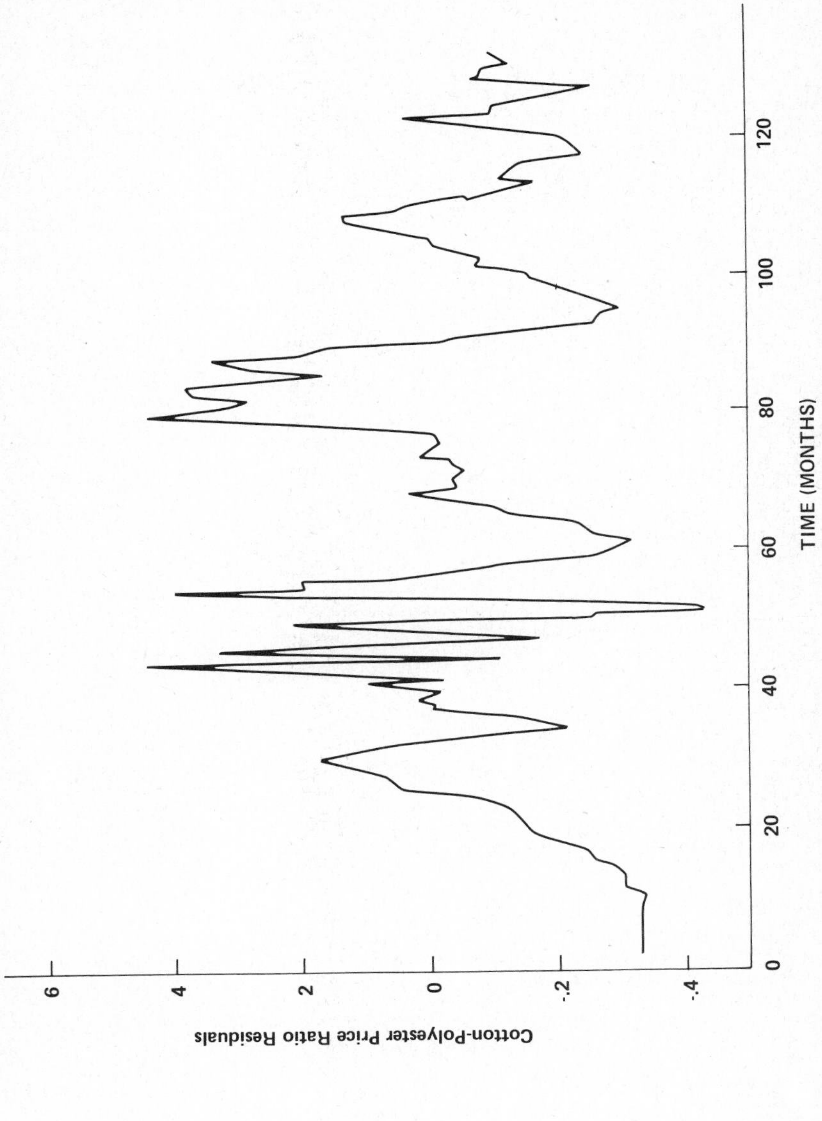

Figure 4 — Plot of Residuals for the Deterministic Model for Price Ratio

$$r_{2t} = -0.027\, r_{1t-1} - 0.158\, r_{2t-1} - 0.727\, r_{1t-2} + 0.653\, r_{2t-2} + a_{2t} - 0.475\, a_{1t-1} + 0.894\, a_{2t-1} \, . \tag{8}$$

The adequacy of the fitted ARMAV(2,1) model was checked by examining the crosscorrelation function between the a_{1t} and a_{2t} series and the autocorrelation functions for the a_{1t} and a_{2t} series. Figure 5 shows a plot of both of these functions. Since all of the values of the autocorrrelation and crosscorrelation functions fall within the $\pm\, 2/N^{1/2}$ limits there appears to be no reason to question the adequacy of the fitted ARMAV(2,1) model. Simultaneous estimation of the deterministic and ARMAV models yielded results in agreement with those shown in equations (5) through (8).

An examination of the r_{1t-i} and r_{2t-i} terms in equations (7) and (8) suggests that there is a carry-over effect which operates within each of the series. This is evidenced by the presence of the r_{1t-1} and r_{1t-2} terms in equation (7) and by the presence of the r_{2t-1} and r_{2t-2} terms in equation (8). The presence of these carry-over effects suggests that, for each series, current values continue to affect future values for two time periods. These carry-over effects represent the manner in which each series gradually adjusts to the shocks which it receives.

Continued examination of the r_{1t-i} and r_{2t-i} terms in equations (7) and (8) suggests that there is a feedback relationship between the stochastic components of the r_{1t} and r_{2t} time series. This feedback is evidenced by the presence of the r_{2t-1} and r_{2t-2} terms in equation (7) and by the r_{1t-1} and r_{1t-2} terms in equation (8). The existence of this feedback suggests that the stochastic components of the percent consumption of cotton and of the cotton-polyester price ratio time series form a closed loop causal system which determines how one part of the system gradually adjusts to shocks which hit the other part of the system. This feedback represents the manner in which each of the series gradually adjusts to the shocks which the other series receives.

An examination of the a_{1t-i} and a_{2t-i} terms in equations (7) and (8) suggests that portions of the effect of a shock which hits either of the series is transmitted to the next period. This is evidenced by the presence of a_{1t-1} in equation (7) and by the presence of a_{2t-1} in equation (8). This transmission of shocks within each of the series from one period to the next suggests that both series continue to react to shocks which hit them during the preceding time period. This intra-series transmission of shocks represents the manner in which each series abruptly adjusts to random shocks which it receives.

Further examination of the a_{1t-i} and a_{2t-i} terms in equations (7) and (8) suggests that portions of the shocks which are experienced by one series are transmitted to the other series. This is evidenced by the presence of a_{2t-1} in equation (7) and by the presence of a_{1t-1} in equation (8). This two-way transmission of shocks between the r_{1t} and r_{2t} time series suggests that the r_{1t} and r_{2t} series each react to random shocks which hit the other series. In both cases these reactions occur one period later. This two-way transmission represents the manner in which each of the series abruptly adjusts to random shocks which the other series receives.

V. OPTIMAL PRICING POLICY

The problem of determining an optimal pricing policy involves specifying how the price of cotton in future time periods should be set in relation to the expected price of polyester so as to maximize the present worth of revenues received by cotton producers. For the sake of simplicity of exposition it will be assumed

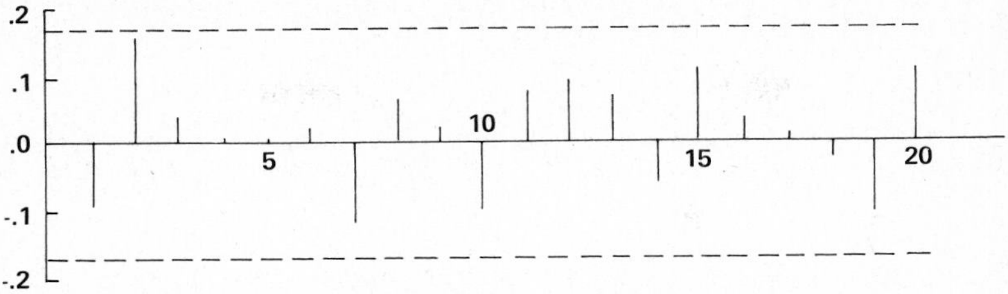

Autocorrelation Function for Cotton Market Share Data Residuals, a_{1t}

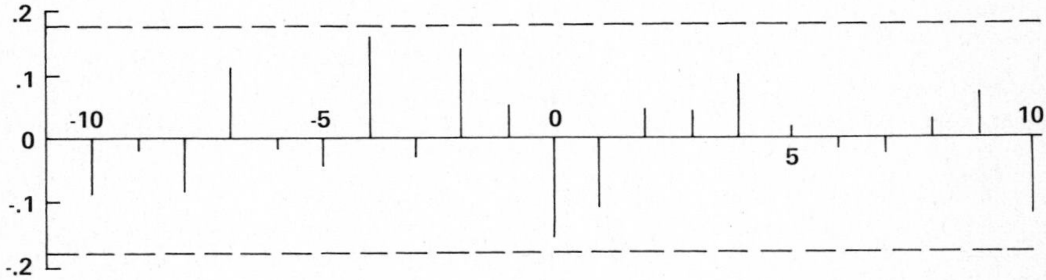

Crosscorrelation Function between Market Share and Price Ratio Data Residuals

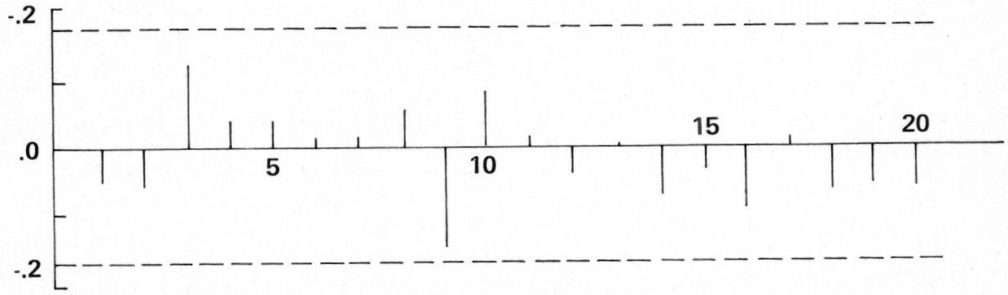

Autocorrelation Function for Price Ratio Data Residuals, a_{2t}

Figure 5 – Autocorrelation and Crosscorrelation Functions for the Residual Series Obtained from the USAMV (2,1) Model

that the size of the market for which cotton and polyester will be competing is known with certainty for each future time period.

The problem of determining how the price of cotton, relative to that of polyester, should be varied as a function of time is a multi-stage optimization problem. One approach for solving such problems is to decompose them into an equivalent sequence of single stage optimization problems and to solve these problems in a recursive fashion.

In order for the above approach to provide an optimal solution it will be necessary for certain conditions to be satisfied. First the system being optimized must be such that knowledge of the current state of the system conveys all of the information about its previous behavior necessary for determining the optimal policy for the remaining stages. Systems with this characteristic are said to satisfy the Markovian property. Second, the function to be optimized must be a monotone function of the solutions to the single stage optimization problems.

Both of these assumptions are satisfied by the optimal pricing policy problem. The Markovian property can be satisfied by employing the approach described in Graupe (1972) to transform the fitted vector time series model into an equivalent form which satisfies the Markovian property. The monotinicity requirement is satisfied since the present worth of expected revenues is equal to the sum of the present worth of the revenue received in each of the future time periods.

The process of decomposing a multi-stage optimization problem into a series of single stage problems which are solved recursively is known as dynamic programming. The basic features which characterize dynamic programming problems are summarized below. (1) The problem can be divided into stages, with a policy decision required at each stage. (2) Each stage has a number of states associated with it. These states represent the various possible conditions in which the system might be at each stage of the problem. The number of states may be either finite or infinite. (3) The effect of the policy decision at each stage is to transform the current state into a state associated with the next stage and to produce a return which contributes to the value of the objective function in a monotone fashion.

In the current case the state of the system is represented by the market share of cotton and the policy decision involves specifying the price of cotton relative to the price of polyester. The Markovian representation of the vector time series model provides the means for determining the state transformation and the return at each stage in the problem.

VI. CONCLUSION

This paper has demonstrated that vector time series analysis can be used to model the price competition in a two product market. It has also suggested that the fitted vector time series model might be transformed into a form which satisfies the Markovian format, and indicated how dynamic programming might be used for determining an optimal pricing policy.

ACKNOWLEDGEMENT

The authors are grateful to the US Department of Agriculture for providing them with price and consumption data for both cotton and polyester.

REFERENCES

DAVIDON, W.C. (1959). Variable Metric Method for Minimization. AEC Research Development Report, ANL-5990.

FLETCHER, R. and POWELL, M.J.D. (1963). A rapidly convergent descent method for minimization. Computer Journal 6, 163-168.

GRAUPE, D. (1972). Identification of Systems. Van Nostrand Reinhold, New York.

KAPOOR, S.G., MADHOK, P. and WU, S.M. (1981). Modeling and forecasting sales data by time series analysis. Journal of Marketing Research 18, 94-100.

PANDIT, S.M. (1973). Data Dependent Systems: Modeling Analysis and Optimal Control via Time Series. Doctoral Dissertation, University of Wisconsin-Madison.

PANDIT, S.M. (1980). Data dependent systems and exponential smoothing. In Analysing Time Series (Proceedings of the International Conference held on Guernsey, Channel Islands, October 1979). Ed: O.D. Anderson, North-Holland, Amsterdam & New York, 217-238.

PANDIT, S.M. and WU, S.M. (1977). Modeling and analysis of closed-loop systems from operating data. Technometrics 19, 477-485.

TERRY, L.M. and TERRY, W.R. (1981). A multiple time series analysis of cotton-polyester market competition. In Applied Time Series Analysis (Proceedings of the International Conference held at Houston, Texas, August 1981). Eds: O.D. Anderson and M.R. Perryman, North-Holland, Amsterdam & New York, 409-413.

TIME SERIES ANALYSIS: Theory and Practice 4
O.D. Anderson (editor)

A TIME SERIES ANALYSIS OF TENSILE STRENGTH IN A DIE CASTING PROCESS

Shiv G. Kapoor

Mechanical and Industrial Engineering, University of Illinois
Urbana, Illinois, USA

W. Robert Terry

Industrial Engineering, University of Toledo
Toledo, Ohio, USA

This paper utilizes a method advanced by Pandit and extended by Kapoor to develop a time series model for predicting tensile strength of cast aluminum engine blocks which were produced by a die casting process. An analysis of the fitted time series model provides a better understanding of the physical phenomena that can be responsible for inadequate tensile strength.

1. INTRODUCTION

A great deal of the current manufacturing technology in the US was developed during an era of artificially low energy prices. These artificially low prices tended to discourage the development of more energy conserving manufacturing processes. However, the end of the era of "cheap energy" has prompted engineers to place more emphasis on increasing the energy efficiency of manufacturing processes.

One strategy for increasing energy efficiency calls for substituting die casting for machining operations in discrete parts production. Doing this would eliminate the energy required to first put unnecessary material on a raw material feedstock and then to remove it. However, substituting die casting for machining can result in a degradation of tensile strengths.

The ultimate objective is to develop a system for controlling the tensile strength of die cast products. Such a system will consist of a mathematical model for predicting tensile strength and a second model which specifies how control variables should be manipulated to optimize performance of the process.

In control engineering the prediction and optimization problems are typically treated separately. The basis for doing this is the separation theorem which states that

> for linear systems with quadratic cost functions
> and subjected to additive white Gaussian
> noise inputs, the optimum stochastic controller
> is realized by cascading an optimal estimator [predictor]
> with a deterministic optimum controller. (Sage, 1968, p. 312)

On the basis of this theorem it was decided to treat the prediction and optimization problems separately. The scope of this paper will be limited to the problem of developing a mathematical model for predicting tensile strength. The problem of developing a model for optimizing the performance of the die casting performance is considered in a forthcoming paper and will not be discussed here.

The remainder of this paper will be organized as follows. The second section will discuss the problems which will typically be encountered in the development of a mathematical model for predicting tensile strength. The third section will describe a systematic approach for overcoming these problems. The fourth section will illustrate the use of this approach by applying it to data obtained from a commercial casting operation. The fifth section will briefly discuss the implications of the resulting model.

2. STATISTICAL PROBLEMS IN MODELING A DIE CASTING PROCESS

There are two approaches for developing a mathematical model for predicting tensile strength. One utilizes theoretical knowledge of the process and deductive logic to derive a model. The other utilizes empirical data and inferential logic to develop a model.

At present, theoretical knowledge of the causal factors which influence tensile strength in a die casting operation is insufficient for deducing a quantitative model for predicting tensile strength. Thus the only viable choice is to infer a model on the basis of empirical data.

The process of inferring a model will be relatively simple if it is possible to control environmental conditions and randomize the order in which the various possible combinations of control variables are run. In most production environments it will not be possible to do either of these. The inability to control experimental conditions precludes eliminating the influence of other causal variables. This opens the possibility that differences between groups of items, that were exposed to different values of the control variables, were caused by differences in the level of other causal variables. The inability to randomize the order that different values of the control variables are run prevents the use of randomization to ensure that the composition of the various experimental groups are probabilistically equivalent. The lack of probabilistically equivalent experimental groups opens the possibility that differences between and among the various groups were caused by differences in the composition of the groups rather than the differences in the treatments which they received.

The inability to conduct controlled experiments creates the possibility that the observed values of tensile strength will not be statistically independent and therefore will not be suitable for classical inferential methods. This can result when the level of an uncontrolled causal factor changes. Such a change constitutes an exogoneous shock which is not accounted for by the system which defines the relationship between the control variables and tensile strength. In a statistical model of this relationship the effects caused by changes in the level of an uncontrolled causal factor will be indistinguishable from the model's error term.

The failure to satisfy the assumption of independence can have a dramatic effect on the level of significance of classical statistical methods. Cochran (1947) showed that a positive correlation between observations in analysis of variance will cause the actual variance of a treatment mean to be larger than would be the case for an independent series, while the estimated variance will be smaller than the variance of an independent series. If the autocorrelation is negative, then the actual variance will be smaller, while the estimated variance will be larger than they would be for an independent series. In a widely discussed paper Box and Newbold (1971) have shown that the failure to correctly account for autocorrelated residuals in a regression analysis can lead to nonsensical results. Beale and Seeley (1979) further discuss this problem and show how a general-purpose linear multiple regression program can be modified to allow for autocorrelated residuals when analysing time series data.

3. STATISTICAL ANALYSIS OF DEPENDENT DATA

3.1 Introduction

The preceding section has noted that the inability to conduct controlled experiments could lead to dependent (or autocorrelated) data. It has also noted that use of statistical procedures, which do not directly account for such dependency, can dramatically increase the risk of making a wrong decision. The purpose of this section is to briefly describe a statistical approach which provides a systematic procedure for accounting for such dependencies.

This approach was advanced by Kapoor (1981). It utilizes Wold's decomposition theorem (Wold, 1954) to extend a procedure developed by Pandit (1973) for modelling stationary time series to the nonstationary case. In particular Wold's decomposition theorem is used to justify splitting the time series into the following components: dynamic deterministic component, d_t, stochastic noise components, s_t; and white noise component, w_t. This approach provides a systematic procedure for determining models for the dynamic and stochastic noise components such that the residuals will be statistically indistinguishable from a white noise process. The purpose of the next two subsections will be to describe systematic procedures for developing models for representing the dynamic deterministic and stochastic noise components.

3.2 Dynamic Deterministic Component

The dynamic deterministic component represents the nonstationary component of the time series. Typically this component results from changes in the levels of uncontrolled variables. The ideal way for dealing with the dynamic deterministic component would be to utilize relevant theory to specify a model which describes how changes in the levels of uncontrolled variables dynamically affect the behavior of the variable of interest and to then use empirical data to estimate the parameters of this model. When this can not be done the analyst should strive to fit the simplist model which will remove the nonstationarity. For noncyclical data an operational procedure might be to fit successively higher order polynomials where t represents time

$$d_t = \sum_{i=0}^{n} b_i t^i$$

and n represents the order of the polynomial. For cyclical data an operational procedure might be to use a regression model with dummy variables to account for the cyclical nonstationarity,

$$d_t = \sum_{i=1}^{m-1} c_i X_i$$

where X_i is a dummy variable which takes a value of 1 for periods corresponding i, $m + i$, $2m + i$, ... and m is the length of the cycle.

3.3 Stochastic Noise Component

The purpose of the stochastic noise component model is to account for the dependency pattern inherent in the residuals from the dynamic deterministic component model. The approach for developing the stochastic noise component model is based on the recognition that there are three classical ways in which a system can respond to the shock created by a change in the level of an unobservable and uncontrolled variable. The first occurs when the strength of the shock is strong enough to permanently alter the structure of the system. In this case the shock will, in the absence of other shocks, cause a permanent change in the system's behavior. The second occurs when the system makes a series of successively smaller partial adjustments to the exogoneous shock. In this case the system will, in the absence of other shocks, more or less gradually return to the pattern of behavior which prevailed prior to the shock. The third occurs when the system requires a nontrivial amount of time to make an adjustment which will return it to its preshock pattern of behavior.

A failure to correctly account for the dependencies in a set of data could cause the levels of significance of the statistical tests, which are used to test hypotheses about a model's parameters, to be distorted. This could cause the investigator to adopt a model which fits historical data well, but which performs poorly when predicting the future.

The risk of adopting such a model can be reduced by utilizing a model which correctly accounts for the dependencies that are inherent in the data. Unfortunately, the knowledge, needed to specify a prior model which accounts for the dependencies in a set of data, is not presently available. This creates the need for a systematic approach which is capable of identifying an appropriate model to account for whatever dependencies exist.

The approach for determining an appropriate model was developed by Pandit (1973) and further discussed by Pandit (1980). It consists of a flexiible set of models and a systematic procedure for determining which member of the set is most appropriate for a particular situation. The general expression for this set of models is

$$r_t = \frac{(1 - \sum_{i=1}^{n-1} \theta_i B^i)}{(1 - \sum_{i=1}^{n} \phi_i B^i)} a_t \tag{1}$$

where r_t represents the residual time series and a_t represents a series of independent identically distributed normal random variables with zero mean and constant variance and the ϕ_i's and the θ_i's are unknown parameters.

The quantity in the denominator, $(1 - \sum_{i=1}^{n} \phi_i B^i)$, is referred to as an autoregressive operator and can be used in the modeling of a system which gradually adjusts to shocks. The quantity in the numerator, $(1 - \sum_{i=1}^{n-1} \theta_i B^i)$ is referred to as a moving average operator and can be used to model a system which abruptly adjusts, after possibly a finite delay, to shocks. The autoregressive and moving average operators together provide the flexibility needed to handle a system which makes both gradual and abrupt adjustments.

The procedure for selecting the most appropriate model consists of fitting a model of the form given in equation (1) for successively higher values of n. This process is continued until the F test below reveals that an increase in n fails to produce a significant improvement in fit.

$$F = \frac{[SS(n) - SS(n+1)]/2}{SS(n+1)/(N-2n-1)}$$

where

$SS(n)$ = sum of squares explained by models of order n,

N = number of observations in time series of interest.

4. ANALYSIS OF DIE CASTING TENSILE STRENGTH DATA

The procedure described above will be used to develop a model for predicting tensile strength for a real world die casting operation. The operation involved making cast aluminum automobile engine blocks.

The manufacturer had established a minimum standard for tensile strength. A sampling program had been established to help ensure that this standard was met. The program consisted of: (1) selecting an engine block from the production line during the first shift every day; (2) cutting out test specimens from each of six prescribed locations; and (3) determining the tensile strength of each specimen.

The data analyzed consisted of 118 observations for each of the six locations. The modeling results are given in Table I. An analysis of these results reveals that the ϕ_1 coefficients in the ARMA (1,0) models for locations 2, 3, and 4 are not significantly different from zero. This indicates that these time series are not significantly different from a white noise process. Both the ϕ_1 and θ_1 coefficients in the ARMA (1, 1) models for locations 1 and 5 are significantly different from zero. In addition the characteristic roots for the AR polynomial operators for both of these locations are a little less than one which indicates that the models are stable. However, one of the characteristic roots for the ARMA (3,3) model for location 6 has an absolute value which is greater than one. This indicated that the model is unstable and prompted a reexamination of the data plot, when it was noticed that tensile strengths at location 6 tended to decrease as a function of time. Thus a model with linear trend for the dynamic deterministic component was fitted. This resulted in a dynamic deterministic component, d_t, of the form

$$d_t = 313.6 - 0.405t$$

and an ARMA (1,0) stochastic noise component

$$r_t = \frac{a_t}{1 + \underset{(.187)}{0.054}B} \quad .$$

Since the ϕ_1 is not significant, the model for location 6 could be simplified by dropping the stochastic noise component. This indicated that residuals for the dynamic deterministic component for location 6 are not significantly different from white noise.

5. DISCUSSION OF FITTED MODELS

If the die casting operation were in control, then the time series for each of the six locations should be six mutually independent white noise processes. The above analysis indicated that only locations 2, 3 and 4 had time series which were indistinguishably different from white noise. This indicated that the operation was not in a state of statistical control and that search for assignable causes should be focused on factors which might cause the time series for locations 1, 5, and 6 to deviate from white noise. The fitted models for these locations can be analyzed to obtain potentially valuable clues as to where these assignable causes might lie. The fitted models for locations 1 and 5, have first order AR components. In section 3 it was noted that an AR component can be

Table I. Model Summary Data

Location	ARMA Order	AR Coefficients (s.e) ϕ_1	ϕ_2	ϕ_3	MA Coefficients (s.e) θ_1	θ_2	θ_3	Q_{cal}	$Q_{.95,25}$
1	(1,1)*	0.912 (.087)			.776 (.133)			38.51	37.66
2	(1,0)	0.137 (.092)						11.90	37.66
3	(1,0)	-0.079 (.093)						16.30	37.66
4	(1,0)	0.102 (.093)						25.72	37.66
5	(1,1)*	0.978 (.042)			.892 (.080)			43.50	37.66
6**	(3,3)*	0.578 (.426)	.986 (.162)	-.536 (.382)	.651 (.389)	.942 (.454)	-.583 (.885)	21.71	37.66

*An ARMA(n,n) model can result from the following procedure to avoid over-fitting: dropping small parameters in an ARMA(n+1,n) model, (2) refitting the simplified model, (3) using the F-criterion and Q-statistic to check the simplified model against the ARMA(n+1,n) model, and (4) selecting the simplified model if the F-value is insignficant.

**The characteristic roots of the autoregressive polynomial are: $\lambda_1 = .5303$, $\lambda_2 = -.982$, $\lambda_3 = 1.634$. Since there is a root greater than one in absolute value, the model is nonstationary.

used to describe a system which gradually adjusts to shocks. This prompted a search for a process which would take place gradually with the passage of time. This search resulted in the following hypothesized causal mechanism for the first order AR component.

The process for manufacturing engine blocks involves using a 10,000 psi pressure differential to inject approximately 300 pounds of molten aluminum through up to seven small openings, known as gates. Each of these gates has a nominal area of 0.7 square inches. The water cooled die rapidly removes heat from the molten aluminum causing it to solidify in approximately 0.35 seconds. This 0.35 seconds solidification time means that the flow rate of the molten aluminum as it passes through the gates is approximately 1550 cubic inches per second.

The enormous velocity of the molten aluminum gradually wears away the gate areas. The enlarged gate causes the entry velocity to decrease. This decrease in entry velocity decreases the inlet pressure of the molten aluminum which makes it more difficult for the molten aluminum to fill the remote sections of the die before solidification occurs. The inability to fill these remote sections can cause the tensile strength in these areas to fall. This problem can be temporarily corrected by welding shut a gate. When low tensile strength problems arise again another gate is welded shut. This process is repeated until only three out of the seven gates are open. When low tensile strength problems arise with only three gates in service, these gates are welded shut and seven new gates are re-drilled in the exact locations as the original gates.

Welding a gate shut can be regarded as inducing a semi-permanent change into the behavior of the time series for the locations that have their flow dynamics affected as a result thereof. Such permanent changes are typically characterized by first order AR coefficients which approach unity as is the case for the models for locations 1 and 5.

The model for location 6 consisted of a deterministic decreasing linear trend. The duration of this trend persisted much longer than the average time between welding shut and re-drilling of all gates. Furthermore, the rate of decrease is much smaller and much more persistent than that which would be accounted for by the reconditioning of the gates. However, such a trend could have been caused by a leak in the cooling system which progressively got worse with the passage of time. This progressive worsening of a leak could have gradually reduced the amount of water available for removing heat from the vicinity of location 6. Failure to remove heat sufficiently fast is known to cause dispersed microporosity problems which can lead to marked reductions in tensile strengths.

6. CONCLUSIONS

This paper has attempted to demonstrate the utility of time series methods for controlling the quality of a die casting operation. Time series models could be utilized to provide an early warning signal for corrective action to be taken. Such early warning could be used to reduce the variability of tensile strength in die casting. This could in turn make it possible for engineers to reduce the weight of engine blocks and thereby help to conserve fuel. Time series models could also be used to obtain information potentially useful for identifying the assignable causes of a quality control problem. This will involve a two phase process. The first phase will utilize the time series model to tentatively identify potential causal mechanisms. The second phase should utilize engineering analysis and experimentation to test the validity of causal mechanisms identified in the first phase.

REFERENCES

BEALE, E.M.L. and SEELEY, R.G. (1979). First order autoregressive regression analysis. In Forecasting (Proceedings of the Institute of Statisticians Annual Conference, Cambridge 1976). Ed: O.D. Anderson, North-Holland, Amsterdam & New York, 167-175.

BOX, G.E.P. and NEWBOLD, P. (1971). Some comments on a paper of Coen, Gomme and Kendall. J. Roy. Stat. Soc. A 134, 229-240.

COCHRAN, W.G. (1974). Some consequences when the assumptions for the analysis of variance are not satisfied, Biometrics 3, 22-38.

KAPOOR, S.G. (1981). Time series modeling of machined surfaces. In Applied Time Series Analysis (Proceedings of the International Conference held at Houston, Texas, August 1981). Eds: O.D. Anderson and M.R. Perryman, North-Holland, Amsterdam & New York, 127-134.

PANDIT, S.M. (1973). Data dependent systems: Modeling analysis and optimal control via time series. PhD Thesis, Department of Mechanical Engineering, University of Wisconsin, Madison.

PANDIT, S.M. (1980). Data dependent systems and exponential smoothing. In Analysing Time Series (Proceedings of the International Conference held on Guernsey, Channel Islands, October 1979). Ed: O.D. Anderson, North-Holland, Amsterdam & New York, 217-238.

SAGE, A.P. (1968). Optimum Systems Control. Prentice-Hall, Englewood Cliffs, NJ.

WOLD, H. (1954). A Study in the Analysis of Stationary Time Series. 2nd ed. Almqvist and Wiksell, Stockholm.

TIME SERIES ANALYSIS: Theory and Practice 4
O.D. Anderson (editor)

A Canonical Correlations Approach to State Vector Analysis of Capital Appropriations and Expenditures

H. D. Vinod,
Fordham University,
Bronx, New York, USA.

B. S. Hui,
Data Resources Inc.,
Lexington, Massachusetts, USA.

State space models using canonical correlations analysis of time series have been recently recommended by Akaike and others. The use of canonical correlations provides initial guesses for unknown parameters for the maximum likelihood iteration in the identification and estimation of multivariate autoregressive moving average models. We apply these methods to Almon's famous data set on capital appropriations and expenditures used by many econometricians. Our multivariate approach reveals the presence of feedbacks ignored in the literature, and provides a new look at the lag structure.

1. Introduction

It is well-known that the method of maximum likelihood estimation can be viewed as the one that minimizes the surprise measured by Kullback's information quantity. When several models are estimated by the method of maximum likelihood, the model with more parameters often fits better. A comparison of models having different numbers of parameters was clarified by Akaike by suggesting the information criterion

$$\text{AIC} = -2 \text{ (Max. of log likelihood)} + 2 \text{ (No. of Independently adjusted parameters)}.$$

This criterion is popular because it addresses statisticians' intuitive preference of parsimonious models as follows. When one chooses the model with the smallest value of AIC, the second term penalizes models with a large number of parameters. The derivation of the second term is readily explained by considering regression models, Sawa(1978). We note that Kullback's information criterion for model selection may be defined in terms of:

$$\int \log g(y) \,/\, f(y|\theta)\; dG(y)$$

where $G(y)$ is the true distribution function of a vector random variable, f is the density of the postulated model, and g is the density of the true model. The AIC criterion gives an approximately unbiased estimate of $-2E(\log f)$, which is related to the Kullback criterion. The second term is designed to correct the bias, and the multiplier 2 is introduced for normalization.

The minimum AIC criterion cannot be readily applied to time series models because the likelihood function becomes complicated when the observations are not mutually independent. Akaike (1976,80) has developed the appropriate criterion for time series models based on canonical correlations. This is a somewhat unconventional use of canonical correlations, since it is used for identification as well as estimation.

In Section 2 we consider a (multivariate) autoregressive moving average vector (ARMAV) model, and indicate the relationship between it and the state space models. In the univariate case ARMA models have been successful in providing parsimonious canonical form representation by a suitable choice of orders for AR and MA portions. In the multivariate case the canonical representation proposed by Akaike (1976) uses the concept of a predictor space. We will not reproduce his intricate geometrical arguments.

2. ARMAV and State Space Representation

Suppose we are given a vector autoregressive moving average ARMAV(p,q) (see Tiao and Box, 1981)

$$\Phi(B)\underline{y}(t) = \theta(B)\underline{u}(t) \,, \tag{1}$$

where

$$\Phi_p(B) = I - \Phi_1 B - \ldots - \Phi_p B^p \,, \tag{2}$$

$$\theta_q(B) = I - \theta_1 B - \ldots - \theta_q B^q \,, \tag{3}$$

where we have a vector of k stationary time series $y(t)$, and $u(t)$ is a k-dimensional vector of random errors (innovations) which are multivariate normal with zero means and covariance matrix Q. Notation B is a backshift operator defined by $By(t) = y(t-1)$. Clearly, the Φ's and θ's are $k \times k$ matrices of autoregressive and moving average coefficients respectively. We assume that the zeros of determinantal polynomials $|\Phi(B)|$ and $|\theta(B)|$ are on or outside the unit circle. When the zeros of $|\Phi(B)|$ are completely outside the unit circle, the series $\underline{y}(t)$ are stationary; and when the zeros of $|\theta(B)|$ are all outside the unit circle, we say the process is invertible (see Tiao and Box 1981). In practice, some nonstationarity can be modelled by simultaneously differencing each of the series, i.e., by allowing the zeroes of $|\phi(B)|$ to be on the unit circle. This may not be satisfactory in some cases.

State vector representation can be written as

$$\underline{x}(t) = F\underline{x}(t-1) + G\underline{u}(t) \,, \tag{4}$$

$$\underline{y}(t) = H\underline{x}(t) \,, \tag{5}$$

where $\underline{x}(t)$ is an n-dimensional state vector. The matrix F is an $n \times n$ transition matrix, G is an $n \times k$ matrix of impulse response coefficients, and where H is a $k \times n$ matrix for selecting the observables in $y(t)$ to go with the state vector.

We build the state vector $\underline{x}(t)$ from the current values and future forecasts as follows:

$$\begin{aligned} \underline{x}(t) = [&y_1(t),\, y_2(t), \ldots, y_k(t)\,; \\ &\hat{y}_1(t+1|t),\, \hat{y}_2(t+1|t), \ldots, \hat{y}_k(t+1|t)\,; \\ &\ldots;\, \hat{y}_1(t+\ell|t), \ldots, \hat{y}_k(t+\ell|t)\,;\, \ldots]' \\ = [&\underline{y}(t)',\, \hat{\underline{y}}(t+1|t)', \ldots, \hat{\underline{y}}(t+\ell|t')', \ldots]' \end{aligned} \tag{6}$$

Here, $\hat{y}_i(t+\ell|t)$ for $i = 1,\ldots,k$ denotes the ℓ period ahead conditional expectation (forecast) of $y_i(t+\ell)$ given all present and past observations $\underline{y}(t)$, $\underline{y}(t-1),\ldots,\underline{y}(1)$.

Selecting a set of linearly independent predictors of $\underline{y}(t)$ from $\underline{x}(t)$ of (6) involves tedious checking to see whether an element in (6) is a linear combination of its antecedents. If so that element and all its future values must be eliminated.

The matrix H in (5) is obtained after one knows which elements of (6) are retained. One simply places a 1 or 0 in the appropriate location for those that are retained and eliminated respectively.

Consider impulse response matrices Ψ_s defined from (1) by the relation

$$\underline{y}(t) = \Phi^{-1}(B)\theta(B)\underline{u}(t) \tag{7a}$$

$$= \sum_{s=0}^{\infty} \Psi_s \underline{u}(t-s) \ , \tag{7}$$

where $\psi_0 = I$.

Letting the summation start at $s = i$ and replacing s by $s - i$ we can write the conditional expectation (by using stationarity)

$$\hat{\underline{y}}(t+i\,|t) = \sum_{s=i}^{\infty} \Psi_s \underline{u}(t-s) \tag{8}$$

Hence, write

$$\hat{\underline{y}}(t+i\,|t+1) = \hat{\underline{y}}(t+i\,|t) + \Psi_{i-1}\underline{u}(t+1) \ , \tag{9}$$

and from conditional expectation methods

$$\hat{\underline{y}}(t+p\,|t) = \Phi_1\hat{\underline{y}}(t+p-1|t) + \Phi_2\hat{\underline{y}}(t+p-2|t) + \ldots + \Phi_p\underline{y}(t) + \Psi_{p-1}\underline{u}(t+1) \tag{10}$$

where we assume $p > q$ for convenience. Hence, combining (9) and (10), we can write the ARMA model (1) in state vector form:

$$\begin{bmatrix} \underline{y}(t+1) \\ \hat{\underline{y}}(t+2|t+1) \\ \cdot \\ \cdot \\ \cdot \\ \hat{\underline{y}}(t+p\,|t+1) \end{bmatrix} = \begin{bmatrix} 0 & I & 0 & \ldots & 0 \\ 0 & 0 & I & \ldots & 0 \\ \cdot & \cdot & \cdot & \ldots & \cdot \\ \cdot & \cdot & \cdot & \ldots & \cdot \\ \cdot & \cdot & \cdot & \ldots & \cdot \\ \Phi_p & \Phi_{p-1} & \cdot & \ldots & \Phi_1 \end{bmatrix} \begin{bmatrix} \underline{y}(t) \\ \hat{\underline{y}}(t+1|t) \\ \cdot \\ \\ \\ \hat{\underline{y}}(t+p-1|t) \end{bmatrix} + \begin{bmatrix} I \\ \Psi_1 \\ \cdot \\ \cdot \\ \cdot \\ \Psi_{p-1} \end{bmatrix} \underline{u}(t+1) \tag{11}$$

or more succinctly (see equation (1))

$$\underline{x}(t+1) = F\underline{x}(t) + G\underline{u}(t+1) \tag{12}$$

We will illustrate this process of writing an ARMA model in an equivalent state vector form with a simple example in Section 4.

We have already noted the determination of the H matrix containing ones and zeroes. We know F and G from (11). Thus we have formulated the problem of obtaining a state space representation (4), (5) of the ARMAV (p,q) model of (1).

3. Canonical Correlation Analysis of Multiple Time Series

In this section we outline Akaike's use of canonical correlation analysis to determine the order and composition of the state vector $\underline{x}_t$ and the F and G matrices. For a more complete discussion, see Akaike (1976).

3.1 Stepwise Autoregression

Given the vector of observed time series $\underline{y}(t)$, we wish to determine the optimal order m such that at time t, all significant past and present information for predicting the future is contained in $\underline{y}(t)$, $\underline{y}(t-1),\ldots,\underline{y}(t-m)$. This is done by successively increasing the order of vector autoregression:

$$\underline{y}(t) = A_1\underline{y}(t-1) + A_2\underline{y}(t-2) + \ldots + \underline{u}(t) \tag{13}$$

and computing the following AIC (Akaike's Information Criterion) after each autoregressive fit:

$$\mathrm{AIC}_i = n \log(|\hat{C}_i|) + 2ik^2 \ , \tag{14}$$

where n is the number of observations $|\hat{C}_i|$ is the determinant of the k-dimensional covariance matrix defined by

$$\hat{C}_i = \frac{1}{n}\sum_{t=1}^{n-i} \underline{y}(t+i)\underline{y}(t) \ , \tag{15}$$

and i is the order of autoregression.

The expression (14) is similar to the general definition in Section 1 since there are k^2 elements or parameters in each of the i matrices when the order of autoregression is i. The first term of (14) evaluates the maximum of an estimate of Kullback's integral.

3.2 Canonical Correlations

Akaike (1976) introduces the use of canonical correlation analysis to determine the state vector $\underline{x}(t)$. For a detailed exposition of canonical correlation analysis, see Anderson (1958). Vinod and Ullah (1981) give more up-to-date discussion and references. Briefly, canonical correlation analysis between two sets of random variables $\{x_1,\ldots,x_m\}$ and $\{y_1,\ldots,y_n\}$ involves finding coefficients $\{a_{11},\ldots,a_{1m}\}$ and $\{b_{11},\ldots,b_{1n}\}$ such that the correlation r_1 between the two linear combinations

$u_1 = \sum_{i=1}^{m} a_{1i}x_i$ and $v_1 = \sum_{j=1}^{n} b_{1j}y_j$ is maximized. Next, we find

$u_2 = \sum_{i=1}^{m} a_{2i}x_i$ and $v_2 = \sum_{j=1}^{n} b_{2j}y_j$ such that u_2 and v_2 are orthogonal to u_1 and v_1, and that $r_2 = r(u_2,v_2) < r_1$ is maximized. This process continues until min(m,n) canonical correlations $(r_1,r_2, \ldots)$ and their corresponding pairs of canonical variates $\{u_1, v_1\}, \{u_2, v_2\}, \ldots$ have been extracted.

3.3 Determination of the State Vector and the F Matrix

Akaike uses a series of canonical correlation analyses between the present and past values of the process:

$$\underset{\sim}{y}^p = \{\underset{\sim}{y}(t)', \underset{\sim}{y}(t-1)', \ldots, \underset{\sim}{y}(t-m)'\}' \tag{16}$$

and the set of present and future values:

$$\underset{\sim}{y}^f = \{y(t)', \underset{\sim}{y}(t+1)', \ldots, \underset{\sim}{y}(t+n)'\}' \tag{17}$$

to determine the state vector $\underset{\sim}{x}(t)$ and the F and G matrices. The order m in (16) is determined by the minimum AIC in stepwise autoregression (see Section 3.1).

Stepwise canonical correlation analyses are performed between the fixed set $\underset{\sim}{y}^p$ and $\underset{\sim}{y}^f$ as the dimension of $\underset{\sim}{y}^f$ increases. At each step the last canonical correlation is tested for significance by the Differenced Information Criterion:

$$\text{DIC} = \text{AIC}(1) - \text{AIC}(2) \tag{18}$$

where AIC(1) is the appropriately redefined AIC of the canonical correlations model when the last canonical correlation is constrained to be zero, and AIC(2) is the corresponding AIC of the unconstrained model. Thus, when DIC is positive, the latest component from $\underset{\sim}{y}^f$ adds significant "new information" to the state vector model. It is therefore included in the state vector $\underset{\sim}{x}(t)$. On the other hand, when DIC is negative, no significant "new information" is added by the latest addition, which can be written as a linear combination of the preceding components in the state vector. The coefficients of this linear combination form a row in the F matrix. We can drop this component and all its future values from the state vector.

This process of stepwise canonical correlation continues until any additional component could be expressed as a linear combination of components already in the state vector.

3.4 Determination of the G Matrix

If none of the components of $\underset{\sim}{y}(t)$ is linearly dependent on the other components, we can write the H matrix in (5):

$$H = [I, 0] \, , \tag{19}$$

where I is a $k \times k$ identity matrix, and if $\underset{\sim}{x}(t)$ is $n \times 1$, 0 is the $k \times (n-k)$ null matrix.

Furthermore, from (11), we see that G consists simply of I and the impulse response matrices, Ψ_i. The Ψ_i's are estimated from the "optimal" order vector autoregression from (13). Let the optimal order by AIC criterion be m. We write

$$\underset{\sim}{y}(t) = A_1\underset{\sim}{y}(t-1) + \ldots + A_m\underset{\sim}{y}(t-m) + \underset{\sim}{u}(t) \, . \tag{20}$$

Equivalently, we can write

$$\begin{aligned} \underline{y}(t) &= (I - A_1 B - \ldots - A_m B^m)^{-1} \underline{u}(t) \\ &= (\Psi_0 + \Psi_1 B + \ldots + \Psi_m B^m) \underline{u}(t) , \end{aligned} \tag{21}$$

where B is the backshift operator as before.

From (21), we get the recursive relations for the impulse response matrices (Ψ_s's) in (7) as follows

$$\begin{aligned} \Psi_0 &= I \\ \Psi_1 &= A_1 \\ \Psi_2 &= A_1\Psi_1 + A_2 \\ \Psi_3 &= A_1\Psi_2 + A_2\Psi_1 + A_3 \\ &\ldots \\ &\ldots \\ \Psi_m &= A_1\Psi_{m-1} + A_2\Psi_{m-2} + \ldots + A_{m-1}\Psi_1 + A_m \end{aligned} \tag{22}$$

We obtain the G matrix from the Ψ_s's (see (11)).

The basic idea in Akaike's (1976, 1980) proposal outlined above is to consider a sequence of impulse response matrices of a system. Next we interpret it as a covariance matrix between y^p, the present and past outputs and y^f, the present and future inputs of the system when the system is driven by a white noise with unit covariance matrix. The relationship of the algorithm with the singular value decomposition of a Hankel matrix is discussed by Akaike (1976, p. 60). For an exposition of singular value decomposition see Vinod and Ullah (1981).

4. Example: Almon Data Feedbacks

Almon (1965) introduced a polynomial distributed lag (impulse response) model and illustrated it by using seasonally adjusted data on capital appropriations and expenditures. These data are used by several authors including Vinod and Pandit (1982), and a slightly updated version of quarterly data from first quarter 1953 to the fourth quarter 1967 is available in Maddala (1977) which is used here to illustrate the state space models. We use a multivariate approach where the feedbacks between capital appropriations and expenditures are explicitly included.

We denote $y(t)$ = capital expenditures and $u(t)$ = capital appropriations. From Box and Pierce's Q statistic we determine that first order differencing may be appropriate for those data. The AIC_i values based on (14) for various AR orders are respectively (0, 1295.0), (1, 1242.6), (2, 1224.7) (3, 1225.3), (4, 1229.9). Thus AIC_i is minimized for AR order 2. The state vector $x(t)$ consists of $[y(t),\ u(t+1),\ y(t+1)]$. The stepwise canonical correlation (ρ) analysis is as follows

$[y(t),\ \rho=1,\ DIC=2992.358]$: keep,

$[u(t+1),\ \rho=0.131,\ \mathrm{DIC}=0.280]$: keep,

$[y(t+1),\ \rho=0.126,\ \mathrm{DIC}=1.972]$: keep,

$[u(t+2),\ \rho=0.029,\ \mathrm{DIC}=-2.267]$: omit,

$[y(t+2),\ \rho=0.008,\ \mathrm{DIC}=-3.544]$: omit.

Our estimates of the various matrices in the state space representation are

$$F = \begin{bmatrix} 0.000 & 0.000 & 1.000 & 0.000 \\ 0.000 & 0.000 & 0.000 & 1.000 \\ 0.106 & -1.152 & -0.098 & 0.821 \\ 0.114 & -0.246 & 0.107 & 1.007 \end{bmatrix},$$

$$G = \begin{bmatrix} 1.000 & 0.000 \\ 0.000 & 1.000 \\ 0.103 & 0.798 \\ 0.075 & 0.604 \end{bmatrix},$$

$$\Phi_0 = I_2 = \theta_0,\ \Phi_1 = \begin{bmatrix} -0.098 & 0.821 \\ 0.107 & 1.007 \end{bmatrix},$$

$$\Phi_2 = \begin{bmatrix} 0.106 & -1.152 \\ 0.114 & 0.246 \end{bmatrix},\quad \theta_1 = \begin{bmatrix} -0.202 & 0.023 \\ 0.232 & 0.404 \end{bmatrix}.$$

For a further discussion of computer programs and forecasting application the reader is referred to Hui and O'Reilly (1982). Akaike(1978) also refers to a program package called TIMSAC which is of interest in this context for model selection based on the method of maximum likelihood. It should be noted that canonical correlation analysis is only for the estimation of initial guess. The method of maximum likelihood should be applied to several possible alternatives.

One of the interesting conclusions is that feedbacks are present, and the lag distribution will change when they are included in the model. The evidence regarding the presence of feedbacks is contained in the fact that both off-diagonal terms of Φ_1 and Φ_2 are non-zero. It stands to reason that the feedback of past year's capital appropriations $u(t-1)$ on capital expenditures $y(t)$ is larger in absolute value (-0.821) than vice versa (0.107), in terms of first differences of both variables. Roughly speaking, one spends more when one has more. The off-diagonal terms of θ_1 may be regarded as the effect of the "residual" from linear least squares forecast (rational expectation) of spending on appropriations or vice versa. We find evidence to support economists' notion that the capital goods producing sector is subject to volatile fluctuations in demand, which in turn contributes to the presence of business cycles, if we evaluate the roots of the AR portion. The reader should remember that direct interpretation of each coefficient remains tricky, since time series models are influenced by noise. This is where economic theory plays an important role.

We feel that most econometricians will find our analysis of Almon's data to be richer and more informative than the traditional lag distributions. Of course, the main practical advantage of state space formulation is that it can be directly used in Kalman filtering algorithms to obtain optimal forecasts of underlying variables.

REFERENCES

Akaike, H. (1976), "Canonical Correlations Analysis of Time Series and the Use of an Information Criterion," in R. Mehra and D. G. Lainiotis (eds), *Advances and Case Studies in System Identification.* New York: Academic Press.

Akaike, H. (1978), "Time Series Analysis and Control Through Parametric Models," in D. F. Findley (ed) *Applied Time Series Analysis.* New York: Academic Press.

Akaike, H. (1980), "On the Identification of State Space Models and Their Use in Control," in D. R. Brillinger and G. C. Tiao (eds), *Directions in Time Series.* Haywood, California:Institute of Mathematical Statistics.

Almon, S. (1965), "The Distributed Lag Between Capital Appropriations and Expenditures," *Econometrica*, 33, 178-196.

Anderson, T. W. (1958), *An Introduction to Multivariate Statistical Analysis.* New York: Wiley.

Chow, G. C. (1975), *Analysis and Control of Dynamic Economic Systems.* New York: Wiley.

Hui, Baldwin and Dan O'Reilly (1982), "State Vector Analysis," in *The DRI Statistician*, 1(no.2),11-23, Lexington, Mass.: Data Resources Inc.

Maddala, G. S. (1977), *Econometrics.* New York: McGraw-Hill.

Tiao, G. C. and Box, G.E.P. (1981), "Modeling Multiple Time Series With Applications," *Journal of American Statistical Association,* 76, 802-816.

Sawa, T. (1978), "Information Criteria for Discriminating Among Alternative Regression Models," *Econometrica*, 46, 1273-1291.

Vinod, H. D. and Ullah, A., (1981), *Recent Advances in Regression Methods.* New York: Marcel Dekker.

Vinod, H. D. and Pandit, S. M. (1982), "Green's Distributed Lag Models and Econometrics of Rational Expectations," unpublished report, Dept. of Economics, Fordham University, Bronx, New York.

TIME SERIES ANALYSIS: Theory and Practice 4
O.D. Anderson (editor)

A THEORETICAL VIEW OF THE USE OF AIC

Ritei Shibata

Mathematical Sciences Research Institute
Berkeley, California
and
Mathematics and Statistics Department
University of Pittsburgh
Pittsburgh, Pennsylvania
U.S.A.

The minimum Akaike's Information Criterion (AIC) has an optimality property for selecting one model from a set of models each specified by many parameters. However, for choosing a model with few parameters the effectiveness of AIC decreases. In such cases, a minimax choice of generalized AIC procedure is suggested. It is also suggested that the inconsistency of the AIC procedure is the inevitable concomitant of balancing underfitting and overfitting risks.

1. INTRODUCTION

This paper aims to summarize recent theoretical work on the problem of model selection from a given family of models, especially that concerning the AIC procedure.

Although there are many applications of the AIC procedure (e.g. see Akaike, 1982), some objections to its use have been raised (Schwarz, 1978, Cox, 1977). A principal objection is to its inconsistency, that is, if the assumed true model has a finite number of parameters, the selection procedure does not always converge to the true model as the sample size increases. In fact, the probability of selecting underfitted models converges to zero, but that of selecting overfitted models does not (Shibata, 1976). It is not difficult to obtain a consistent procedure by modifying the AIC procedure; for example, consider the BIC (Akaike, 1977, Schwarz, 1978) or the ϕ (Hannan and Quinn, 1979).

Is inconsistency really a serious defect of AIC? The concept of consistency is most fundamental in asymptotic theory, and inconsistent estimation or testing procedures are seldom considered. This is true also for our problem, provided that the models being considered are "separated". "Separated" means that the models never overlap each other (see Cox, 1961). For example, an autoregressive model and a moving average model are separated. In such separated cases, the AIC procedure has no problem of inconsistency. Since, a model is selected so as to minimize

$$\text{AIC} = -2\log(\text{max likelihood}) + 2 \times (\text{the number of parameters}),$$

and log maximum likelihoods, for each separated model, differ by an order of magnitude for sample size n, it follows that the procedure is consistent.

The AIC procedure is inconsistent only when models are not separated. An example is a set of autoregressive models, in which the model of order k includes models of lower order $\ell < k$. Then, log maximum likelihoods for overfitted models are the same to order n, and the AIC procedure is inconsistent. Our main concern here is for such non-separated models, in particular, for a set of nested models. One should note that, then, overfitting is not as serious a fault as underfitting. Even if the model is overfit, consistency of estimates or of estimated predictors is retained even though the variances are increased. On the other hand, underfitting sometimes leads to a significant increase in bias.

Although the AIC procedure is inconsistent, it satisfactorily balances both underfitting and overfitting risks, and is asymptotically efficient for selecting one model from a family of models, each specified by many parameters (Shibata, 1981). On the other hand, the balancing achieved by consistent procedures is not satisfactory. In fact, it will be shown in Section 2 that they are not asymptotically efficient, with asymptotic effiencies sometimes zero. In Section 3, we will show that similar results hold true even when selecting a model with few parameters. For some parameter values, the underfitting risk connected with consistent selection procedures diverges to infinity.

Therefore, if one wishes to have a good estimator or predictor, it is not necessarily wise to insist on the consistency of the selection procedure. Inconsistency does not imply a defect in the selection procedure, but rather the inevitable concomitant of balancing underfitting and overfitting risks.

To simplify the discussion in this paper, attention is restricted to the selection of the order of an autoregressive model. It is not difficult to extend the discussion to more general problems, such as subset selection, and selection of the orders of autoregressive moving average models in which both autoregressive and moving average operators have no approximately common roots (Shibata, 1981, Taniguchi, 1980).

Consider the autoregressive model of order k, AR(k),

$$Z_t = \phi_1 Z_{t-1} + \cdots + \phi_k Z_{t-k} + \epsilon_t ,$$

where the ϵ_t's are independent normally distributed random variables with mean 0 and variance $\sigma^2 \leqslant \infty$, and $\phi(z) = 1 - \phi_1 z - \cdots - \phi_k z^k$ is not zero for $|z| \leqslant 1$. An order k is selected from $1 \leqslant k \leqslant K$, where K is a given number. Let $Z_1, \ldots, Z_n$ be observed from an AR(k_0) with parameters $\phi_1, \ldots, \phi_{k_0}$ and σ^2. Here k_0 is not necessarily finite. Under the model of order k, one may obtain least squares estimates $\hat{\phi}_1(k), \ldots, \hat{\phi}_k(k)$ and $\hat{\sigma}^2(k)$, putting $Z_1, \ldots, Z_K$ as initial conditions which don't depend on each k.

Consider an estimated one step ahead predictor

$$\hat{Z}^*_{t+1} = \sum_{\ell=1}^{k} \hat{\phi}_\ell(k)\, Z^*_{t+1-\ell}$$

for another process $\{Z^*_t\}$ which is independent of $\{Z_t\}$ but which has the same probabilistic structure as $\{Z_t\}$. Then its prediction error is

$$(\hat{\underset{\sim}{\phi}}(k) - \underset{\sim}{\phi})' \, \Gamma \, (\hat{\underset{\sim}{\phi}}(k) - \underset{\sim}{\phi}) + \sigma^2,$$

where $\hat{\underset{\sim}{\phi}}(k)' = (\hat{\phi}_1(k),\ldots,\hat{\phi}_k(k),0,\ldots)$, $\underset{\sim}{\phi}' = (\phi_1,\ldots,\phi_{k_0},0,\ldots)$, and $\boldsymbol{\Gamma}$ is the $\infty \times \infty$ autocovariance matrix. Then a natural loss function for a selection $\hat{k}$ is,

$$L(\hat{k}) = n(\hat{\underset{\sim}{\phi}}(\hat{k}) - \underset{\sim}{\phi})' \boldsymbol{\Gamma} (\hat{\underset{\sim}{\phi}}(\hat{k}) - \underset{\sim}{\phi}),$$

and the risk is $R(\hat{k}) = E[L(\hat{k})]$.

Under this loss function, the minimum risk order, k^*, is defined as that k which minimizes the risk R(k) in $1 \leqslant k \leqslant K$. The selection $\hat{k} \equiv k^*$ is the best one in the non-random selection $\hat{k} \equiv k$, $1 \leqslant k \leqslant K$. The k^*, of course, depends on the unknown parameters and also upon the sample size n. For large enough n, $k^* = k_0$, but for small sample size n, $k^* \leqslant k_0$ even if k_0 is finite and $1 \leqslant k_0 \leqslant K$.

In this paper, the generalized AIC criterion is considered, that is, the order $\hat{k}_\alpha$ is selected so as to minimize

$$AIC_\alpha(k) = n \log \hat{\sigma}^2(k) + \alpha k.$$

If $\alpha = \log n$, one obtains the BIC criterion, and if $\alpha = 2c \log \log n$, for some $c > 1$, one obtains the ϕ criterion.

2. THE CASE WHEN k^* IS LARGE

The result proved in Shibata (1980a) is that a lower bound for the loss $L(\hat{k})$ of any $\hat{k}$ is given asymptotically in probability by $L(k^*)$ and that

$$p - \lim_{n\to\infty} \frac{L(\hat{k}_2)}{L(k^*)} = 1 . \qquad (2.1)$$

A similiar result holds true also for the risk $R(\hat{k})$: a lower bound is given by $R(k^*)$ and

$$\lim_{n\to\infty} \frac{R(\hat{k}_2)}{R(k^*)} = 1, \qquad (2.2)$$

(Shibata, 1982a). Furthermore, (2.1) and (2.2) hold for any possible values of parameters $\phi_1,\ldots,\phi_{k_0}$ and σ^2, if and only if $\alpha = 2$, which is the case for the AIC procedure. The key assumption is that $k^* = k^*(n)$ diverges to infinity as $n \to \infty$, together with K. If one assumes that observations $Z_1,\ldots,Z_n$ come from an AR(∞) whose order is outside $1 \leqslant k \leqslant K$, then the above key assumption is easily justified. Even if one assumes a finite order AR(k_0), the above results hold approximately for large k_0; i.e., for large k^* (Shibata, 1980a).

An essential point in the proof is that

$$n \log \{ \frac{\hat{\sigma}^2(k)}{\sigma^2(k)} \} - n \log \{ \frac{\hat{\sigma}^2(\ell)}{\sigma^2(\ell)} \}$$

is asymptotically χ^2-distributed with degree of freedom $\ell-k$, for $\ell > k$, so that

$$\frac{1}{\ell-k} [n \log \{ \frac{\hat{\sigma}^2(k)}{\sigma^2(k)} \} - n \log \{ \frac{\hat{\sigma}^2(\ell)}{\sigma^2(\ell)} \}]$$

can be approximated by 1 for large enough $\ell-k$. Here $\sigma^2(k)$ and $\sigma^2(\ell)$ are the asymptotic values of $\hat{\sigma}^2(k)$ and $\hat{\sigma}^2(\ell)$, respectively. Roughly speaking, if k^* is large, it is enough to consider just ℓ and k in the neighbourhood of k^*, so that the above approximation is applicable.

Computer simulations (Shibata, 1980b, 1982a) suggest that k^* must be very large for (2.1) to hold; but, for (2.2) to hold, $k^* \geqslant 10$ is sufficient for practical purposes. In other words, when the selected number $\hat{k}_2$ is larger than 10, one can trust the above optimality property of AIC, even taking into consideration the overfitting behaviour of the AIC procedure. Of course, it should be noted that the optimality property is only in the sense of the loss function $L(\hat{k})$ or, equivalently, in the sense of the Kullback-Leibler information measure.

Some applications of AIC show that $\hat{k}_2 = 65$ (Akaike, 1973), $\hat{k}_2 = 15$ (Akaike, 1969), $\hat{k}_2 = 11$ (Tong, 1977).

3. THE CASE WHEN k^* IS SMALL

We have just considered k^* large, but often k^* is very small, for example 2 or 3. For such cases, one cannot expect an optimality property as in Section 2. However, the essential point is the same, namely, that a good selection procedure is one which satisfactorily balances underfitting and overfitting risks.

We now assume that observations $Z_1,\ldots,Z_n$ come from an AR(k_0), $1 \leqslant k_0 \leqslant K$, with k_0 fixed. The sample size n needs to be fairly large, say, $n \geqslant 100$. Without loss of generality we can assume $\sigma^2 = 1$. If the nonzero parameters $\phi_1,\ldots,\phi_{k_0}$ are not too small (more precisely, $n^{1/2}\phi_1,\ldots,n^{1/2}\phi_{k_0}$ are significantly large relative to α), then underfitting scarcely occurs by $\hat{k}_\alpha$. We call such a case a "sharp fit"; otherwise it is a "dull fit", when the underfitting risk is not negligible.

The overfitting risk, in the presence of a sharp fit, is evaluated using random walk theory (Shibata, 1982a),

$$\lim_{n\to\infty} R(\hat{k}_\alpha) = \sum_{m=1}^{K-k_0} P(\chi^2_{m+2} > \alpha m) + k_0. \tag{3.1}$$

As is easily seen, a lower bound for the above asymptotic risk is k_0, so that one may define an approximate mean efficiency, in the presence of a sharp fit, by the quotient

$$k_0 \Big/ \{ \sum_{m=1}^{K-k_0} P(\chi^2_{m+2} > \alpha m) + k_0 \} .$$

Some numerical examples are shown in Table 1.

Table 1. Approximate mean efficiency in the presence of a sharp fit, K = 10.

α	k_0= 1	2	3	4	5	6	7	8	9	10
1.0	.141	.267	.401	.487	.583	.673	.758	.839	.918	1.000
2.0	.313	.490	.602	.682	.745	.798	.845	.891	.940	1.000
4.0	.688	.801	.858	.890	.910	.925	.948	.958	.972	1.000

The values in Table 1 show that AIC with α = 2 yields low efficiency when k_0 is small. As is seen from (3.1), the larger α the better. For example, BIC with α = log n, or ϕ with α = 2c log log n, c > 1, is always better than AIC with α = 2. However, in the event of a "dull fit", the above is not true.

To illustrate this, consider the case when the only possible underfitting is $\hat{k}_\alpha = k_0 - 1$. (This occurs when ϕ_{k_0} is close to zero, but ϕ_{k_0-1} is not.) Reparametrizing $\phi_1,\ldots,\phi_{k_0}$ by partial autocorrelations $\psi_1,\ldots,\psi_{k_0}$, in which ψ_{k_0} is close to zero, one can obtain the following result, (Shibata, 1982b): the risk $R(\hat{k}_\alpha)$ is asymptotically equal to

$$E[\{n \psi^2_{k_0} - (W_{k_0} - n^{1/2}\psi_{k_0})^2\} I_{(M_{K-k_0} < W,\ W > 0)}]$$

$$+ E\{ (M_{K-k_0} + \alpha T_{K-k_0}) I_{(M_{K-k_0} > W)}\} + k_0. \qquad (3.2)$$

Here $M_{K-k_0} = \max_{k_0 \leqslant k \leqslant K} (S_k)$ is the maximum of the random walk $S_k = \sum_{m=k_0+1}^{k} (W_m^2 - \alpha)$ from time k_0 to K with the provision that $S_{k_0} = 0$. Each W_m^2, $k_0+1 \leqslant m \leqslant K$ is an independent χ^2-distributed random variable with one degree of freedom, and T_{K-k_0} is the waiting time for the maximum M_{K-k_0}. W_{k_0} is normally distributed with mean $n^{1/2}\psi_{k_0}$ and variance 1, and $W = -(W^2_{k_0} - \alpha)$.

In (3.2), the first and second terms essentially define, respectively, the underfitting and overfitting risks. As α becomes large, the random walk S_k drifts negative, so that the second term in (3.2) approaches zero. On the other hand, the first term is bounded away from

$$n \psi^2_{k_0} P(M_{K-k_0} < W,\ W > 0) - 1. \qquad (3.3)$$

As before, when α becomes large, (3.3) approaches

$$n\psi_{k_0}^2 P(W > 0) - 1 = n\psi_{k_0}^2 P(W_{k_0}^2 < \alpha) - 1 .$$

Here, for $\delta\alpha \leqslant n\psi_{k_0}^2 \leqslant \alpha$, the following inequality holds,

$$n\,\psi_{k_0}^2 P(W_{k_0}^2 < \alpha) \geqslant \delta\alpha \{\Phi(\alpha^{1/2}) - 1/2\},$$

where $\Phi(\chi)$ is the normal distribution function. Therefore, for large n and α,

$$R(\hat{k}_\alpha) \geqslant \delta\alpha \{\Phi(\alpha^{1/2}) - 1/2\} + k_0 - 1,$$

when $\delta\alpha \leqslant n\psi_{k_0}^2 \leqslant \alpha$.

Consequently, we see that the risk $R(\hat{k}_\alpha)$ when $(\delta\alpha/n)^{1/2} \leqslant |\psi_{k_0}| \leqslant (\alpha/n)^{1/2}$, will diverge to infinity if α is a divergent sequence in n, for example α = log n or α = 2c log log n. Thinking of the risk function, $R(\hat{k}_\alpha)$, α should be bounded with respect to the sample size n.

Next, we shall obtain an appropriate α using the minimax principle. As is easily seen, the minimax solution for the risk $R(\hat{k}_\alpha)$, itself, is a trivial selection $\hat{k} \equiv K$. Hence, consider, instead of $R(\hat{k}_\alpha)$, a "regret",

$$\delta\, R(\hat{k}_\alpha) = R(\hat{k}_\alpha) - R(k_0).$$

This regret indicates how much the risk increases by selecting $\hat{k}_\alpha$, rather than $\hat{k} \equiv k_0$, when the true order k_0 is known.

In Shibata (1982b), a minimax solution for $\delta R(\hat{k}_\alpha)$ is obtained using computer simulations under the constraint $k_0 - 1 \leqslant \hat{k}_\alpha \leqslant K$. The solutions are shown in Table 2. The solution without such constraint will be smaller than the solution obtained here.

Table 2. Minimax solution of α

K	2	3	4	5	6	7	8	≥9
α	2.0	2.0	2.0	2.8	2.8	2.8	2.8	3.5
minimax value	0.60	0.98	1.47	1.60	1.74	1.74	1.75	1.84

Therefore, for instance, if we use $\hat{k}_{3.5}$ for $K \geqslant 9$, the prediction error is bounded above by

$$\sigma^2(1 + \frac{k_0}{n} + \frac{1.84}{n}) \qquad (3.4)$$

and below by

$$\sigma^2(1 + \frac{k_0}{n}) .$$

If another $\hat{k}_\alpha$ is used, the upper bound is larger than that given by (3.4). For example: 12.23 for α = 1.0, 2.75 for α = 2.0 and 2.29 for α = 5.0, instead of 1.84 as given in (3.4).

As a practical application, the values of $AIC_{3.5}(k)$ for the Canadian Lynx data are shown in Table 3.

Table 3. $AIC_{3.5}(k)$ for Canadian Lynx data

k	$AIC_{3.5}$	k	$AIC_{3.5}$	k	$AIC_{3.5}$	k	$AIC_{3.5}$
1	-238.39	6	-312.17	11	-318.40*	16	-305.23
2	-316.00	7	-314.18	12	-316.13	17	-301.74
3	-315.09	8	-312.16	13	-313.85	18	-300.44
4	-316.93	9	-309.99	14	-310.50	19	-297.35
5	-315.08	10	-310.13	15	-307.05	20	-294.43

The asterisk * shows the minimum of $AIC_{3.5}(k)$. All values are calculated from the values reported by Tong (1977). The selected number is 11 which is the same given by the AIC procedure.

As a conclusion, the use of AIC_α with a divergent sequence $\alpha = \alpha_n$ is not recommendable in terms of the mean squared error or the prediction error. The use of AIC can be recommended if the k^* is expected to be large. Otherwise the minimax choice of α seems to yield a reasonable selection procedure.

This research was partly supported by the National Science Foundation, Grant MCS-812-0790.

REFERENCES

AKAIKE, H. (1970). A method of statistical identification of discrete time parameter linear systems. Ann. Inst. Statist. Math. 21, 225-242.

AKAIKE, H. (1973). Information theory and an extension of the maximum likelihood principle. In 2nd International Symposium on Information Theory, Eds. B.N. Petrov and F. Csaki, Akademia Kiado, Budapest, 267-81.

AKAIKE, H. (1977). An objective use of Bayesian models. Ann. Inst. Statist. Math., 29 A 9-20.

AKAIKE, H. (1982). Citation classic. Current Contents; Engineering, Technology and Applied Sciences 12, No. 51, 22.

COX, D.R. (1961). Tests of separate families of hypotheses. Proc. 4th Berkeley Symp 1, 105-23.

COX, D.R. (1977). Discussion of the papers by Campbell and Walker, and Morris and Tong. J.R. Statist. Soc. A 140, 453.

HANNAN, E.J. and QUINN, B.G. (1979). The determination of the order of autoregression. J.R. Statist. Soc. B41, 190-5.

SCHWARZ, G. (1978). Estimating the dimension of a model. Ann. Statist. 6, 461-4.

SHIBATA, R. (1976). Selection of the order of an autoregressive model by Akaike's information criterion. Biometrika 63, 117-26.

SHIBATA, R. (1980a). Asymptotically efficient selection of the order of the model for estimating parameters of a linear process. Ann. Statist. 8, 147-164.

SHIBATA, R. (1980b). Selection of the number of regression parameters in small sample cases. In Statistical Climatology; Developments in Atmospheric Science 13, Eds.: S. Ikeda et al, Elsevier, 137-148.

SHIBATA, R. (1981). An optimal selection of regression variables. Biometrika 68, 45-54.

SHIBATA, R. (1982a). Selection of regression variables; asymptotic mean efficiency and a modification of FPE statistic. Tech. Rept. Univ. of Pittsburgh, 82-16.

SHIBATA, R. (1982b). Selection of the number of regression variables; underfitting and overfitting risks. Tech. Rept. Univ. of Pittsburgh, 82-22.

TANIGUCHI, M. (1980). On selection of the order of the spectral density model for a stationary process. Ann. Inst. Statist. Math. 32 A, 401-419.

TONG, H. (1977). Some comments on the Canadian Lynx Data. J.R. Statist. Soc. A 140, 432-436.

TIME SERIES ANALYSIS: Theory and Practice 4
O.D. Anderson (editor)

ORDER SELECTION FOR AUTOREGRESSION WITH APPLICATION TO X-RAY PHOTOELECTRON SPECTROSCOPY DATA

William C. Torrez

Department of Statistics
University of California
Riverside, California 92521
U.S.A.

An observed process X(t) is assumed to be the result of an input positive real signal S(t) degraded by broadening and the superposition of white noise. The broadening of S(t) is modeled by convolution so that we may write X(t) = D*S(t) + N(t), where D(t) is the broadening function (which is assumed known) and N(t) represents white noise. An empirical study is carried out of the autoregressive order selection problem for the autocovariance function $\hat{S}(f)$ for the modeling of X-ray photoelectron spectroscopy (XPS) data. Monte Carlo simulations show that maximum likelihood estimation of the order is inconsistent and, in fact, overestimates the true order of the model if certain broadening components are removed. Alternative estimation procedures based on information theoretic methods are discussed.

1. INTRODUCTION

An observed process X(t) is assumed to be the result of a positive real input signal S(t) degraded by "broadening" and the superposition of white noise. Degradation of S(t) by broadening is modeled by convolution so that we may write

$$X(t) = D * S(t) + N(t) \tag{1.1}$$

where D is the broadening function and N(t) represents white noise. Let '^' denote Fourier transform. The signal S(t) (defined on values in the energy domain) and its Fourier transform $\hat{S}(f)$ (defined on values in the frequency domain) can be considered as a spectral density and its autocovariance function, respectively. This correspondence is valid since S(t) is real and positive and was first applied to a model like (1.1) in Clayton and Ulrych (1977). When only a sample of the autocovariance function is known, it has been shown (cf. van den Bos 1971; Torrez, 1980) that extension of the autocovariance function by maximizing the entropy of a stationary random process whose power spectral density corresponds to the desired extracted signal S(t) is equivalent to the (least squares) fitting of an autoregressive model to the known autocovariance sequence.

Taking the Fourier transform of Equation (1.1) gives $\hat{X}(f) = \hat{D} \cdot \hat{S}(f) + \hat{N}(f)$.

If we wish to recover the signal S(t), we can reasonably attempt Fourier division by D ($\neq 0$) and write

$$\hat{X}/\hat{D} = \hat{S} + (\hat{N}/\hat{D}). \tag{1.2}$$

Now the "deconvolved" autocovariance function $\hat{S}(f)$ exhibits reasonable behavior in the low-frequency regions (where D has significant spectral energy) but for higher frequencies f, this function gives little useful information due to the noise enhancement caused by spectral division by D. By fitting an autoregressive model of order p to $\hat{S}$ and using the Levinson-Durbin recursion procedure to solve the Yule-Walker equations for the calculation of the estimates of the autoregressive coefficients, we may hope to extract the signal S(t) as described above. The choice of the order p remains a problem. In this article, we consider this order selection problem and carry out an empirical study of an order selection criterion proposed by Vasquez et al (1981) for the resolution of data arising in the chemical analysis of silicon-dioxide interface phenomena by the use of X-ray photoelectron spectroscopy (XPS). Monte Carlo simulations show that this criterion overestimates the model order p and is especially sensitive to the broadening component D(t). Alternative procedures based on information theoretic methods are also discussed.

Early work on the analysis of time series with superposed error was done by Walker (1960) and Hannan (1963). However, they were not concerned with the model order selection problem. Parzen (1967) examined the problem of fitting a model of signal plus white noise to an observed discrete parameter time series X(t). Later Pagano (1974) considered the same model fitting problem where the signal is an AR process of known order q and, using Whittle's (1963) representation of such processes as ARMA models, he was able to estimate the parameters of the process in a strongly consistent and efficient way. Tong (1975) using this ARMA representation of AR signal plus white noise uses Akaike's (1973) AIC procedure to select model order. The work of Akaike (1974) and Graupe et al (1975) considered the suitability of applying the AIC procedure to the ARMA model but no results were given for asymptotic optimality. Assuming an infinite order AR process, Shibata (1980) derives an asymptotic lower bound for the mean square error of prediction when the order of the model is selected from the data. Ulrych and Clayton (1976) noted that, for short realizations of the observed process, the AIC criterion is difficult to apply.

Referring to model (1.1), we assume that the broadening function D(t) is known. Clayton and Ulrych (1977) modeled the signal S(t) as a sum of phase shifted Dirac impulses

$$S(t) = \sum_{j=1}^{n} A_j \, \delta(t-t_j) \qquad (1.3)$$

where A_j is the amplitude of the jth impulse and t_j is the offset in t of the signal S(t). They noted that the AIC criterion is not directly applicable to purely harmonic sequences of the type expressed in equation (1.3) and they reported that this criterion generally underestimates the true AR model order. However, they do not propose any method of AR model order selection.

2. THE XPS DATA MODEL

The data y generated from XPS experiments are reasonably modeled (Clayton and Ulrych, 1977; Vasquez et al, 1981) as the convolution of the desired output S with a disturbing (or broadening) function D, plus the addition of white noise, i.e., $X = D * S + N$. This is the model (1.1): where S represents the chemical distribution function of the various chemical states within the sampled volume; D, the broadening function, is composed of the convolution of Lorentzian and Gaussian components (i.e. terms which have spectral density functional forms $\exp(-|x|)$ and $\exp(-x^2)$, respectively) representing various degrading characteristics due to instrument broadening, atomic line splitting, and solid state specific broadening phenomena; and N represents white noise due to the accumulation process of electron arrivals in the spectrometer.

Write $D = g * h$, where h represents a Voight profile (the convolution of instrument broadening with Lorentzian and Gaussian components) and g is a natural line-shape function due to a spin-orbit doublet of well-defined splitting and intensity ratio.

For XPS spectra (and generally for any experimental spectra obtained by photo emission techniques) the signal dominates the low frequency part of the spectrum, while the noise lies primarily in the high frequency region. In many cases, for example in the study of the electronic structure of oxide/gallium arsenide and silicon dioxide/silicon interfaces (cf. Grunthaner et al, 1979), much chemical information is contained in the high frequency ranges, so that a central problem in the resolution enhancement of spectra of such samples is to extract the chemical information S in the high frequency regions as accurately as possible from the spectrometer output X.

A standard procedure for this model is first to "deconvolve" from X certain broadening characteristics, e.g., we can estimate h * S by the inverse Fourier transform of $\hat{X}/\hat{g}$. (Of course, due to noise enhancement, we perform this division only over the frequency region where $\hat{S}(k)$ has significant spectral energy). By further such "deconvolution," we can remove all or some of the Voight component h. In practice some of the Lorentzian broadening is retained to enhance numerical stability of the computational algorithms. We will examine this point later.

There are a variety of linear methods which we may utilize to restore enhancement, but they suffer in their application to data sets with additive noise effects such as the XPS model, and in many cases are unable to resolve closely situated peaks. Linear methods of spectral resolution (and spectral density estimation) such as truncation and Fourier filtering simply do not take into consideration the possibility of periodicities in the autocovariances of the desired function S. When smoothing by simple truncation is performed, the line-shapes are resolved somewhat, but annoying Gibbs oscillations result; while filtering the Fourier transform with a roll-off function, like sin x/x, may fail to resolve the closely neighboring peaks.

The periodicity evident in the low frequency terms of the autocovariances suggests that these low frequency terms are likely to give us valuable information about the function S even at the high frequency ranges. Thus extrapolation to high frequencies is feasible.

3. ORDER SELECTION AND XPS DATA

As mentioned in Section 1, in the model (1.1) and its deconvolutional form (1.2), we consider the function S and its Fourier transform $\hat{S}$ as a spectral density and its autocovariance function, respectively. The covariance matrix generated by the empirical covariance function $\hat{S}$ will, in general, not be positive definite due to the effect of the noise term $\hat{N}/\hat{D}$. The Yule-Walker equations for $\hat{S}$ become

$$\begin{aligned}
\pi_{N0}\,\hat{S}(0) + \pi_{N1}\,\hat{S}(1) + \dots + \pi_{NN}\,\hat{S}(N) &= P_N \\
\pi_{N0}\,\hat{S}(1) + \pi_{N1}\,\hat{S}(0) + \dots + \pi_{NN}\,\hat{S}(N-1) &= 0 \\
\vdots \qquad\qquad & \quad \vdots \\
\pi_{N0}\,\hat{S}(N) + \pi_{N1}\,\hat{S}(N-1) + \dots + \pi_{NN}\,\hat{S}(0) &= 0,
\end{aligned}$$

or in matrix notation

$$\Gamma_N \,\underset{\sim}{\pi}_N = \underset{\sim}{P}_N \tag{3.1}$$

where Γ_N is the (N+1)x(N+1) covariance matrix consisting of the N+1 observed $\hat{S}(k)$'s, k=0,1,...,N, and $\underset{\sim}{\pi}_N=(\pi_{N0},\dots,\pi_{NN})'$ is the (N+1)-point prediction error filter. Note that $\pi_{N0} = 1$.

Here $\underset{\sim}{P}_N = (P_N, 0, \ldots, 0)'$ where the real valued quantity P_N is known as the prediction error power (PEP). The extrapolation of $\hat{S}(k)$, $k>N$, is then accomplished by the application of the Levinson-Durbin recursive procedure. Now a necessary and sufficient condition for the covariance matrix Γ_N to be positive definite is that the PEP sequence up to and including P_N be positive (cf. Smylie et al, 1973). Note that the PEP sequence should decrease monotonically to a positive value as the order N increases, and then flatten out (cf. Makhoul, 1975). However, due to the effects of noise, what is actually observed is a monotonic decrease in the value of the PEP, a partial flattening, then an accelerated rate of decline until the PEP goes negative (see Figure 2).

It is of importance here to mention the standard methods of order selection.

(1) Final Prediction-Error (FPE) Criterion: (Akaike, 1969, 1974). This criterion estimates the mean-square error in prediction expected when a predictive filter, calculated from one realization of the process, is applied to another independent realization of the process. For a filter of order M, the FPE is defined by, in the case of a zero mean process,

$$\mathrm{FPE}(M) = \frac{N+M+1}{N-M-1} P_M$$

where P_M is the output error power of the filter. Since P_M decreases as M increases, while the scale factor increases, the FPE(M) will have a minimum at some value $M=M_o$. This value defines the optimal value for the order of the prediction error filter.

(2) Information Theoretic Criterion (AIC). This method, developed by Akaike (1974), is based on the minimization of the log-likelihood of the prediction-error variance as a function of the filter order M. The AIC is defined by

$$\mathrm{AIC}(M) = \ell n(P_M) + \frac{2M}{N}.$$

Here AIC(M) may be calculated at each recursion of the filter design procedure, and M_o is the value for which the AIC(M), expressed as a function of M is minimized. It can be shown (Ulrych and Clayton, 1976) that FPE(M) and AIC(M) are asymptotically equivalent, i.e.

$$\lim_{N\to\infty} \left\{\ell n \; [\mathrm{FPE}(M)]\right\} = \mathrm{AIC}(M).$$

(3) Autoregressive Transfer Function Criterion (CAT). According to this criterion (proposed by Parzen, 1974), the optimal filter order, M_o, is obtained when the estimate of the difference in the mean-square errors between the true filter, which exactly gives the prediction error, and the estimated filter, is minimum. Parzen has shown that this difference can be calculated, without explicitly knowing the exact filter, by using the formula

$$\mathrm{CAT}(M) = \frac{1}{N} \sum_{m=1}^{M} \frac{N-m}{NP_m} - \frac{N-M}{NP_M}$$

where P_m, $1 \leq m \leq M$, is the output error power of the filter at the mth iteration.

Any (or all) of the criteria described above may be calculated in each iteration of the algorithm used to design the prediction-error filter. The value thus obtained for the pertinent objective criterion is compared with the previously calculated value. The iteration terminates when the new value is larger than the previous one, and the optimal filter order is thereby attained. The application of any of the above criteria yields a value M_o which is, in most cases, the best compromise between the optimal precision and minimum bias of the spectral estimate. It has been demonstrated (Landers and Lacoss, 1977) that for one period of a complex exponential signal, for example, all three criteria give the same value for M_o.

Ulrych and Clayton (1976) and Clayton and Ulrych (1977) discuss the applicability of the FPE and AIC criteria to sinusodial processes, of which XPS data is an example (Vasquez et al, 1981). They concluded after careful study that these criteria are not directly applicable to purely haromonic sequences; and, in fact, for data with sharp spectral lines, these criteria exhibit no clear minimum. Vasquez et al (1981) have developed an alternative criterion for XPS data. This criterion chooses the optimal value to be the order M corresponding to the maximum positive deviation $P_M - P_{M+1}$. This criterion is a variation of the maximum likelihood procedure where M is chosen as the largest order for which $1 - P_{M+1}/P_M < \delta$, for some prescribed value δ. This method is described in some detail in e.g., Anderson (1971) and Wold (1938).

Our interest in this procedure was to check its sensitivity to the deconvolution of broadening components. We used a test spectrum (Figure 1(A)) consisting of a

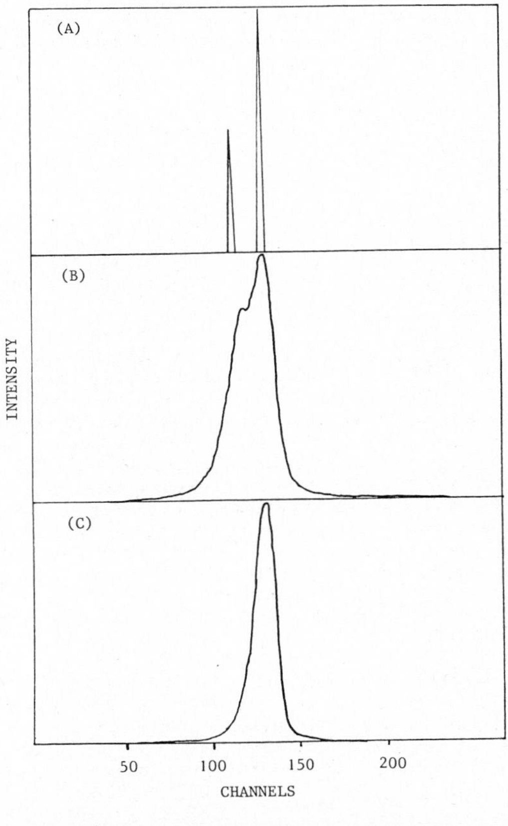

FIGURE 1. THE DATA MODEL: (A) THE CHEMICAL DISTRIBUTION FUNCTION (CDF); (B) THE CDF BROADENED WITH A VOIGHT PROFILE WITH NOISE ADDED TO SIMULATE SPECTROMETER OUTPUT; (C) THE FUNCTION TO BE DECONVOLVED.

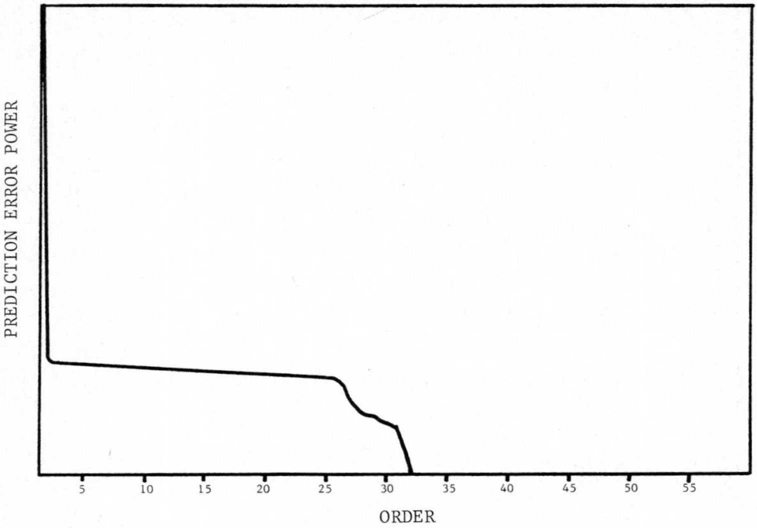

FIGURE 2. PLOT OF THE PREDICTION ERROR POWER (PEP) FOR THE DATA MODEL IN FIGURE 1(B).

line shape of intensities 0.5 and 1.0 at channels 113 and 128, respectively. This line shape was broadened by a Voight profile composed of a Gaussian spectrum of 12 channels FWHM (full width at half maximum) convolved with a Lorenztian spectrum of 6 channels FWHM. White noise (signal to noise ratio of 1000:1) was superimposed to obtain the test data (Figure 1(B)). A component consisting of a Gaussian spectrum of 12 channels FWHM convolved with a Lorenztian spectrum of 3 channels FWHM was deconvolved from the broadened signal (Figure 1(C)). Note that the natural lineshape of deconvolution results is Lorenztian, thus extrapolation by fitting an AR model is best suited to a Lorenztian spectrum S. It is for this reason that we choose not to deconvolve all of the Lorenztian component from the Voight broadening. Using this particular test spectrum, we performed 100 simulations of the order estimator described above. The simulation was performed on a Varian computer (series 72) at the Jet Propulsion Laboratory's electron spectroscopy laboratory for chemical analysis. A typical PEP sequence for the test spectrum is shown in Figure 2. The results of the simulation gave a mean optimal order of 28.21 with a standard deviation of 2.07. Figure 3 shows the extracted signal using order 28 and one notes the somewhat poorer resolution as compared with the extracted signal using order 22 (Figure 4). This overestimation was evident in numerous other test cases with signal to noise ratios varying from 10:1 to 10,000:1 (see Table 1).

If, however, a Gausian spectrum of 12 channels FWHM convolved with a Lorenztian spectrum of 5 channels FWHM was deconvolved from the broadened signal, then the resultant extracted signal resolved the original line-shape extremely well (Figure 5). This held true in repeated applications. As we mentioned above, the numerical stability of the Levinson-Durbin procedure is enhanced when some Lorenztian broadening components are retained and not deconvolved. These results show, however, that if too much Lorenztian broadening is retained, then the order selection procedure will definitely overestimate the optimal order.

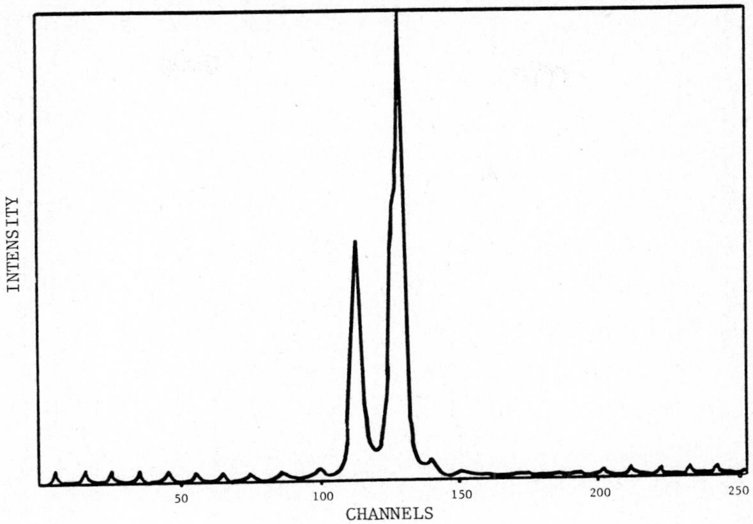

FIGURE 3. SIGNAL EXTRACTED SPECTROMETER IN FIGURE 1(B) USING ORDER = 28.

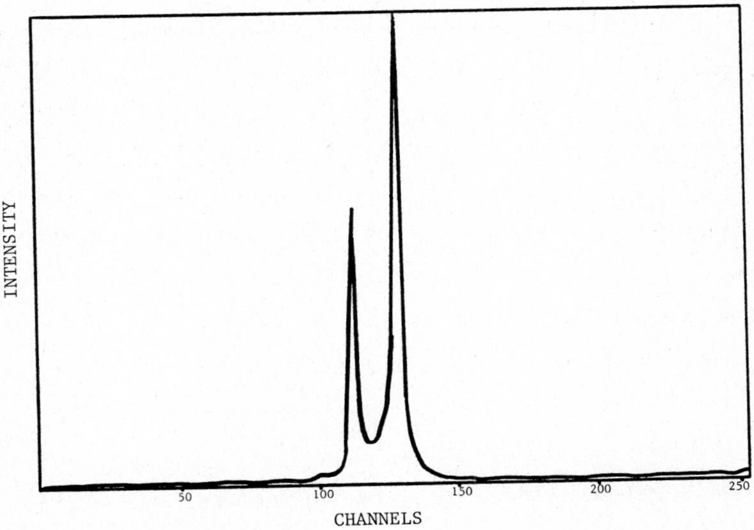

FIGURE 4. SIGNAL EXTRACTED FROM SPECTROMETER OUTPUT IN FIGURE 1(B) USING ORDER = 22.

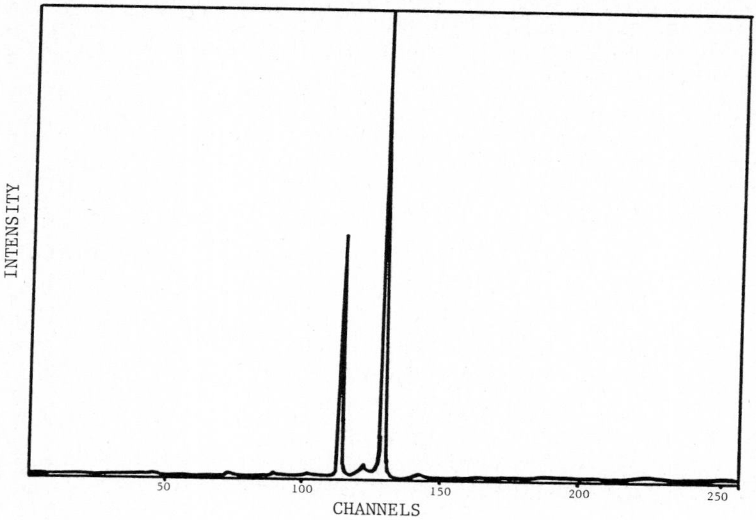

FIGURE 5. SIGNAL EXTRACTED FROM SPECTROMETER OUTPUT DEPICTED IN FIGURE 1(B) WITH DECONVOLUTION OF 5 CHANNELS FWHM AND ORDER = 21.

NUMBER OF SIMULATIONS	SIGNAL TO NOISE RATIO	MEAN OPTIMAL ORDER	STANDARD DEVIATION
500	10:1	16.13	2.95
100	100:1	20.35	2.74
100	1000:1	28.21	2.07
500	10,000:1	39.70	2.05

FIGURE 6. MEAN AND STANDARD DEVIATION OF OPTIMAL ORDER ESTIMATE FOR VARYING SIGNAL TO NOISE RATIO.

4. ALTERNATIVE PROCEDURES

Katz (1981) presents results which show that the AIC procedure has a substantial probability of overestimating the true order of a Markov chain. This inconsistency of the AIC procedure has previously been established in the case of estimating the order of AR processes by Shibata (1976). Katz considered the question as to whether the AIC estimator of a Markov chain's order can be modified to produce a consistent estimator. He defined a Bayesian information criterion (BIC) (also considered previously by Akaike, 1978), and under similar assumptions to those of Schwarz (1978), Katz extends the asymptotic optimality of the BIC estimator to the Markov chain case. Under more restrictive conditions, the BIC procedure is not only consistent, but also asymptotically optimal in the sense of minimizing the expected loss. The derivation of this approximate Bayesian criterion, for selecting a Markov chain's order, formally requires the specification of a prior distribution.

Let $p_k>0$, $k=0,1,\ldots,N$, be the prior probability that the AR model is of kth order, and let μ_k be the conditional prior distribution of the AR vector parameter $\underline{a}_k$ in

(3.1), given that the model is of order k. For simplicity, we take μ_k to be a diffuse prior for the vector π_k, i.e., each vector π_k is assumed to have an independent Dirichlet distribution with all parameters unity. If a fixed penalty is assumed for choosing the wrong model, the Bayes procedure consists of selecting the model whose posterior probability, say $BP_k(X(1),...,X(N))$, is greatest. However these procedures are not without some drawbacks. Atkinson (1980) states that the Bayesian procedure overlooks a "philosophical" problem, stating that the "consistent" procedures of Schwarz (1978) and Katz (1981), pick only the simplest of the true models. In particular, it is not clear that the true model will remain constant as the number of observations increases.

REFERENCES

AKAIKE, H. (1969). Fitting autoregressive models for prediction. Ann. Inst. Statist. Math. 21, 243-247.

AKAIKE, H. (1973). Information theory and an extension of the maximum likelihood principle. In Second International Symposium on Information Theory. Eds: B. N. Petrov and F. Csaki, Akademia Kiado, Budapest, 267-281.

AKAIKE, H. (1974). A new look at the statistical model identification. IEEE Trans. Autom. Cont., AC-19, 716-723.

AKAIKE, H. (1978). A Bayesian analysis of the minimum AIC procedure. Ann. Inst. Statist. Math. 30A, 9-14.

ANDERSON, T. W. (1971). The Statistical Analysis of Time Series. Wiley, New York.

ATKINSON, A. C. (1980). A note on the generalized information criterion for choice of a model. Biometrika 67, 413-418.

CLAYTON, R. W. and ULRYCH, T. J. (1977). A restorative method for impulsive functions. IEEE Trans. Inform. Theory IT-23, 262-264.

GRAUPE, D., KRAUSE, D. J., and MOORE, J. B. (1975). Identification of autoregressive moving-average parameters of time series. IEEE Trans. Autom. Cont. AC-20, 104-107.

GRUNTHANER, F. J., GRUNTHANER, P. J. VASQUEZ, R. P., LEWIS, B. F., MASERJIAN, J., and MADHUKAR, A. (1979). Local atomic and electronic structure of oxide/GaAs and SiO_2/Si interfaces using high resolution XPS. J. Vac. Sci. Technol. 16(5), 1443-1454.

HANNAN, E. J. (1963). Regression for time series with errors of measurement. Biometrika 50, 293-302.

KATZ, R. W. (1981). On some criteria for estimating the order of a Markov chain. Technometrics 23, 243-249.

LANDERS, T. E., and LACOSS, R. T. (1977). IEEE Trans., CE-15(1), 26-32.

MAKHOUL, J. (1975). Linear prediction: a tutorial review. Proc. IEEE 63, 561-580.

PAGANO, M. (1974). Estimation of models of autoregressive signal plus white noise. Ann. Statist. 21, 274-281.

PARZEN, E. (1967). Time series analysis for models of signal plus white noise. In Advanced Seminar on Spectral Analysis of Time Series. Ed: B. Harris, Wiley, New York.

PARZEN, E. (1974). Some recent advances in time series modeling. IEEE Trans. Autom. Cont., AC-19, 723-730.

SCHWARZ, G. (1978). Estimating the dimension of a model. Ann. Statist. 6, 461-464.

SHIBATA, R. (1976). Selection of the order of an autoregressive model by Akaike's information criterion. Biometrika 63, 117-126.

SHIBATA, R. (1980). Asymptotically efficient selection of the order of the model for estimating parameters of a linear process. Ann. Statist., 8, 147-164.

SMYLIE, O. E., CLARKE, G. K. C., and ULRYCH, T. J. (1973). Analysis of irregularities in the earth's rotation. In Methods in Computational Physics, 13, 391-430, Academic Press, New York.

TONG, H. (1975). Autoregressive model fitting with noisy data by Akaike's information criterion. IEEE Trans. Inform. Theory IT-21, 476-480.

TORREZ, W. (1980). Autoregressive spectral estimation: the maximum entropy method. Univ. of Calif. Riverside, Dept. of Stat. Tech. Report No. 69.

ULRYCH, T. J. and BISHOP, T. N. (1975). Maximum entropy spectral analysis and autoregressive decomposition. Rev. Geophys. 13, 183-200.

ULRYCH, T. J. and CLAYTON, R. W. (1976). Time series modeling and maximum entropy. Phys. Earth Planetary Interiors 12, 188-200.

VAN DEN BOS, A. (1971). Alternative interpretation of maximum entropy spectral analysis. IEEE Trans. Inf. Th., IT-17, 493-494.

VASQUEZ, R. P., KLEIN, J. D., BARTON, J. J., and GRUNTHANER, F. J. (1981). Application of maximum entropy spectral estimation to deconvolution of XPS data, J. Elect. Spectros. Rel. Phen. 23, 63-81.

WALKER, A. M. (1960). Some consequences of superimposed error in time series analysis. Biometrika 47, 33-43.

WHITTLE, P. (1963). Prediction and Regulation. Van Nostrand, Princeton.

WOLD, H. (1938). A Study in the Analysis of Stationary Time Series. Almqvist and Wiksell, Stockholm. Second ed. with Appendix by P. Whittle, 1954.

TIME SERIES ANALYSIS: Theory and Practice 4
O.D. Anderson (editor)

SUBSET TRANSFER FUNCTION MODEL FITTING

O.B. Oyetunji

Department of Statistics
University of Ibadan
Ibadan, Nigeria.

In this paper, a Transfer Function Model of order k,m, TF(k,m) is defined as a linear regression equation between $\{Y_t, Y_{t-1}, \ldots, Y_{t-k}\}$ and $\{X_{t-1}, X_{t-2}, \ldots, X_{t-m}\}$, where $\{Y_t\}$ and $\{X_t\}$ are assumed to be two observable, stationary time series. A method of fitting subset TF models, that is where some coefficients are constrained to zero, is given. Using some animal population data, the usefulness of TF models for investigating causal relationship is indicated. It is also shown that when subset TF models are fitted, these are more parsimonious than models obtained by fitting each series separately. An application of subset TF modelling for a nonlinear series is also given.

1. INTRODUCTION

In classical statistics, external analysis of given data is well known. By external analysis, we mean that we have a variable of interest, Y say, and we search for some other (explanatory) variables, $\{X_1, X_2, \ldots, X_k\}$ say, to explain its variation. The most commonly used form of external analysis is simple linear regression, when we have one explanatory variable X, and fit the linear equation

$$Y = aX + e \tag{1.1}$$

where $e \sim N(0,\sigma^2)$. If there is more than one explanatory variable, the resulting equation is referred to as multiple linear regression.

It is therefore reasonable to consider a time series model where a stationary process $\{Y_t\}$ does not only depend on its past history but also on the past history of a number of other stationary processes. The study of simultaneous dependence between more than one process is referred to as multiple time series analysis. The basic theoretical properties of multiple time series are well established (Quenouille, 1957 and Hannan, 1970) although, until recently, there were few published applications. This was probably due to computational difficulties inherent in such models.

Suppose we have two stochastic processes, $\{X_t\}$ and $\{Y_t\}$; a transfer function model of order (k,m), abbreviated to TF(k,m), is defined by

$$(1 - \delta_1 B - \ldots - \delta_k B^k)Y_t = (\omega_o + \omega_1 B + \ldots + \omega_m B^m)X_{t-b} + N_t \tag{1.2}$$

where B is the backshift operator, b is the delay in the response of Y to a change in X, and N_t (the error process) is assumed to follow an ARMA process. The identification and fitting of a TF model defined by (1.2) is discussed in Box and Jenkins (1970).

Transfer function models could be useful in investigating series with causal relationship, especially in ecology where the prey-predator phenomenon is well known. Some authors, Bulmer (1974), Jenkins (1975), Chan and Wallis (1978), Tong and Lim (1980) and Jenkins and Alavi (1981), have studied such interaction in animal populations. The TF model could also be useful in economic situations where there is a direct effect, such as milk/coffee prices and more recently price index/oil prices.

In this paper, we consider a particular class of TF models in which the error process is assumed to be white noise, and define a TF(k,m) as

$$Y_t - \delta_1 Y_{t-1} - \dots - \delta_k Y_{t-k} = \omega_1 X_{t-1} + \omega_2 X_{t-2} + \dots + \omega_m X_{t-m} + \varepsilon_t \quad (1.3)$$

where $\{\varepsilon_t\}$ is assumed to be a white noise process, with variance σ^2, and $\{X_t\}$ and $\{Y_t\}$ to be marginally and jointly stationary. This model is analogous to the simple linear regression model (1.1). As in linear regression, an extension whereby more than one explanatory process, say $\{X_{1t}, X_{2t}, \dots, X_{pt}\}$, is used, is straightforward.

With k + m parameters in the model, we are likely to have more parameters than in the corresponding univariate autoregressive models. Therefore, to obtain a parsimonious model, we need to reduce the number of parameters in the TF model. Hence our interest is in fitting subset TF models. A subset TF(k,m) is defined as a TF(k,m) in which some δ's and some ω's are constrained to zero such that $\delta_k \neq 0$ and $\omega_m \neq 0$. A subset model that contains p non-zero coefficients is referred to as a p-variate model.

2. SUBSET TRANSFER FUNCTION MODEL SELECTION

We shall assume that $\{Y_t\}$ and $\{X_t\}$ are zero-mean stationary processes with respective autocovariance functions, $\gamma_1(r)$ and $\gamma_2(r)$, and cross-covariance function $\gamma_{12}(r)$. We rewrite the TF(k,m) model (1.3) as

$$Y_t = \delta_1 Y_{t-1} + \delta_2 Y_{t-2} + \dots + \delta_k Y_{t-k} + \omega_1 X_{t-1} + \dots + \omega_m X_{t-m} + \varepsilon_t \,. \quad (2.1)$$

Multiplying (2.1) by Y_{t-r} and taking expectations, for r = 1,2,...,k, and also by X_{t-r} and taking expectations, for r = 1,2,...,m, we obtain a set of k + m linear simultaneous equations which, using the relationship $\gamma_{12}(-r) = \gamma_{21}(r)$, can be written in matrix form as

$$\underline{\gamma} = \underline{\Sigma}\ \underline{\theta} \quad (2.2)$$

where $\underline{\gamma} = \begin{bmatrix} \underline{\gamma}_1 \\ \hline \underline{\gamma}_{21} \end{bmatrix}$, $\underline{\Sigma} = \begin{bmatrix} \underline{\Sigma}_{11} & \underline{\Sigma}_{12} \\ \underline{\Sigma}_{12}^T & \underline{\Sigma}_{22} \end{bmatrix}$, $\underline{\theta} = \begin{bmatrix} \underline{\delta} \\ \hline \underline{\omega} \end{bmatrix}$,

$$\underline{\gamma}_1 = [\gamma_1(1)\ \gamma_1(2)\ \ldots\ \gamma_1(k)]^T \ , \quad \underline{\gamma}_{21} = [\gamma_{21}(1)\ \gamma_{21}(2)\ \ldots\ \gamma_{21}(m)]^T \ ,$$

$$\underline{\delta} = [\delta_1\ \delta_2\ \ldots\ \delta_k]^T \ , \quad \underline{\omega} = [\omega_1\ \omega_2\ \ldots\ \omega_m]^T .$$

$\underline{\Sigma}_{11}$ is the (k × k) autocovariance matrix of $\{Y_t\}$, $\underline{\Sigma}_{22}$ the (m × m) autocovariance matrix of $\{X_t\}$ and $\underline{\Sigma}_{12}$ is the (k × m) covariance matrix of $\{Y_t, X_t\}$. Let $c_i(r)$ and $c_{ij}(r)$ denote the sample estimates of $\gamma_i(r)$ and $\gamma_{ij}(r)$ respectively, i = 1,2, j = 1,2. Substituting these sample estimates in (2.2), we obtain sample estimates of $\underline{\gamma}$ and $\underline{\Sigma}$, say $\underline{c}$ and $\underline{S}$ respectively. $\underline{S}$ is a positive definite symmetric matrix and we obtain estimates of the coefficients from

$$\underline{\theta} = \underline{S}^{-1}\ \underline{c} \tag{2.3}$$

where $\hat{\underline{\theta}} = \begin{bmatrix} \hat{\underline{\delta}} \\ \hline \hat{\underline{\omega}} \end{bmatrix}$.

With the advent of criteria which avoid hypothesis testing in the selection of the optimum model, subset selection has become a practical proposition. McClave (1975) has given a method of fitting subset autoregressive models based on Hocking and Leslie's (1967) algorithm. Suppose there are N observations on a linear multiple regression given by

$$Y_i = a_1X_{1i} + a_2X_{2i} + \ldots + a_kX_{ki} + e_i \qquad (i = 1,2,\ldots,N) \tag{2.4}$$

where $e_i \sim N(0,\sigma^2)$ for all i, and $\text{cov}(e_i,e_j) = 0$ for all $i \neq j$. (2.4) may be written in matrix form as $\underline{Y} = \underline{X}\ \underline{a} + \underline{e}$. Hocking and Leslie have given an algorithm which identifies the p-variate model ($p \leq k$) with minimum residual variance without necessarily having to evaluate all the possible $2^k - 1$ subsets. Note that the autoregressive model is a form of linear multiple regression, except that the X's are no longer fixed. However, it is well known that, asymptotically, results obtained by assuming the X's to be deterministic can be extended to time series models. Therefore results obtained for multiple regression may be extended to time series model fitting.

Furnival (1971), describing a method of evaluating all possible $2^k - 1$ subset regressions, has given an algorithm for including and excluding variables in a multiple linear regression. By applying matrix operations to an augmented matrix, $\begin{bmatrix} \underline{A} & \underline{d} \\ \underline{d}^T & s \end{bmatrix}$, where $\underline{A} = \underline{X}^T\underline{X}$, $\underline{d} = \underline{X}^T\underline{Y}$ and $s = \underline{Y}^T\underline{Y}$, estimates for the regression coefficients and residual variance of the corresponding regression model are obtained. Furnival pointed out that the application of Gaussian elimination to the augmented matrix has the advantage of saving a great deal of computer time, although it yields only the maximum likelihood estimate of the residual variance and not the coefficients in the model. This is not a disadvantage since coefficients for any model of interest may easily be recomputed, as is seen below.

In view of the similarity between models (2.1) and (2.4), it is proposed to extend Furnival's algorithm to the case of subset TF model fitting. Suppose (k,m) is the maximum order to be fitted. Then

$$\underline{P} = \left[\begin{array}{c|c} \underline{S} & \underline{c} \\ \hline \underline{c}^T & c_1(0) \end{array}\right] \tag{2.5}$$

has exactly the same algebraic structure as the augmented matrix associated with multiple regression. Applying Gaussian elimination to $\underline{P}$, on any subset p ($p \leq k + m$) of its first $(k + m)$ pivotal elements, will yield the approximate maximum likelihood estimate of the residual variance after fitting the corresponding p-variate subset TF model.

Since we are comparing models that differ appreciably in numbers of parameters, we define the best subset model as that with minimum BIC, where BIC for a p-variate model is as defined by Akaike (1977),

$$BIC(p) = (N-p)\ \ln \tilde{\sigma}_p^2 - (N-p)\ \ln\left(1 - \frac{p}{N}\right) + p\ \ln\left(\frac{\tilde{\sigma}_o^2 - \tilde{\sigma}_p^2}{p}\right) + p\ \ln N$$

where $\tilde{\sigma}_o^2$, the data variance, is the mean-corrected sum of squares divided by N (the number of obervations), and $\tilde{\sigma}_p^2$ is the approximate maximum likelihood estimate of the residual variance after fitting a p-variate model. The use of BIC as a criterion for subset selection is discussed in Oyetunji (1979).

For example, let $k = 1$ and $m = 1$. The full order TF(1,1) is given by $Y_t = \delta_1 T_{t-1} + \omega_1 X_{t-1} + \varepsilon_t$, and the corresponding augmented matrix is

$$\underline{P} = \left[\begin{array}{cc|c} c_1(0) & c_{12}(0) & c_1(1) \\ c_{12}(0) & c_2(0) & c_{21}(1) \\ \hline c_1(1) & c_{21}(1) & c_1(0) \end{array}\right] .$$

There are $2^{1+1} - 1 = 3$ possible subsets. Using Furnival's algorithm to decide the order in which lags are included in the model, the first step corresponds to fitting $Y_t = \delta_1 Y_{t-1} + \varepsilon_t$, the second $Y_t = \omega_1 X_{t-1} + \varepsilon_t$, while the third $Y_t = \delta_1 Y_{t-1} + \omega_1 X_{t-1} + \varepsilon_t$.

At the second step, Gaussian elimination is applied to $\underline{P}$ using the second diagonal element as the pivot. This will yield the approximate maximum likelihood estimate of the residual variance, after fitting $Y_t = \omega_1 X_{t-1} + \varepsilon_t$, as follows:

$$\underline{P}^{(2)} = \left[\begin{array}{cc|c} c_1(0) - \dfrac{c_{12}^2(0)}{c_2(0)} & 0 & c_1(1) - \dfrac{c_{12}(0)c_{21}(1)}{c_2(0)} \\ c_{12}(0) & c_2(0) & c_{21}(1) \\ \hline c_1(1) - \dfrac{c_{12}(0)c_{21}(1)}{c_2(0)} & 0 & c_1(0) - \dfrac{c_{21}^2(1)}{c_2(0)} \end{array}\right].$$

It can be verified that $P^{(2)}(3,3) = c_1(0) - c_{21}^2(1)/c_2(0)$ is the approximate maximum likelihood estimate of the residual variance after fitting $Y_t = \omega_1 X_{t-1} + \varepsilon_t$.

At the third step, Gaussian elimination is applied to $\underline{P}^{(2)}$, using the first diagonal element of $\underline{P}^{(2)}$ as the pivot. If the resulting matrix is denoted by $\underline{P}^{(3)}$, it can be verified that

$$P^{(3)}(3,3) = c_1(0) - \frac{c_1^2(1)c_2(0) - 2c_1(1)c_{21}(1)c_{12}(0) + c_{21}^2(1)c_1(0)}{c_1(0)c_2(0) - c_{12}^2(0)}$$

which is the approximate maximum likelihood estimate of the residual variance after fitting $Y_t = \delta_1 Y_{t-1} + \omega_1 X_{t-1} + \varepsilon_t$.

For a general order (k,m), applying Furnival's algorithm with Gaussian elimination to the matrix (2.5), the estimated residual variance is obtained for each of the $2^{k+1} - 1$ possible subset TF models. The BIC value is computed for each model, together with BIC(o) = N ln $\tilde{\sigma}_o^2$, where $\tilde{\sigma}_o^2 = c_1(0)$ is the data variance of $\{Y_t\}$. BIC(o) is the BIC value associated with the case where no time series model is fitted to the data. The models with smallest BIC values, together with the lags for coefficients contained in those models, are noted.

From this information, it is possible to obtain the approximate maximum likelihood estimates for the coefficients of the required models. For any particular model, setting to zero those rows and columns of $\underline{S}$ which correspond to lags not contained in the model (except for diagonal elements) yields $\underline{S}^*$. Setting to zero the elements of $\underline{c}$ corresponding to those lags not in the model gives a new vector $\underline{c}^*$. The approximate maximum likelihood estimates of the coefficients of this TF model are given by $\underline{\theta}^* = \underline{S}^{*-1}\, \underline{c}^*$, c.f. (2.3).

3. EXAMPLES

100 independent series of full order TF(3,3) each with 500 observations, given by $Z_t - 0.6Z_{t-1} - 0.79Z_{t-2} + 0.504Z_{t-3} = V_{t-1} - 0.32V_{t-2} - 0.38V_{t-3} + \varepsilon_t$ where $\{\varepsilon_t\}$ is white noise with variance $\sigma^2=1$, and $\{V_t\}$ is a zero-mean stationary process, were simulated. Maximum orders were set at k = 8 and m = 8. Applying subset TF model selection to the simulated data, the correct full order TF(3,3) was identified as the best model 96 times.

Also, 100 independent series, each with 500 observations, of the subset TF model given by $Z_t - 1.5Z_{t-2} + 0.56Z_{t-4} = 1.2V_{t-2} + 0.36V_{t-4} + \varepsilon_t$ were simulated, and the method, with maximum orders k = 8 and m = 8, was applied to the simulated data. The correct model was selected as the best model (best in the sense of minimum BIC) 62 times, as the second best model 14 times, and as the third best model 8 times. These results demonstrate the feasibility of the proposed approach.

As an application of TF models in investigating causal relationships, we look at the prey-predator phenomenon. The data source is the records of fur sales for the Hudson's Bay Company in North Canada. It is assumed that trapping effort is approximately constant each year, so that the number of animals caught is a constant proportion of the population. Animal populations analysed are those of Lynx, Hare, Mink and Muskrat. There are 90 observations on the Lynx-Hare populations (1844-1933), and 83 observations on Muskrat-Mink populations (1767-1849).

To stabilize the variances in the series, and to make the series stationary, some transformation of data was necessary. Therefore models for these animal populations are fitted to the following transformed series: $L_t = \log_{10}$ Lynx(t), $H_t = \log_{10}$ Hare(t), $MK_t = \ell n$ Muskrat(t) - ℓn Muskrat(t-1), and $M_t = \ell n$ Mink(t) - ℓn Mink(t-1); where, for example, Lynx(t) denotes the Lynx population at time t.

First, we fitted full order univariate autoregressive models to these series and obtained:

$$H_t - 0.66H_{t-1} - 0.003H_{t-2} + 0.05H_{t-3} + 0.24H_{t-4} = \varepsilon_t, \ \tilde{\sigma}^2 = 0.1217 \quad (3.1)$$

$$M_t + 0.5M_{t-1} + 0.27M_{t-2} = \varepsilon_t, \ \tilde{\sigma}^2 = 0.2388 \quad (3.2)$$

$$MK_t + 0.51MK_{t-1} + 0.32MK_{t-2} = \varepsilon_t, \ \tilde{\sigma}^2 = 0.4458. \quad (3.3)$$

For the Lynx data, we fitted AR(11), with $\tilde{\sigma}^2 = 0.0401$ and coefficients given in the following table:

1	2	3	4	5	6	7	8	9	10	11
-1.168	0.556	-0.259	0.315	-0.151	0.140	-0.042	0.020	-0.120	-0.201	0.326.

In fitting a full order TF model between H_t and L_t, we set the maximum orders as k = 6 and m = 12, and obtained a TF(1,10) with $\tilde{\sigma}^2 = 0.0898$. This seems to imply that the Hare population is heavily dependent on the Lynx population. Although the variance of this TF model is less than that of the AR(4) given by (3.1), it has many more parameters. However, in fitting a full order TF model between L_t and H_t, setting the maximum orders at k = 12 and m = 6, we obtained TF(9,3) with $\tilde{\sigma}^2 = 0.0434$, which has only one more parameter than the AR(11) fitted to the Lynx population, but a bigger residual variance. This seems to imply that the Hare is the prey of Lynx (because the TF(1,10) fitted between H_t and L_t suggests H_t is heavily dependent on L_t), which is reasonable.

To obtain a parsimonious TF model, we searched for the best (in the sense of minimum BIC) subset TF model between H_t and L_t. We set the maximum orders as k = 4 and m = 12 and obtained

$$H_t - 0.51H_{t-1} = 0.4L_{t-7} - 0.35L_{t-8} + 0.62L_{t-9} - 0.32L_{t-10} + \varepsilon_t, \quad \tilde{\sigma}^2 = 0.0914. \quad (3.4)$$

Comparing this with the AR(4) model (3.1), we see that there is a reduction in the residual variance and the two models have about the same number of parameters.

The results for the Muskrat-Mink interaction, with maximum orders as k = 6 and m = 6, are as follows:

$$M_t + 0.5M_{t-1} + 0.27M_{t-2} = 0.03MK_{t-1} + \varepsilon_t, \quad \tilde{\sigma}^2 = 0.2382 \quad (3.5)$$

$$MK_t + 0.44MK_{t-1} + 0.18MK_{t-2} = 0.25M_{t-1} + 0.39M_{t-2} + \varepsilon_t, \quad \tilde{\sigma}^2 = 0.4135. \quad (3.6)$$

The subset models obtained are the same as the full order models given by (3.5) and (3.6), respectively.

Comparing (3.5) with (3.2), we see that the dependence of Mink on Muskrat is negligible - there is little reduction in residual variance and the coefficient of MK_{t-1} in (3.5) is only 0.03. However, there is a reduction in residual variance for the TF(2,2), given by (3.6), fitted between Muskrat and Mink. These models therefore suggest that the Muskrat population is affected by Mink population, but not vice-versa.

Another application of TF modelling can be found in Haggan and Ozaki (1979) for amplitude dependent nonlinear random vibrations. The model, referred to as the exponential autoregressive model, is defined for the pth order [EAR(p)] by

$$X_t = (\phi_1 + \pi_1 e^{-\gamma X_{t-1}^2})X_{t-1} + \dots + (\phi_p + \pi_p e^{-\gamma X_{t-1}^2})X_{t-p} + \varepsilon_t \quad (3.7)$$

where ε_t is a white noise process with variance σ^2. The conditions for (3.7) to exhibit limit cycle behaviour are given as:

(i) the roots of $\lambda^p - \phi_1\lambda^{p-1} - \dots - \phi_p = 0$ lie inside the unit circle

(ii) the roots of $\lambda^p - (\phi_1 + \pi_1)\lambda^{p-1} - \dots - (\phi_p + \pi_p) = 0$ do not all lie inside the unit circle and

(iii) $(1 - \sum_{i=1}^{p} \phi_i)/\sum_{i=1}^{p} \pi_i \quad < 0 \text{ or } > 1$.

From (3.7) we see that for a given value of γ, we need to estimate two coefficients for each lag. That is for an EAR(p), we estimate 2p parameters. For example, using the criterion AIC, Akaike (1974), Haggan and Ozaki (1979) fitted an EAR(11) to the Lynx population, with $\hat{\gamma} = 3.89$, $\hat{\sigma}^2 = 0.0321$ and coefficients given by:

i	1	2	3	4	5	6	7	8	9	10	11
$\hat{\phi}_i$	1.09	-0.28	0.27	-0.45	0.41	-0.36	0.22	-0.10	0.22	-0.07	-0.38
$\hat{\pi}_i$	0.01	-0.49	-0.06	0.30	0.54	0.61	-0.53	0.30	-0.18	0.18	0.16

It was therefore necessary to estimate 22 parameters from 103 observations. Obviously, there are too many parameters and it is desirable to reduce them. (3.7) may be rewritten as

$$X_t - \phi_1 X_{t-1} - \ldots - \phi_p X_{t-p} = \pi_1 e^{-\gamma X_{t-1}^2} X_{t-1} + \ldots + \pi_p e^{-\gamma X_{t-1}^2} X_{t-p} + \varepsilon_t . \quad (3.8)$$

Hence, for a given value of γ, EAR(p) may be viewed as a TF(p,p). To identify γ, we need to perform a grid search. That is, for various values of γ, fit EAR; and $\hat{\gamma}$ is then that value which gives the EAR with smallest AIC value. Using this grid search, and fitting full order models for the Lynx data, $\hat{\gamma}$ was found to be 3.89. We then set $\gamma = 3.89$ in the search for the best subset model.

Applying the subset TF selection method to the model given by (3.8), we obtained the best (in the sense of minimum BIC) subset as

$$X_t = (1.04 + 0.03e^{-\hat{\gamma}X_{t-1}^2})X_{t-1} - (0.1 + 0.5e^{-\hat{\gamma}X_{t-1}^2})X_{t-2} - (0.08 + 0.06e^{-\hat{\gamma}X_{t-1}^2})X_{t-4}$$
$$+ (0.24 - 0.03e^{-\hat{\gamma}X_{t-1}^2})X_{t-9} - (0.36 + 0.29e^{-\hat{\gamma}X_{t-1}^2})X_{t-11} + \varepsilon_t$$

with $\hat{\gamma} = 3.89$ and $\hat{\sigma}^2 = 0.0341$. The moduli of the roots of the equations in conditions (i) and (ii) for limit cycle behaviour are given below.

	1	2	3	4	5	6	7	8	9	10	11
(i)	.924	.924	.841	.841	.998	.998	.897	.897	.793	.969	.969
(ii)	.821	.821	1.020*	1.020*	.879	.879	.719	.719	.760	.543	.610

Asterisks denote roots outside the unit circle satisfying (ii) and $(1 - \sum_{i=1}^{11} \hat{\phi}_i)/\sum_{i=1}^{11} \hat{\pi}_i = -0.963$.

Hence all conditions for limit cycle behaviour are satisfied. With 10 parameters in the subset model, we have achieved a great reduction in the number of parameters, and comparison with the AR(11) fitted to the Lynx data is more straightforward.

4. CONCLUSION

We have defined a special class of Transfer Function model as an example of multiple time series, and have shown that TF models can be used in investigating one-way dependence between two stationary processes.

Using full order TF models, in prey-predator relationships in animal populations, it can be determined which is the prey and which the predator. This is achieved by looking at the order of the fitted TF model. If m is much greater than k, this will suggest the series $\{Y_t\}$ is heavily dependent on $\{X_t\}$. This, in prey-predator relationships, will imply that X_t is the predator and Y_t the prey. From our results, the Muskrat-Mink relationship is such that Mink are predators of Muskrat, a conclusion also arrived at by Bulmer (1974) and Jenkins (1975). We also found that the Hare population is heavily dependent on the Lynx population, which indicates Lynx are predators of Hare. Hence, using TF models, prey-predator relationships can be determined and the order identified.

Another application is the representation of the EAR model as a TF model; which makes fitting EAR models easier, as EAR is nonlinear.

With the application of the subset TF model selection method a great reduction in number of parameters is achieved. Comparison of subset TF models with the corresponding univariate models is more straightforward, since they both have about the same number of parameters. However, the fitted subset TF model of Hare on Lynx gives a curious result. It indicates a delay of seven years before the Hare population responds to changes in the Lynx population.

REFERENCES

AKAIKE, H. (1974). A new look at statistical model identification. IEEE Transactions Automatic Control 19, 716-723.

AKAIKE, H. (1977). On entropy maximisation principle. Proceedings of Symposium on Application of Statistics. Dayton, Ohio; June 1976. Ed: P.R. Krishnaiah, Amsterdam: North-Holland.

BOX, G.E.P. and JENKINS, G.M. (1970). Time Series Analysis, Forecasting and Control, San Francisco: Holden-Day.

BULMER, M.G. (1974). A statistical analysis of the 10-year cycle in Canada. Journal of Animal Ecology 43, 707-718.

CHAN, W.Y.T. and WALLIS, K.F. (1978). Multiple time series modelling: another look at the Mink-Muskrat interaction. Applied Statistics 27, 1-48.

FURNIVAL, G.M. (1971). All possible regressions with less computation. Technometrics 13, 403-408.

HAGGAN, V. and OZAKI, T. (1978). Amplitude dependent AR model fitting for nonlinear random vibrations. In Time Series (Proceedings of the International Conference held at Nottingham University, March 1979). Ed: O.D. Anderson, North-Holland, Amsterdam and New York, 57-71.

HOCKING, R.R. and LESLIE, R.N. (1967). Selection of best subset in regression analysis. Technometrics 19, 531-540.

JENKINS, G.M. (1975). Interaction between Muskrat and Mink-cycles in North Canada. In Proceedings of 8th International Biometric Conference, Constanta, Romania, August 1974. Eds: L.C.A. Corsten and T. Postelnicu.

JENKINS, G.M. and ALAVI, A.S. (1981). Some aspects of modelling and forecasting multivariate time series. Journal of Time Series Analysis 2, 1-48.

McCLAVE, J. (1975). Subset autoregression, Technometrics 17, 213-218.

OYETUNJI, O.B. (1979). PhD thesis. Mathematics Department, University of Manchester, Institute of Science and Technology, UK.

QUENOUILLE, M.H. (1975). The Analysis of Multiple Time Series. London: Griffin.

TONG, H. and LIM, K.S. (1980). Threshold autoregression, limit cycles and cyclical data. Journal of Royal Statistical Society B 42, 425-292.

TIME SERIES ANALYSIS: Theory and Practice 4
O.D. Anderson (editor)

A TIME SERIES APPROACH TO THE PREDICTION OF OIL DISCOVERIES

Phillip A. Cartwright
and
Paul Newbold

University of Illinois
Department of Economics
Champaign, Illinois 61820
U.S.A.

A nonlinear, two state (outlier/no-outlier) forecasting model with time varying coefficients is presented and applied to North Sea oil discovery data. The research indicates that extensive gains in forecast performance may be obtained from the application of broader and more flexible classes of time series models.

I. Introduction

In this paper, a nonlinear, two state (outlier/no-outlier) forecasting model with time varying coefficients is presented and applied to North Sea oil discovery data. Conditional on the first 99 discoveries declared in the North Sea, the production potential of the geographical region lying between 56^{0} and 62^{0} north latitude and east of the Shetland Islands is analyzed. Forecasts of future discoveries are generated and the results are compared with alternative assessments.

II. Models of Oil and Gas Discovery

In this section the principal approaches to predicting quantities of undiscovered oil and gas are described. These principal approaches may be classified into six groups: (1) life-cycle; (2) rate-of-effort; (3) geologic-volumetric; (4) subjective probability; (5) discovery process; and (6) econometric.

Life-cycle models are based on the assumption that there is a simple functional relationship between time and amounts of undiscovered hydrocarbons. This assumption is motivated by the observation that production curves tend to rise from zero during the initial phase of the discovery process, peak and then decline toward zero as the population of possible discoveries from a region or field is depleted over time. These models, e.g., Hubbert (1962), do not introduce geological or physical aspects of the discovery process, nor are they responsive to changes in economic variables.

Rate-of-effort models are quite similar to life-cycle models. Both of these models assume that discoveries of hydrocarbons can be produced by defining a simple functional relationship between discoveries and an exogenous variable. Rate-of-effort models take the amount of cumulative exploratory effort rather than time as an independent variable.

Among the more advanced of the rate-of-effort models is that proposed by Bromberg and Hartigan (1975). In their research, future additions to reserves are projected from new discoveries and extensions and revisions. The model explicitly characterizes random fluctuations about an exponential trend, and introduces uncertainty about the parameters of the model by introducing diffuse prior distributions.

Geologic-volumetric models such as that developed by Mallory (1975) are used to appraise mineral potential of large regions. Beginning with an analysis of geochemical, geological and geophysical data, analysts predict (1) yield in barrels per unit area or volume of unexplored sediment in the basin and (2) the volume of productive sediment remaining to be explored. On the basis of these predictions a point forecast of the undiscovered mineral remaining is computed by multiplying (1) x (2).

Subjective probability models are best described in terms of an evaluative process. Following the description provided by the U.S.G.S. (1975), at the outset of the evaluative process, individual petroleum provinces are cited to be analyzed. The geologic framework within which selected provinces are located must be defined. Based upon planimetering of tectonic maps, point estimates of sedimentary rock areas and volume are developed. The models involve the quantitication of expert judgments in the form of probabilities. These probabilities are often derived from consensus estimates, which may in turn, be supported by the use of any of the procedures mentioned in this section.

Discovery process models are developed from assumptions that describe physical characteristics of depositions of individual fields and the pattern of discovery. A probability law governing the discovery process is proposed based on historical data. Although parameters of the probability law may not be known with certainty, a functional form for the density function characterizing the law may be assumed to be known with certainty. Unlike those models mentioned to this point, the discovery process models are intended for application to units consisting of similar deposits, and are generally employed at the micro (play) level as opposed to the regional level. Examples of these models have been developed by Barouch and Kaufman (1976), Smith (1980), and by Meisner and Demirmen (1981).

Econometric analyses involve the application of statistical procedures to empirically test theories concerning economic relationships. Among the earliest attempts to apply econometric modelling methodology to the problem of predicting new oil discoveries is that of Fisher (1966). Fisher's study addresses the question of the sensitivity of wildcat drilling and new oil discoveries to economic incentives. Several studies have been directed toward estimating the price responsiveness of new discoveries of natural gas under regulatory policies. The best known of these studies are those by Erickson and Spann (1971), Khazzoom (1971) and the much broader modelling effort by Pindyck and MacAvoy (1975).

III. The Model

Priestley (1980) has developed a class of nonlinear time series models called state dependent models (S.D.M.). This broad class of models includes the conventional ARMA model as well as bilinear, exponential and threshold autoregressive models. The principal advantage of the S.D.M. is that it allows a general form of nonlinearity, i.e., the model provides a nonlinear description which does not require an investigator to impose a specific nonlinear form a priori.

Consider a simple linear autoregressive model of order p,

$$X_t + \phi_1 X_{t-1} + \dots + \phi_p X_{t-p} = \mu + \varepsilon_t \tag{1}$$

where ε_t denotes strict white noise and μ, ϕ_1, ..., ϕ_p are constants. At time (t-1) the future evolution of the process $\{X_t\}$ is determined by the values of X_{t-1}, ..., X_{t-p}, along with future values of the ε_t process. The vector x_{t-1} = $\{X_{t-1}, \dots, X_{t-p}\}'$ may be regarded as the state vector of the process defined by (1). This vector contains all past information on the process which is relevant to its own future.

Following Priestley, suppose that the process $\{X_t\}$ is characterized by a nonlinear autoregressive model of order p. A model is sought which approximates the behavior of the series by a linear AR(p) model, while permitting a description of its behavior about small deviations of the process from its current state. In the state dependent model all of the coefficients in the AR(p) model are functions of x_{t-1}, the state vector at time (t-1). The state dependent AR(p) model is given by

$$X_t + \phi_1(x_{t-1})X_{t-1} + \dots + \phi_p(x_{t-1})X_{t-p} = \mu(x_{t-1}) + \varepsilon_t \ . \tag{2}$$

Provided that the coefficients of the S.D.M. are smooth functions of the state vector, the S.D.M. may be fit to data without imposing any further conditions on the specific functions which relate the coefficients to the state vector. A general model for $\mu(x_t)$ and the autoregressive coefficients may be written as

(3) $$\mu(x_t) = \mu(x_{t-1}) + \Delta X_t' \, \alpha^{(t)}$$

(4) $$\phi_n(x_t) = \phi_n(x_{t-1}) + \Delta X_t' \, \gamma_n^{(t)},$$

where $\{\alpha^{(t)}\}$ and $\{\gamma^{(t)}\}$ are gradients defined in terms of small departures of the process from the local state. In fitting the model, the gradients are permitted to wander in the form of random walk. Thus, the model is allowed to bend so as to minimize the discrepancy between the observation X_t and its predicted value. The random walk model for the gradients of the AR model is given by

(5) $$B_t = B_{t-1} + V_t$$

where $B_t = \{\alpha^{(t)}; \gamma_p^{(t)}, \ldots, \gamma_1^{(t)}\}$ and V_t is a sequence of independent matrix valued random variables. It may be assumed that the elements of each matrix are distributed MVN $\sim (0,\Sigma_v)$. Priestley has shown that the extension of the autoregressive model to the ARMA case is straightforward. In keeping with the application, however, attention has been restricted to the autoregressive model in this paper.

If any time series model is to be successfully applied to the problem of forecasting oil discoveries, it is imperative that the model capture the discovery decline phenomenon. As described by Root and Attanasi (n.d.), the discovery decline or depletion phenomenon is a reflection of the decreasing effectiveness of exploration effort over the mature phase of a finite geological structure, e.g., a petroleum play. Exploratory effort which takes place during the early life of a play may result in a limited number of small discoveries reflecting a limited amount of knowledge about a particular structure. Once the larger discoveries have been taken, the general declining trend of discovery size occurs. Having drilled that portion of the total area containing the larger hydrocarbon deposits, the likelihood of locating additional prospects tends to decrease with each additional discovery.

In order to capture the discovery decline phenomenon in the structure of the model, the model for the constant term given by Priestley and described by equation (3) is respecified. In particular, the model is written in the form

(6) $$\mu(x_t) = \mu(x_{t-1}) + \Delta X_t' \, \alpha^{(t)} + \delta^{(t)}$$

where the drift term $\delta^{(t)}$ is given a random walk.

In order to confine attention to a specific model of interest and establish the recursive algorithm for updating the coefficients an AR(1) model is formulated below. The equations of interest are

(7) $$X_t = H_t \, \theta_t + \varepsilon_t$$

and

(8) $$\theta_t = F_{t-1}\,\theta_{t-1} + W_t$$

where (7) and (8) give the observation and state equations, respectively. Specifically, H_t is a (1x5) row vector given by

(9) $$H_t = (1, X_{t-1}, 0, 0, 0)$$

θ_t is a (5x1) column vector given by

(10) $$\theta_t = (\mu_t, \phi_{1t}, \alpha^{(t)}, \gamma_1{}^{(t)}, \delta^{(t)})' ,$$

and F_{t-1} is the (5x5) transition matrix given by

(11) $$F_{t-1} = \left[\begin{array}{cc|ccc} 1 & 0 & \Delta X'_{t-1} & 0 & 1 \\ 0 & 1 & 0 & \Delta X'_{t-1} & 0 \\ \hline 0 & 0 & 1 & 0 & 0 \\ 0 & 0 & 0 & 1 & 0 \\ 0 & 0 & 0 & 0 & 1 \end{array}\right] .$$

The elements $\Delta X'_{t-1} = (X_{t-1} - X_{t-2})$ and the vector W_t is given by

(12) $$W_t = (0, 0, V_{1t}, V_{2t}, V_{3t})$$

where V_{1t}, V_{2t} and V_{3t} are the columns of V_t from (5).

The algorithm developed by Kalman (1963) may be applied directly to equations (7) and (8) as shown below.

(13) $$\hat{C}_t = \hat{\phi}_t - \hat{K}_t[H_t\,\hat{\phi}_t\,H'_t + \hat{\sigma}^2_\varepsilon]\hat{K}'_t$$

(14) $$\hat{\phi}_t = F_{t-1}\,\hat{C}_{t-1}\,F'_{t-1} + \hat{\Sigma}_w$$

(15) $$\hat{K}_t = \hat{\phi}_t\,H'_t[H_t\,\hat{\phi}_t\,H'_t + \hat{\sigma}^2_\varepsilon]^{-1}$$

(16) $$\hat{\theta}_t = F_{t-1}\,\hat{\theta}_{t-1} + \hat{K}_t[X_t - H_t\,F_{t-1}\,\hat{\theta}_{t-1}]$$

(17) $$\hat{X}_t = H_t\,\hat{\theta}_t .$$

In the recursion (13)-(17), ϕ_t is the variance-covariance matrix of the one-step prediction errors of θ_t, i.e., $\phi_t = E[(\theta_t - F_{t-1}\,\hat{\theta}_{t-1})(\theta_t - F_{t-1}\,\hat{\theta}_{t-1})']$, $C_t = E[(\theta_t - \hat{\theta}_t)(\theta_t - \hat{\theta}_t)']$, the variance-covariance matrix of $(\theta_t - \hat{\theta}_t)$, and K_t is the Kalman gain vector given by

(18) $$\hat{K}_t = \hat{\phi}_t\,H'_t\,\hat{\sigma}^{-2}_e ,$$

where $\hat{\sigma}^2_e$ is the variance of the prediction errors of X_t. At time $t = t_0$ the system requires start-up values denoted by the double circumflex for

(19) $$\hat{\theta}_{t0} = (\hat{\hat{\mu}}, \hat{\hat{\phi}}_1, 0, 0, 0)$$

as well as the variance-covariance matrix, C_{t0}, and $\hat{\hat{\sigma}}_{\varepsilon}^{2}$. These may be obtained by estimating a standard OLS AR(1) model using an initial stretch of the data. The actual number of observations chosen may vary depending on the length of the series being analyzed. The recursion should be started midway into the initial stretch. The initial values of the gradients may be set to zero. Additionally, values must be chosen for Σ_W. The methodology employed to estimate these values will be discussed in Section III of this paper.

The problem of sensitivity of parameter estimates to outlying observations must be taken into account in the development of a statistical model appropriate for forecasting future oil discoveries. While it is assumed that the largest discoveries are taken first, this is a probabilistic statement and it does not preclude the possibility that unusually large or small discoveries may occur at an atypical point along a so-called discovery decline curve. In order to deal with the outlier problem, a Bayesian forecasting procedure developed by Harrison and Stevens (1976) is integrated into the framework of the state-dependent model.

The occurrence of outliers is modelled as follows. Recall the system given by (13)-(17) and define

$$N = \text{number of states of nature}$$

π_j = prior probability of the occurrence of state j, (j = 1, ..., N). In order to compute forecasts of the process of oil discovery, a two state model is required, i.e., outlier and no-outlier. Using notation consistent with (13)-(17), the two-state system is written as

$$\hat{C}_{t,j} = \hat{\phi}_{t,j} - \hat{K}_{t,j}[H_t \hat{\phi}_{t,j} H_t' + \hat{\sigma}^2_{\varepsilon,j}]\hat{K}_{t,j}' \tag{20}$$

$$\hat{\phi}_{t,j} = F_{t-1} \hat{C}_{t-1,j} F_{t-1}' + \hat{\Sigma}_{W,j} \tag{21}$$

$$\hat{K}_{t,j} = \hat{\phi}_{t,j} H_t'[H_t \hat{\phi}_{t,j} H_t' + \hat{\sigma}^2_{\varepsilon,j}]^{-1} \tag{22}$$

$$\hat{\theta}_{t,j} = F_{t-1} \hat{\theta}_{t-1,j} + \hat{K}_{t,j}[X_t - H_t F_{t-1} \hat{\theta}_{t-1,j}] \tag{23}$$

$$\hat{X}_{t,j} = H_t \hat{\theta}_{t,j} , \qquad (j = 1,2) . \tag{24}$$

Generation of forecasts of the process X_t (t = 1,2, ...), requires that the analyst specify prior probabilities for each of the N states and compute posterior probabilities $p^i_{t,j}$, i.e., the probability posterior to observation X_t that the process was in the ith state at t-1 and is at present in the jth state. Specifically, for the AR(1) specification

$$p^i_{t,j} = \frac{k}{\sqrt{2\pi V^i_{e,j}}} \exp - \frac{[X_t - \hat{X}^i_t]^2}{2 V^i_{e,j}} \pi_j b^i_{t-1} \tag{25}$$

where

$$V^i_{e,j} = \Omega^{2^i}_{\mu_{t-1}} + 2\Omega^i_{\mu_{t-1}\phi_{1,t-1}} + \Omega^{2^i}_{\phi_{1,\ t-1}} + \sigma^2_{\varepsilon,j}$$

$$b^i_{t-1} = \sum_{j=1}^{N} p^i_{t-1,j}$$

and k is a normalizing constant. The variance and covariance terms $\Omega^{2^i}_{\mu_{t-1}}$, $\Omega^i_{\mu_{t-1}\phi_{1,t-1}}$ and $\Omega^{2^i}_{\phi_{1,\ t-1}}$ are defined below.

The condensing procedure described by Harrison and Stevens (1976) is dependent upon ordered approximations which permit the summary of model information relevant to future times as N normal distributions and N posterior probabilities. Having computed the posterior probabilities, the procedure for computing weighted forecasts is quite straightforward. A practical difficulty arises, however. Beginning with an N component prior results in the generation of an N^2 component posterior. Thus, for N = 2, estimation of the system given by (20)-(24) results in twenty expressions, i.e., N^2 = 4 equations for each term, $\hat{C}^i_{t,j}$, $\hat{\phi}^i_{t,j}$, $\hat{K}^i_{t,j}$, $\hat{\theta}^i_{t,j}$ and $\hat{X}^i_{t,j}$. Again, the i superscript identifies the state prevailing in the period t-1.

To generate single forecasts for a given period it is necessary to condense the expressions. This may be accomplished by computing weighted averages of the state-dependent terms based upon values of the posterior probabilities. First of all it is necessary to compute

$$b_{t,j} = \sum_{i=1}^{N} p^i_{t,j} \tag{26}$$

which permits computation of the weighted values of the state vector. For example, the level of the process can be written in the form

$$\mu^*_{t,j} = \sum_{i=1}^{2} p^i_{t,j}\ \mu^i_{t,j}/b_{t,j}\ . \tag{27}$$

Similarly, for the AR(1) process,

$$\phi^*_{1t,j} = \sum_{i=1}^{2} p^i_{t,j}\ \phi^i_{1t,j}/b_{t,j}\ . \tag{28}$$

Second moments are computed using the same general procedure. For example,

$$\Omega^2_{\mu_{t,j}} = \sum_{i=1}^{2} p^i_{t,j}[\sigma^{2^i}_{\mu_{t,j}} + (\mu^i_{t,j} - \mu^*_{t,j})^2]/b_{t,j}\ . \tag{29}$$

Having condensed the N^2 components, the probabilities given in (26) are computed as are the posterior probabilities given by (25) and the next iteration proceeds.

IV. Model Identification and Estimation

In order to fit the Bayesian model, start-up values are needed for the state vector, the variance-covariance matrix $C_{0,j}$ and the residual variances $\hat{\hat{\sigma}}^2_{\varepsilon,j}$ $(j = 1,2)$. In addition, values for the prior probabilities π_j are required. In order to compute start-up values the first several observations are used to compute OLS estimates. The minimization of the residual variance over the remaining observations is taken as the model selection criterion. This is approximately maximum likelihood. The recursion is started half-way into the stretch of data used for computing start-up values.

In order to locate the minimum residual variance, a grid-type search is performed over elements of the Σ_W matrix, i.e., for the AR(1) model, $\hat{\hat{\sigma}}^2$, $\hat{\hat{\sigma}}^2_{\gamma_1}$, $\hat{\hat{\sigma}}^2_{\delta}$. Additionally, the Bayesian form of the model requires that a search be conducted over the residual variance of the atypical (outlier) state. So as to minimize the complexity of the grid search, initially the non-Bayesian form of the model is fit to the data. This permits the researcher to identify the order of the process utilizing a grid search of lower dimension than that required by the Bayesian model.

The data are the logarithms of the first 99 discoveries declared in the North Sea. The chronologically ordered discoveries are located in the region lying between 56° and 62° north latitude and to the east of the Shetland Islands. Based upon work by Barouch and Kaufman (1977), logarithms of the data are taken on the basis of evidence indicating that deposition data are lognormally distributed.

The start-up values for the OLS AR(1) model are computed using the first 20 observations. The start-up values are shown in Table 1.

The recursion is started half-way into the stretch of twenty observations used to obtain start-up values. A grid-type search was performed over the elements of the Σ_W matrix. The minimum value of the standard error over the last 79 periods is $\hat{\sigma}^2_{\varepsilon} = .5356$. It is of interest to note that the grid search indicates $\hat{\hat{\sigma}}^2_{V\alpha} = \hat{\hat{\sigma}}^2_{V\gamma_1} = .5356$. Thus, conditional on the model selection criterion, the correct specification of the model parameters μ_t and $\phi_{1,t}$ does not involve state-dependency as captured by the term $\Delta X'_{t-1} = (X_{t-1} - X_{t-2})$.

The actual data are shown along with the points fit by the model in Chart 1. The smooth pattern exhibited by the fit data points, particularly, over the last several observations is considered desirable inasmuch as a substantial level of autocorrelation is not anticipated over the remaining life of the play. The behavior of the parameter estimates is shown by the plotted values in Chart 2 and Chart 3. Note, the estimates of the level of the process, $\hat{\mu}_t$, follow a pattern consistent with the process of discovery decline. The behavior of the AR(1)

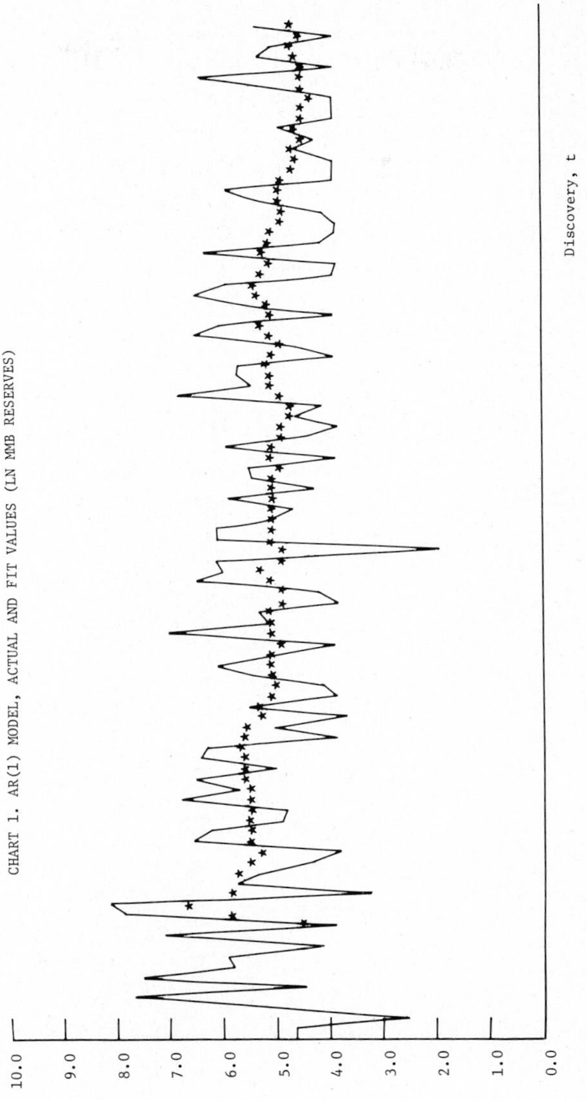

CHART 1. AR(1) MODEL, ACTUAL AND FIT VALUES (LN MMB RESERVES)

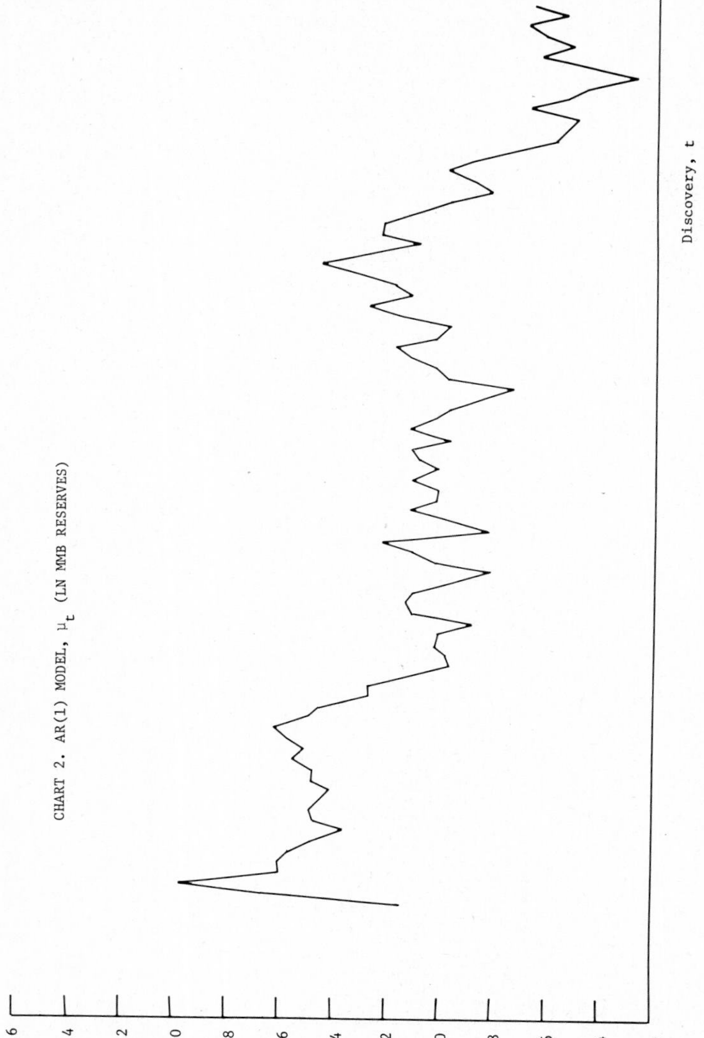

CHART 2. AR(1) MODEL, μ_t (LN MMB RESERVES)

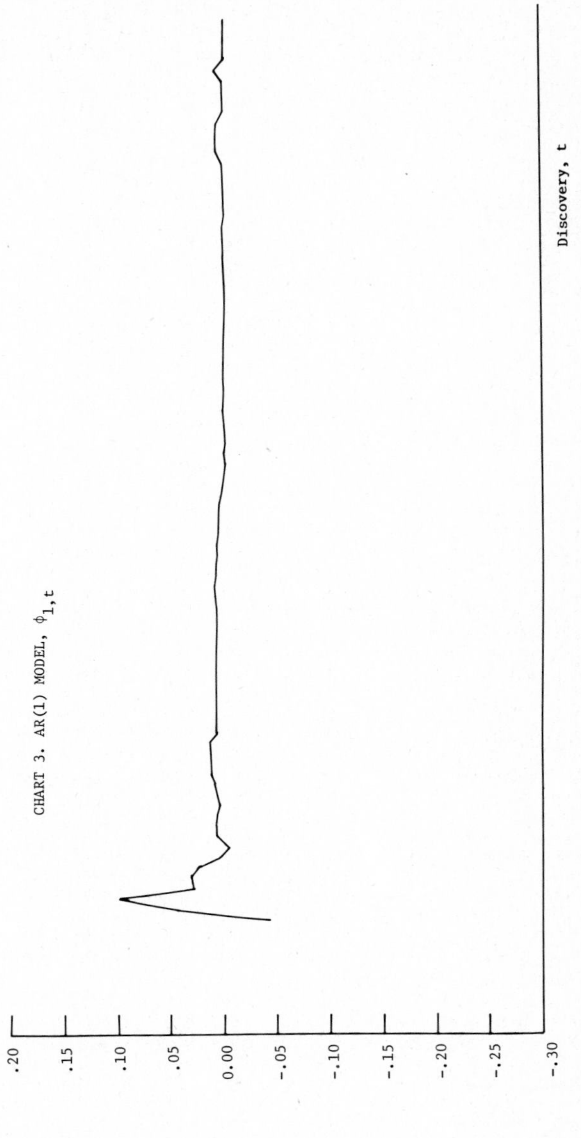

CHART 3. AR(1) MODEL, $\phi_{1,t}$

TABLE 1

START-UP VALUES FOR THE AR(1) MODEL

Coefficient	Variance-Covariance	
$\hat{\hat{\mu}} = 6.50$	$\hat{\hat{\sigma}}^2_{\mu} = 1.79$	$\hat{\hat{\sigma}}^2_{v\delta} = .0002$
$\hat{\hat{\phi}}_1 = -.19$	$\hat{\hat{\sigma}}^2_{\phi_1} = .06$	$\hat{\hat{\sigma}}_{v\alpha\gamma_1} = 0.00$
	$\hat{\hat{\sigma}}_{\mu\phi_1} = -.31$	$\hat{\hat{\sigma}}_{v\alpha\delta} = 0.00$
	$\hat{\hat{\sigma}}^2_{\varepsilon} = 2.79$	$\hat{\hat{\sigma}}_{v\gamma_1\delta} = 0.00$
	$\hat{\hat{\sigma}}^2_{v\alpha} = 0.00$	
	$\hat{\hat{\sigma}}^2_{v\gamma_1} = 0.00$	

parameter estimates is quite stable about zero. This is consistent with the expectation that a low level of autocorrelation is anticipated over the sequence of remaining discoveries.

As a check on the AR(1) model specification, an AR(2) model is fit to the logarithms of the North Sea data. Following the general form given by equations (4) and (6), the parameter models for the AR(2) case are written as

$$\mu_t = \mu_{t-1} + \alpha_1^{(t)} \Delta X_t^{(1)} + \alpha_2^{(t)} \Delta X_2^{(2)} + \delta^{(t)} \tag{30}$$

$$\phi_{1,t} = \phi_{1,t-1} + \gamma_{11}^{(t)} \Delta X_t^{(1)} + \gamma_{12}^{(t)} \Delta X_t^{(2)} \tag{31}$$

$$\phi_{2,t} = \phi_{2,t-1} + \gamma_{21}^{(t)} \Delta X_t^{(1)} + \gamma_{22}^{(t)} \Delta X_t^{(t)} \tag{32}$$

where

$$\Delta X_t^{(1)} = (X_t - X_{t-1}) \text{ and } \Delta X_t^{(2)} = (X_{t-1} - X_{t-2}).$$

The start-up values based upon the application of OLS using the first 30 observations of the data are given in Table 2.

The standard error of the residuals over the last 79 periods is minimized conditional upon a grid-type search over the parameters in the Σ_w matrix. Consistent with the methodology adopted in fitting the AR(1) model, the off-diagonal terms are set to zero. The minimum value of the standard error of the residuals is located where the values of $\hat{\hat{\sigma}}^2_{v\alpha_1}$, $\hat{\hat{\sigma}}^2_{v\gamma_{21}}$ and $\hat{\hat{\sigma}}^2_{v\delta}$ are non-zero. Thus,

TABLE 2

START-UP VALUES FOR AR(2) MODEL

Coefficient	Variance-Covariance	
$\hat{\mu} = 6.68$	$\hat{\sigma}^2_{\mu} = 2.76$	$\hat{\sigma}^2_{v\alpha_1} = .0001$
$\hat{\phi}_1 = -.16$	$\hat{\sigma}^2_{\phi_1} = .04$	$\hat{\sigma}^2_{v\alpha_2} = 0.00$
$\hat{\phi}_2 = -.05$	$\hat{\sigma}^2_{\phi_2} = .04$	$\hat{\sigma}^2_{v\gamma_{11}} = 0.00$
	$\hat{\sigma}_{\mu\phi_1} = -.25$	$\hat{\sigma}^2_{v\gamma_{12}} = 0.00$
	$\hat{\sigma}_{\mu\phi_2} = -.24$	$\hat{\sigma}^2_{v\gamma_{21}} = 9x10^{-6}$
	$\hat{\sigma}_{\mu_1\phi_2} = .01$	$\hat{\sigma}^2_{v\gamma_{22}} = 0.00$
	$\hat{\sigma}^2_{\varepsilon} = 2.17$	$\hat{\sigma}^2_{v\delta} = .002$

the random walk terms $\alpha_1^{(t)}$, $\gamma_{21}^{(t)}$ and $\delta^{(t)}$ are non-zero. Alternatively, the parameters μ_t and $\phi_{2,t}$ are a function of $(X_{t-1} - X_{t-2})$ through the terms $\alpha_1^{(t)}$, and $\gamma_{21}^{(t)}$, respectively.

The plot of actual and fitted values is shown in Chart 4. Comparison with the fit of the estimated values generated by the AR(1) model shown in Chart 1 does not reveal any significant differences. The behavior of the parameter estimates is plotted in Charts 5, 6 and 7. Chart 5 plots the estimates of μ_t. Notice that the pattern exhibited by the estimates is quite similar to the pattern of the estimates generated by the AR(1) model shown in Chart 2. As is the case under the AR(1) specification, the constant term appears to capture the discovery decline process. In contrast to the estimates of the autocorrelations under the AR(1) specification, the estimates of the ϕ_{1t} and ϕ_{2t} parameters for the AR(2) model plotted in Chart 6 and Chart 7 are quite variable.

The minimum value of the standard error of the residuals is $\hat{\sigma}_{\varepsilon 79} = .5365$. This is not as low as the value achieved under the AR(1) specification. Thus, it is concluded the autoregressive process of order 1 is probably adequate.

Having identified the order of the process, the Bayesian form of the model is fit to the data. As interest in applying the multi-state model is due to concern that the parameter estimates are overly sensitive to transient observations, state-dependent values must be assigned to the residual variances

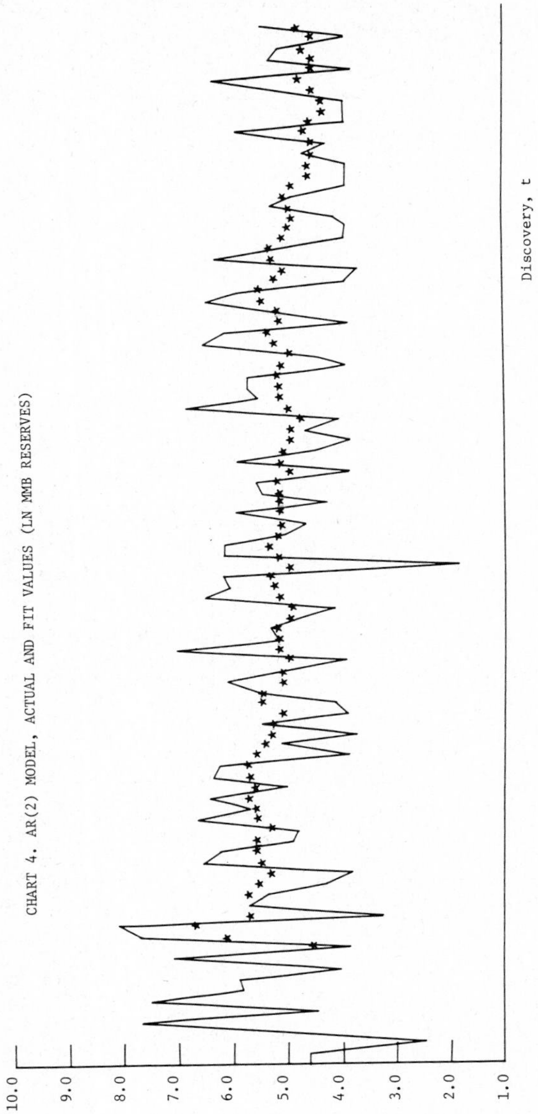

CHART 4. AR(2) MODEL, ACTUAL AND FIT VALUES (LN MMB RESERVES)

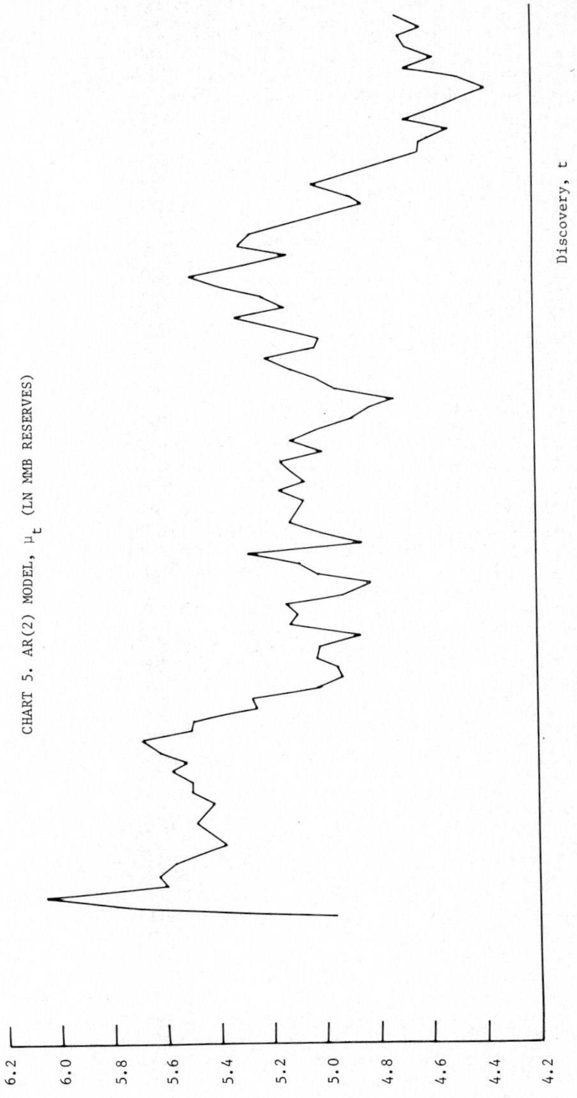

CHART 5. AR(2) MODEL, μ_t (LN MMB RESERVES)

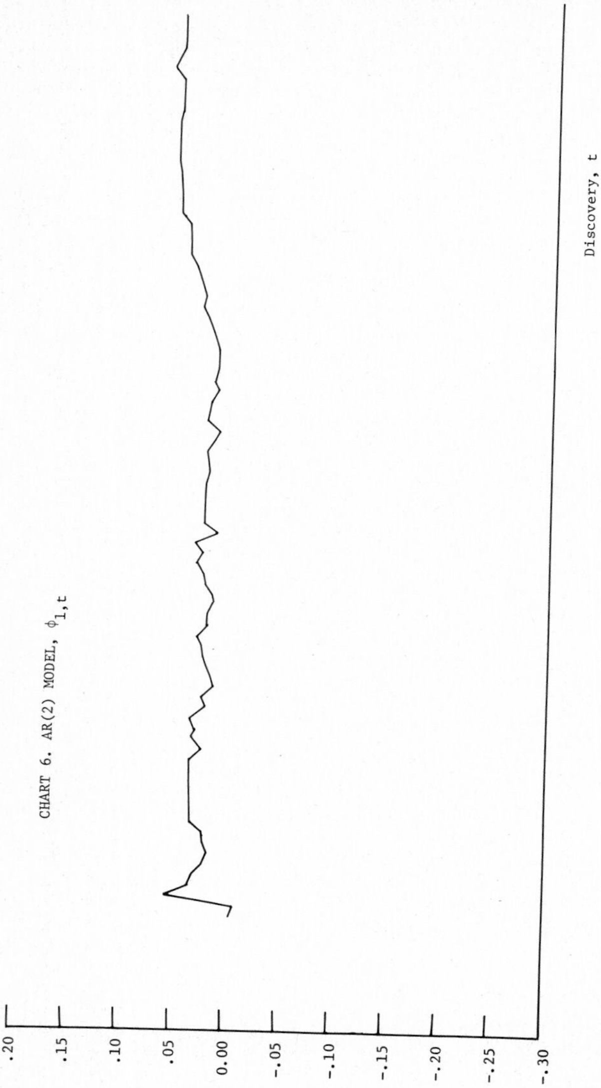

CHART 6. AR(2) MODEL, $\phi_{1,t}$

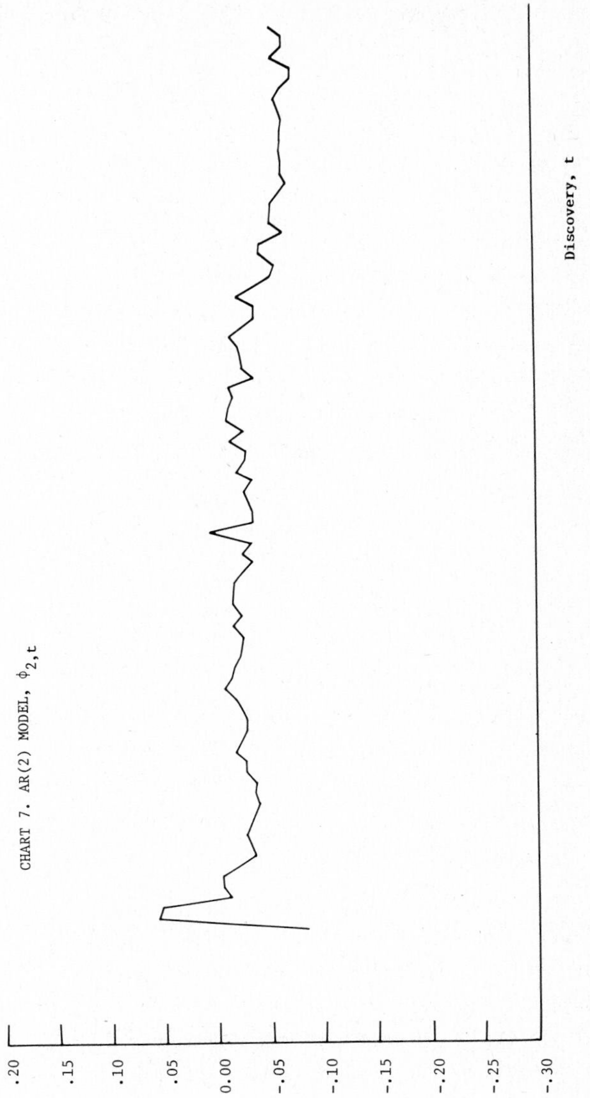

CHART 7. AR(2) MODEL, $\phi_{2,t}$

$\hat{\sigma}_{\varepsilon,j}$, (j = 1,2). Additionally, the prior probabilities must be assigned to the states. The assignment of prior probabilities may be based upon evaluation of the discovery history of the play. In addition to observing past behavior of the data points, the researcher may base the assignment of prior probabilities on knowledge obtained from sources other than the time series of discoveries, e.g., geological estimates of the production potential of the geological play.

The assignment of values to the residual variances may be achieved by using the OLS start-up value to characterize the variance in the prevalent state (no-outlier), and by searching over alternative values in the alternate state (outlier) until the value which results in the best model fit is identified. While this is not the only possibility, provided that there are not an excessive number of parameters in the model so that the search procedure becomes too complicated, this procedure seems reasonable. The complexity of the search procedure may be reduced somewhat by permitting the variance-covariance matrix for the random walk parameters to be state independent. In practice, this procedure is adopted.

On the basis of past analyses of the North Sea data, the prior probabilities assigned to the outlier and no-outlier states are π_1 = .99 and π_2 = .01. The start-up values for the AR(1) model are given in Table 1. The residual variance is assigned to state 1, thus, $\hat{\sigma}^2_{\varepsilon,1} = 2.79$. A grid-type search is implemented over values of $\hat{\sigma}^2_{v\alpha}$, $\hat{\sigma}^2_{v\gamma_1}$, $\hat{\sigma}^2_{v\delta}$ and $\hat{\sigma}^2_{\varepsilon,2}$.

The minimum standard error of the residuals over the last 79 periods is $\hat{\sigma}_{\varepsilon 79} = 1.028$ and this value is located where $\hat{\sigma}^2_{v\alpha} = .002$, $\hat{\sigma}^2_{v\gamma_1} = 0.00$, $\hat{\sigma}^2_{v\delta} = 5 \times 10^{-6}$ and $\hat{\sigma}^2_{\varepsilon,2} = 3.40$. Note, that the fit is not as good as that achieved under the non-Bayesian specification, however, this is not disturbing given the weighted nature of the results.

The actual and fit values are shown in Chart 8. Again, the fit values exhibit a reasonably smooth pattern. The parameter estimates are plotted in Chart 9 and Chart 10. The estimates of the constant term reflect the discovery decline phenomenon. The pattern of the autocorrelations in both states is plotted in Chart 10. It is interesting that the estimates of $\phi_{1t,1}$ are consistently below those of $\phi_{1t,2}$. Both sets of estimates are quite stable over the last 50 discoveries.

V. Forecasting North Sea Oil Discoveries

The output of the fitting algorithm gives estimated values of $\hat{\theta}_t$ based upon information given by the value of X at point t. From the estimates of the parameters in θ_t, and given the observation of X at point (t-1), estimates of X_t are computed. Clearly, the algorithm must be altered for the purpose of

CHART 8. AR(1) BAYESIAN MODEL, ACTUAL AND FIT VALUES (LN MMB RESERVES)

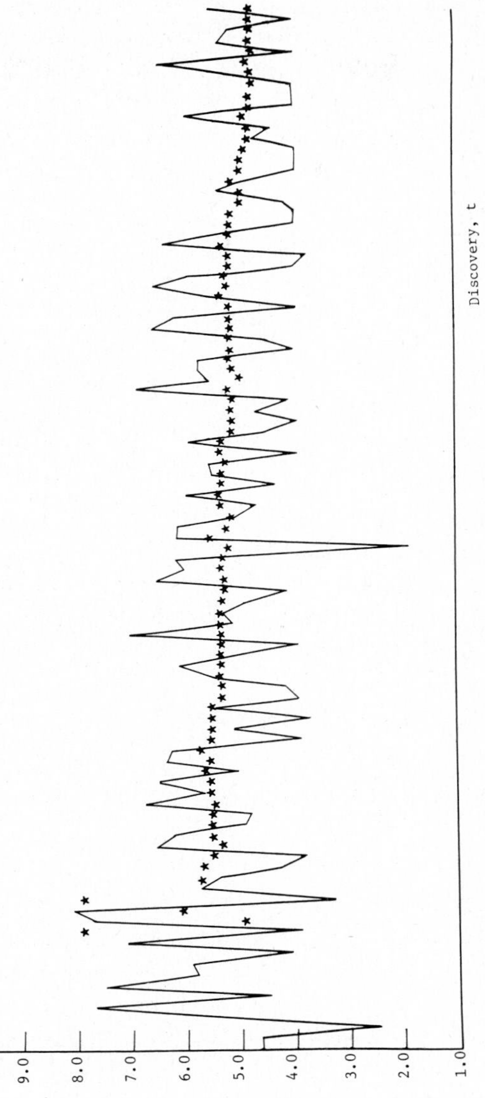

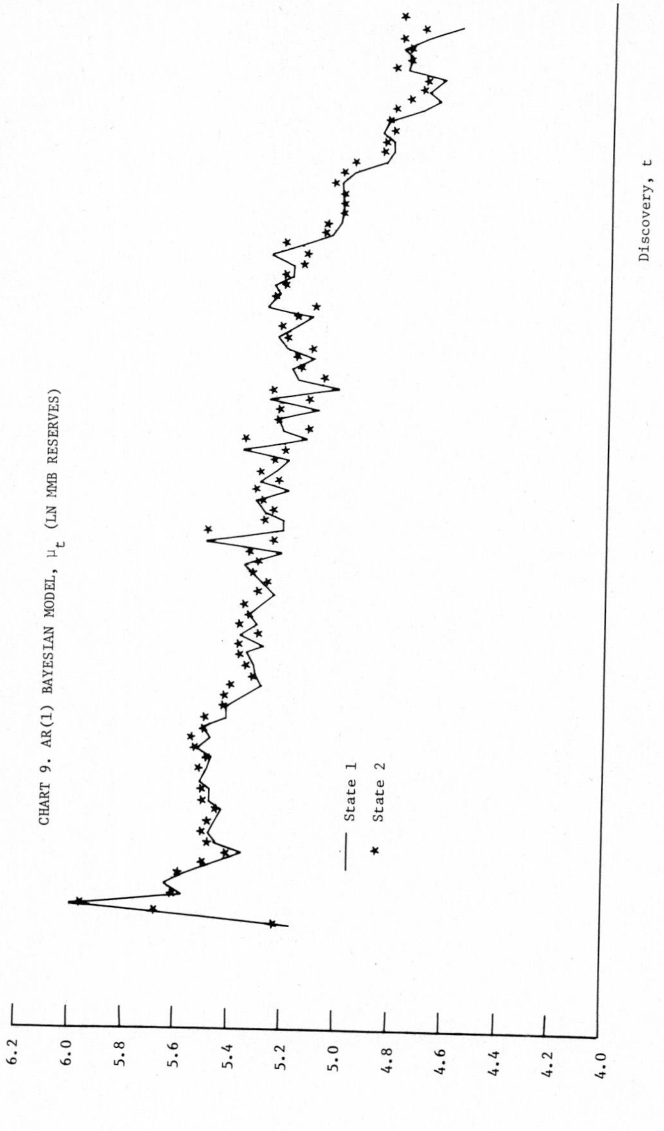

CHART 9. AR(1) BAYESIAN MODEL, μ_t (LN MMB RESERVES)

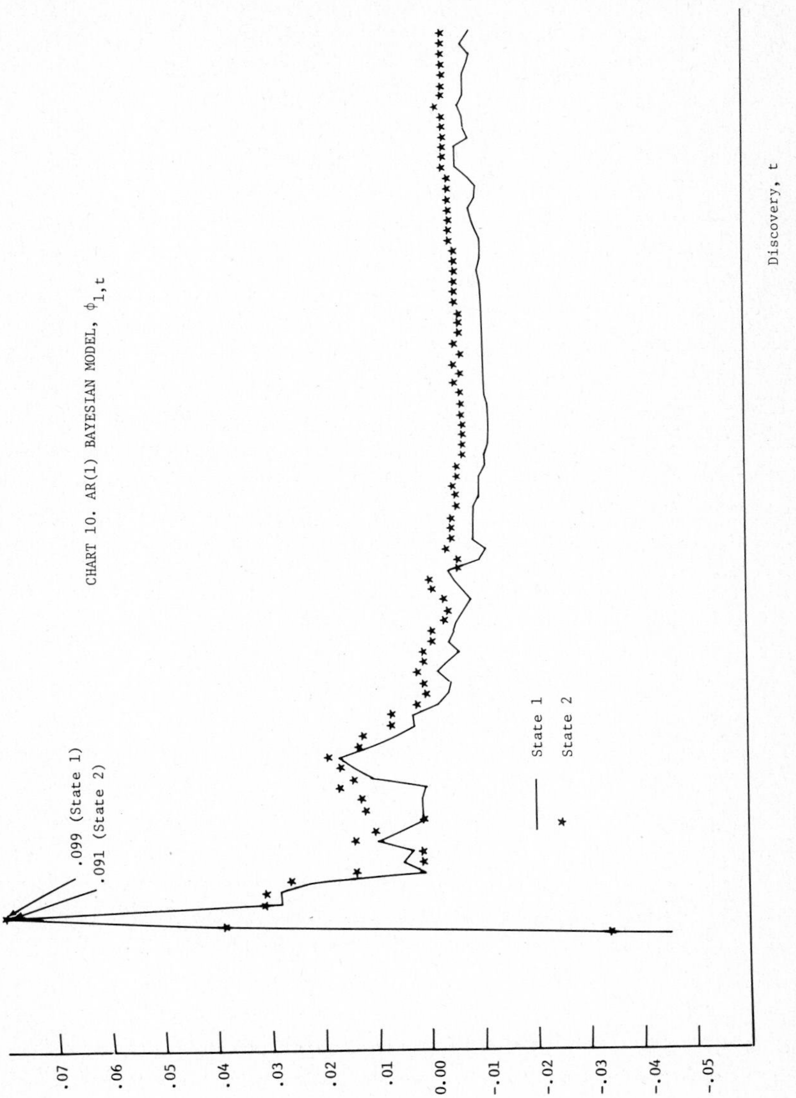

CHART 10. AR(1) BAYESIAN MODEL, $\phi_{1,t}$

forecasting. The system is altered by using the parameter vector $\hat{\theta}_t$ and knowledge of the most recent observation X_t to give the forecasting equation

$$\hat{X}_{t+1} = \hat{\mu}_t + \hat{\phi}_{1,t} X_t \, . \tag{33}$$

Following the procedure indicated by equation (33), one-step forecasts are generated by the AR(1) model. The plot of the forecasts is shown in Chart 11. The smooth pattern of the forecasts is considered desirable inasmuch as large autocorrelations are not anticipated over the mature phase of the play. The standard error of the residuals over the last 20 periods is $\hat{\sigma}_{\varepsilon 20}$ = .8373. The performance of the model over the last 20 periods is considered as it is believed that decisions relating to the usefulness of the forecasts must be based upon knowledge of the performance of the model over the recent past. A standard OLS AR(1) model has been estimated. The value of the standard error of the residuals over the last 20 periods is $\hat{\sigma}_{\varepsilon 20}$ = .9717. Furtherfore, use of the AR(1) non-Bayesian S.D.M. results in a standard error $\hat{\sigma}_{\varepsilon 20}$ = .9165. Thus, the Bayesian model outperforms a conventional OLS model as well as a non-Bayesian AR(1) S.D.M. Table 3 records ten-step forecasts generated by the model. Notice that the model does seem to track the discovery decline phenomenon.

TABLE 3

TEN-STEP FORECASTS GENERATED BY THE BAYESIAN S.D.M. (MMB RESERVES)

Discovery, t	Forecast, t+1
99	108.91
100	114.55
101	110.65
102	111.30
103	110.29
104	109.53
105	108.70
106	107.91
107	107.13
108	106.35

Smith (1980) has recently published estimates of remaining reserves in seven designated size classes for the petroleum play under analysis in this research. Smith reports that the total volume of ultimately recoverable reserves is approximately 43.2 billion barrels of oil and gas equivalent. Smith's estimates indicate most of the remaining 221 discoveries are in size classes bounded from 0 to 50 and 50 to 100 million barrels. As a check on the projections generated by the time series model, 221 discoveries were generated beyond the 99th. The results indicate that approximately 13.4 billion barrels of oil and gas

CHART 11. AR(1) BAYESIAN MODEL, ACTUAL VALUES AND ONE-STEP FORECASTS (LN MMB RESERVES)

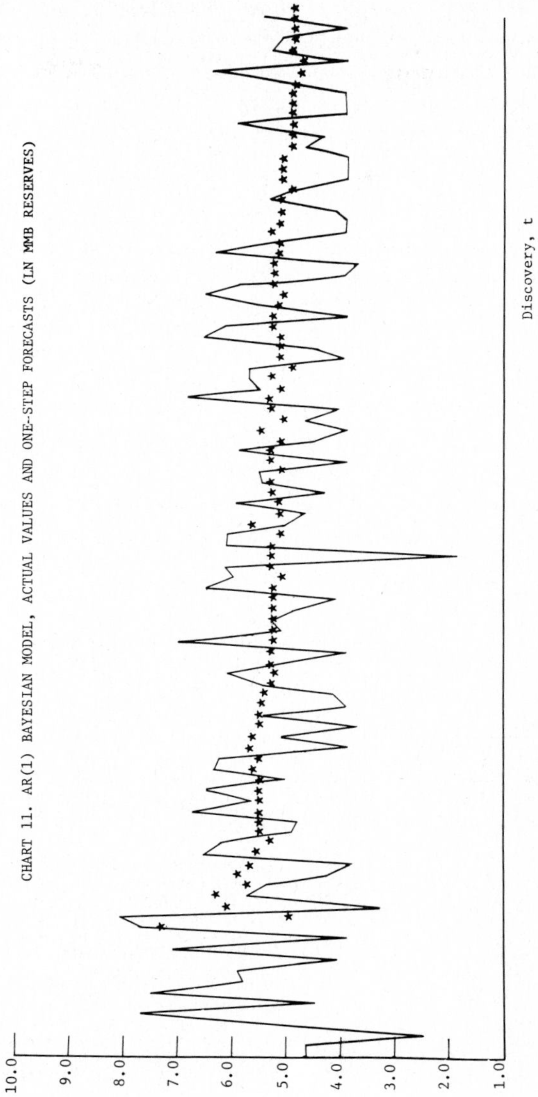

equivalent remain to be discovered. Adding this projection of the 32.6 billion barrels reported as having been discovered by the first 99 discoveries yields an estimate of the total volume amounting to approximately 46.0 billion barrels. This estimate is quite close to that published by Smith, and it is well within the bounds established by industry estimates which range from 35 to 67 billion barrels.

VI. Conclusions

This research presents the application of a nonlinear, state dependent, Bayesian time series model to the problem of forecasting sizes of future oil discoveries at the play level. The research indicates that there are likely to be substantial gains obtained as the result of consideration of broader classes of time series models beyond those belonging to the conventional linear ARMA class. This research lends support to the claim that more complex models are likely to provide a better fit to the data than that provided by standard linear models, and that the gains in modelling flexibility achieved by applying a broader class of models may be substantial.

ACKNOWLEDGEMENT

This research was supported by the National Science Foundation, under Grant No. SES-8011175.

REFERENCES

BAROUGH, E. and KAUFMAN, E. M. (1976). Probabilistic Modelling of Oil and Gas Discovery. In Energy: Mathematics and Models. SIAM Institute for Mathematics and Society.

__________. (1977). Estimation of Undiscovered Oil and Gas. In Proceedings of Symposia in Applied Mathematics 21, American Mathematical Society, 77-91.

BROMBERG, L. and HARTIGAN, J. A. (1975). Report to the Federal Energy Administration: United States Reserves of Oil and Gas. Department of Statistics, Yale University.

ERIKSON, E. W. and SPANN, R. W. (1971). Supply Price in a Regulated Industry: The Case of Natural Gas. Bell Journal of Economics and Management Science 2, 94-121.

FISHER, F. M. (1964). Supply and Cost in the U.S. Petroleum Industry: Two Econometric Studies. John Hopkins Press.

HARRISON, P. J. and STEVENS, C. F. (1976). Bayesian Forecasting. Journal of Royal Statistical Society (B) 38, 205-247.

HUBBERT, M. K. (1962). Energy Resources, A Report to the Committee on Natural Resources. National Academy of Sciences, National Research Council, Publication 1000D.

KALMAN, R. E. (1963). New Methods in Wiener Filtering Theory. In Proceedings of the First Symposium on Engineering Applications of Random Function Theory and Probability. Eds: J. L. Bogdanoff and F. Kozin, Wiley, 270-388.

KHAZZOOM, J. D. (1971). The F.P.C. Staff's Econometric Model of Natural Gas Supply in the United States. Bell Journal of Economics and Management Science 2, 51-93.

MACAVOY, P. W. and PINDYCK, R. S. (1975). The Economics of the Natural Gas Shortage (1960-1980). North-Holland.

MALLORY, PETER F. (1975). Accelerated Natural Gas Resource Appraisal (ANGORE). In Methods of Estimating the Volume of Undiscovered Oil and Gas Resources. Ed: J. D. Haun, American Association of Petroleum Geologists, 23-30.

MEISNER, J. and DEMIRMEN, F. (1981). The Creaming Method: A Bayesian Procedure to Forecast Future Oil and Gas Discoveries in Mature Exploration Provinces. Journal of Royal Statistical Society (A) 144, 1-31.

PRIESTLEY, M. B. (1980). State-Dependent Models: A General Approach to Non-Linear Time Series Analysis. Journal of Time Series Analysis 1, 47-71.

ROOT, D. H. and ATTANASI, E. D. (n.d.). An Analysis of Petroleum Discovery Data and a Forecast of the Rate of Peak Production. Paper prepared for U.S. Geological Survey, Reston, Virginia.

SMITH, JAMES L. (1980). A Probabilistic Model of Oil Discovery. Review of Economics and Statistics 62, 587-594.

U.S. Department of the Interior. (1975). Geological Estimates of Undiscovered Oil and Gas Reserves in the United States. In Geological Survey Circular 75, Reston, Virginia.

TIME SERIES ANALYSIS: Theory and Practice 4
O.D. Anderson (editor)

AN INTEGRATED TIME SERIES ANALYSIS COMPUTER PROGRAM: THE SCA STATISTICAL SYSTEM

Lon-Mu Liu

Department of Quantitative Methods
University of Illinois at Chicago
Box 4348, Chicago, Illinois 60680
U.S.A.

Gregory B. Hudak

Scientific Computing Associates
P.O. Box 625
DeKalb, Illinois 60115
U.S.A.

This paper briefly describes comprehensive time series analysis capabilities encompassing both the time and frequency (spectral) domains in the SCA statistical system. The time domain analysis includes univariate ARIMA, transfer function, intervention, multivariate ARMA, and simultaneous transfer function models. Spectral analysis may be based on periodograms, auto and cross covariances, or parametric models. The system provides the user an English-like language to more easily implement an analysis. Hence, although the system is capable of handling complicated models, the user language makes model specification easy. In addition, matrices can be manipulated using analytic formulas.

1. INTRODUCTION

Newly developed statistical software to facilitate time series analysis is described in this paper. Special attention is given to explain facilities that simplify model identification, estimation, diagnostic checking, and forecasting. The SCA (Scientific Computing Associates) system is designed as integrated software for a wide variety of statistical analyses. It is suitable for both routine data analysis and statistical research. Its current capabilities include descriptive statistics, data plotting, data manipulation and editing, matrix algebra, linear and nonlinear regression, and time series analysis. The system also provides macro procedure capabilities to increase flexibility and ease of use, especially in interactive mode. Furthermore, the workspace for each computing session can be saved to a disk file permitting more convenient and efficient computing.

The SCA system's user control language has two basic types of statements, analytic and English-like. Most unambiguous vector and matrix manipulations are expressed as analytic statements. More complicated tasks are expressed in English-like statements. For example, the computation

$$Z_t = (Y_t + \mu)^{\lambda} \qquad t=1,2,\ldots,n$$

can be expressed as the analytic statement

```
Z = (Y+MU)**LAMBDA
```

where Z and Y are vectors, and MU and LAMBDA scalars. The solution to the normal equations

$$\beta = (X'X)^{-1}X'Y$$

can be expressed in this system as

```
BETA = INV(T(X)#X)#T(X)#Y
```

where T and INV are matrix transposition and inversion functions and "#" is the matrix multiplication operator. The user is not limited to references to a whole matrix or vector, but may also specify a column, a row, or an element of a matrix or a vector in an analytic statement. The analytic statements also include control statements such as IF-THEN-ELSE, DO-loop, GO FORWARD and GO BACKWARD.

The SCA system has a number of English-like statements for data input, output, manipulation, editing, and analyses. An English-like statement reads like a paragraph of text. Therefore, it is also referred to as a paragraph. Each paragraph begins with a unique paragraph name and as many simple sentences of instructions as required. Sentences are separated by periods (.). As an example, the next two SCA paragraphs access two variables to be identified as "TAX" and "INCOME" from a file identified as number 10 and then displays them:

```
INPUT VARIABLES ARE TAX, INCOME.  FILE IS 10.
PRINT VARIABLES ARE TAX, INCOME.
```

Although a paragraph may consist of many sentences, the sentences usually possess default options or values. Hence it is not necessary to specify all sentences of a paragraph. In the interactive environment, if any important information is omitted, the program will prompt meaningful messages and allow the user to provide the information.

Both analytic and English-like statements can be blended in any order within a system session as long as the order is logical. Further, the system provides extensive on-line HELP information for both types of statements, facilitating interactive use.

The time series capabilities of the SCA system include both time domain and frequency (spectral) analyses. The time domain analysis handles several general classes of parametric models and will be described below. Spectral analysis (not discussed here) may be based on periodograms, auto and cross covariances, or parametric models.

2. TIME SERIES MODELS IN THE SCA SYSTEM

The SCA system handles very general forms of univariate ARIMA, transfer function, intervention, multivariate ARMA, and simultaneous transfer function (also referred to as rational distributed lag structural form models, see Wall 1976) models. The system provides convenient tools for tentative model identification, specification (writing an identified model in the SCA language), estimation, diagnostic checking, and forecasting.

Tentative model identification usually is the most sophisticated step in time series analysis. Several useful tools may be employed in this step. For univariate ARIMA models, we provide paragraphs to compute the sample estimates of the ACF, PACF, and EACF. The ACF, autocorrelation function (Box and Jenkins 1976), is useful for identifying MA models; PACF, partial autocorrelation function (Box and Jenkins 1976), for AR models; and EACF, extended autocorrelation function (Tiao and Tsay 1981), for mixed ARMA models. In addition, the IACF (inverse autocorrelation function) and the Corner Method (Beguin, Gourieroux, and Monfort 1980) are also available. Univariate and simultaneous transfer function model identification is based primarily on the procedure proposed by Liu and Hanssens (1982) and Hanssens and Liu (1982). For multivariate ARMA models, the procedures developed by Tiao and Box (1981) are followed.

Although the SCA system is capable of handling very complicated models, the user language makes model specification easy. For example, the following two-input transfer function model

$$Y_t=((\omega_1B^2+\omega_2B^3)/(1-\delta_1B))X_{1t}+((\omega_3B^2)/(1-\delta_2B))X_{2t}+((1-\theta_1B^4)(1-\theta_2B^{12})/(1-\phi B))a_t$$

can be specified as

```
     UTSMODEL  NAME IS EXAMPLE1.                                              @
               MODEL IS Y=(W1*B**2+W2*B**3)/(1-D1*B)X1+(W3*B**2)/(1-D2*B)X2   @
                      + (1-TH1*B**4)(1-TH2*B**12)/(1-PHI*B)NOISE.
```

The paragraph name UTSMODEL invokes univariate model specification, the NAME sentence uniquely specifies the model name, which is "EXAMPLE1" (since a user may have many models in a computing session), B is the backward shift operator, and W1, W2, W3, D1, D2, TH1, TH2, and PHI are parameter names specified by the user. The parameters can be set to fixed values or equal to other parameters. For example, D1, D2, and PHI may be constrained to be the same value. A model can be subsequently modified by adding, deleting, or changing some components. The "@" symbol above signifies a continuation of a statement.

A more condensed version of the UTSMODEL paragraph is available to more experienced users. An operator such as $(1-\phi_1B-\phi_3B^3)$ may be reduced to only its polynomial orders, i.e., (1,3). Consequently, the above model could be abbreviated to

```
     UTSMODEL NAME IS EXAMPLE1.                                  @
              MODEL IS Y=(2,3)/(1)X1+(2)/(1)X2+(4)(12)/(1)NOISE.
```

Model specification for multivariate ARMA models is similar to univariate ARIMA. For example, the model

$$(1-\Phi_1B-\Phi_2B^2)(1-\Phi_3B^{12})Y_t = (1-\Theta_1B)(1-\Theta_2B^{12})a_t ,$$

where Φ_1, Φ_2, Φ_3, Θ_1, and Θ_2 are matrices, can be specified as

```
     MTSMODEL NAME IS EXAMPLE2.                                  @
              MODEL IS (1-PHI1*B-PHI2*B**2)(1-PHI3*B**12)SERIES  @
                     =(1-TH1*B)(1-TH2*B**12)NOISE.
```

where PHI1, PHI2, PHI3, TH1, TH2 are parameter matrices storing the initial values or final estimates. The elements in the parameter matrices can be fixed to specific values, or set to be equal among them. Furthermore, the user may also specify the covariance structure for the error term, a(t), in order to express endogenous and exogenous variables in the system, or deterministic series.

Parameter estimation for the time series models in the SCA system are based on both conditional least squares, CLS (Box and Jenkins 1976), and maximum likelihood estimaton, MLE (Hillmer and Tiao 1979), methods. The CLS method usually requires much less computing time, but the MLE method provides more efficient parameter estimates.

The SCA system allows storage of useful information at each step of an analysis in user specified variables. Thus the residuals from an estimated model may be stored in some designated variables for use in checking model adequacy.

Once an adequate model is obtained for a set of data, the user may forecast by means of the appropriate SCA paragraphs.

3. ANNOTATED COMPUTER EXAMPLES

In this section, we use two annotated examples to illustrate some capabilities in the SCA system. The examples were run interactively on the IBM 4341 at the University of Illinois at Chicago. The output has been slightly edited for briefness and "~~" is added to the beginning of each input line for clarity.

Example 1: Series A of Box and Jenkins (1970), chemical process concentration readings, is used. In this example, we illustrate model identification, estimation, diagnostic checking, and forecasting for univariate ARIMA modelling using the SCA system.

```
~~INPUT VARI IS SERIESA.
```

Data is transmitted to the system and held under the variable name "SERIESA".

```
REAL      VARIABLE SERIESA ,  A  197  BY    1 MATRIX, IS STORED IN THE WORKSPACE
--

~~UIDEN  VARIABLE IS SERIESA.  OUTPUT NOPRINT(PLOT).
```

The sample ACF and PACF of the series will be examined. The plot of these statistics will be supressed.

```
TIME PERIOD ANALYZED . . . . . . . . . . 1  TO    197
EFFECTIVE NUMBER OF OBSERVATIONS . . .            197
NAME OF THE SERIES . . . . . . . . . .        SERIESA
STANDARD DEVIATION OF THE SERIES . . .         0.3992
MEAN OF THE (DIFFERENCED) SERIES . . .        17.0624
STANDARD DEVIATION OF THE MEAN . . . .         0.0284
T-VALUE OF MEAN (AGAINST ZERO) . . . .       599.8357

AUTOCORRELATIONS

  1- 12     .57  .50  .40  .36  .33  .35  .39  .32  .30  .25  .19  .16
  ST.E.     .07  .09  .10  .11  .12  .12  .13  .13  .14  .14  .14  .14
    Q      65.0  114  146  172  194  219  251  272  291  305  312  318

 13- 24     .19  .24  .14  .18  .20  .20  .14  .18  .10  .13  .11  .14
  ST.E.     .14  .15  .15  .15  .15  .15  .15  .15  .15  .15  .15  .16
    Q       326  338  342  349  358  367  371  378  381  384  387  391

 25- 36     .13  .13  .15  .05  .09  .02  .09  .12  .11  .06 -.06 -.03
  ST.E.     .16  .16  .16  .16  .16  .16  .16  .16  .16  .16  .16  .16
    Q       395  399  404  405  407  407  408  412  415  416  417  417

PARTIAL AUTOCORRELATIONS

  1- 12     .57  .25  .07  .07  .07  .12  .16 -.03  .01 -.02 -.07 -.02
  ST.E.     .07  .07  .07  .07  .07  .07  .07  .07  .07  .07  .07  .07

 13- 24     .06  .09 -.12  .05  .10  .07 -.07  .05 -.11  .05 -.03  .05
  ST.E.     .07  .07  .07  .07  .07  .07  .07  .07  .07  .07  .07  .07
--

~~UIDEN   VARIABLE IS SERIESA.  DFORDER IS 1. @
~~        MAXLAG IS 12.  OUTPUT NOPRINT(PLOT).
```

It is possible that the series is nonstationary. The sample ACF and PACF of the series $(1-B)Z_t$ will be examined as above, but only 12 lags will be calculated.

```
        .
        .
        .
AUTOCORRELATIONS

  1- 12    -.41  .02 -.07 -.01 -.07 -.02  .15 -.07  .04  .02 -.05 -.06
  ST.E.     .07  .08  .08  .08  .08  .08  .08  .08  .08  .08  .08  .09
    Q      33.9 34.0 34.9 34.9 35.9 35.9 40.3 41.2 41.5 41.6 42.1 42.9
```

```
PARTIAL AUTOCORRELATIONS

 1- 12    -.41 -.18 -.17 -.14 -.19 -.21 -.00 -.05 -.02  .04 -.01 -.08
 ST.E.     .07  .07  .07  .07  .07  .07  .07  .07  .07  .07  .07  .07
--
```

The identification stage leads us to the model $(1-B)Z_t = (1-\theta B)a_t$. The UTSMODEL paragraph will specify this. The model will be internally held under the label "IMA11".

Note (1-B) appearing to the right of "SERIESA" below is a modifier of the series to be analyzed. The variable "THETA1" is used to specify the initial value and later store the estimate of θ. Since "THETA1" has not been previously defined, the value 0.1 will be used as an initial value.

```
~~UTSMODEL  NAME IS IMA11.                                    @
~~          MODEL IS SERIESA((1-B)) = (1-THETA1*B)NOISE.
```

A summary of the model is now printed.

```
SUMMARY FOR UNIVARIATE TIME SERIES MODEL --  IMA11

----------------------------------------------------------------------------
VARIABLE  TYPE OF   PARAMETER-   ORIGINAL    DIFFERENCING
          VARIABLE   IZATION    OR CENTERED
                                                  1
SERIESA   RANDOM     OUTPUT      ORIGINAL    (1-B  )
----------------------------------------------------------------------------

 PARAMETER    VARIABLE   NUM./  FACTOR  ORDER  CONS-     VALUE     STD      T
   LABEL        NAME     DENOM.                TRAINT             ERROR   VALUE

  1  THETA1    SERIESA    MA       1      1     NONE      .1000
--
```

The model labelled "IMA11" is now estimated using CLS. Residuals will be retained in the variable "RESI".

```
~~UESTIM  MODEL IS IMA11.  HOLD RESIDUALS(RESI).

ITERATION TERMINATED DUE TO:
RELATIVE CHANGE IN THE OBJECTIVE FUNCTION LESS THAN 0.1000D-03

TOTAL NUMBER OF ITERATIONS . . . . . . . . . . .              6
RELATIVE CHANGE IN THE OBJECTIVE FUNCTION. . . . 0.2638D-04
MAXIMUM RELATIVE CHANGE IN THE ESTIMATES . . . . 0.6165D-02

SUMMARY FOR UNIVARIATE TIME SERIES MODEL --  IMA11

TIME SPAN FOR MODEL ESTIMATION :   1 TO  197

----------------------------------------------------------------------------
VARIABLE  TYPE OF   PARAMETER-  ORIGINAL    DIFFERENCING
          VARIABLE   IZATION   OR CENTERED
                                                 1
SERIESA   RANDOM     OUTPUT     ORIGINAL    (1-B  )
----------------------------------------------------------------------------

 PARAMETER    VARIABLE   NUM./  FACTOR  ORDER  CONS-     VALUE     STD      T
   LABEL        NAME     DENOM.                TRAINT             ERROR   VALUE

  1  THETA1    SERIESA    MA       1      1     NONE      .7015   .0511   13.73

TOTAL SUM OF SQUARES . . . . . . . .  0.312420D+02
```

```
TOTAL NUMBER OF OBSERVATIONS . . . .              197
RESIDUAL SUM OF SQUARES. . . . . . .      0.198854D+02
EFFECTIVE NUMBER OF OBSERVATION. . .              196
RESIDUAL VARIANCE ESTIMATE . . . . .      0.101456D+00
RESIDUAL STANDARD ERROR. . . . . . .      0.318522D+00
--
```

Diagnostic checking of the residuals, "RESI", will now be performed. We examine only the sample ACF.

```
~~ACF  VARIABLE IS RESI.  MAXLAG IS 12.  OUTPUT NOPRINT(PLOT).
     .
     .
     .
```

AUTOCORRELATIONS

1- 12	.10	.01	-.11	-.12	-.13	-.01	.14	.02	.04	-.01	-.11	-.12
ST.E.	.07	.07	.07	.07	.07	.08	.08	.08	.08	.08	.08	.08
Q	2.1	2.2	4.7	7.7	10.9	11.0	14.9	15.0	15.2	15.3	17.6	20.9

--

As an alternative to the "ordinary" means to identify the model, we employ the sample extended autocorrelation function proposed by Tiao and Tsay (1981). We also specify the maximum values of p and q to examine; here 7 and 12 respectively.

```
~~EACF  VARI IS SERIESA. MAXLAG IS AR(7),MA(12).

TIME PERIOD ANALYZED . . . . . . . . . .   1  TO   197
EFFECTIVE NUMBER OF OBSERVATIONS . . .              197
NAME OF THE SERIES . . . . . . . . . .          SERIESA
STANDARD DEVIATION OF THE SERIES . . .           0.3992
MEAN OF THE (DIFFERENCED) SERIES . . .          17.0624
STANDARD DEVIATION OF THE MEAN . . . .           0.0284
T-VALUE OF MEAN (AGAINST ZERO) . . . .         599.8357
```

THE EXTENDED ACF TABLE

(Q-->)	0	1	2	3	4	5	6	7	8	9	10	11	12
(P= 0)	1.00	.57	.50	.40	.36	.33	.35	.39	.32	.30	.25	.19	.16
(P= 1)	.57	-.39	.04	-.06	-.01	-.07	-.01	.16	-.07	.04	.04	-.04	-.06
(P= 2)	.25	-.29	-.27	-.04	.01	-.05	-.01	.17	.03	.04	.07	-.02	-.05
(P= 3)	.08	-.50	-.01	.09	-.01	-.01	-.03	.16	-.03	.11	-.02	-.01	.01
(P= 4)	.09	-.48	-.02	.08	-.02	-.01	-.04	.14	.03	.09	-.03	-.02	.00
(P= 5)	.07	-.39	-.41	-.17	.01	-.17	-.02	.10	-.01	.06	.07	-.01	.01
(P= 6)	.14	-.49	.15	-.18	-.00	-.26	-.06	.09	-.10	.05	.02	-.02	-.03
(P= 7)	.19	.19	-.01	.04	.34	.26	-.08	-.23	.03	.01	.03	-.05	-.07

SIMPLIFIED EXTENDED ACF TABLE

(Q-->)	0	1	2	3	4	5	6	7	8	9	10	11	12
(P= 0)	X	X	X	X	X	X	X	X	X	X	X	X	X
(P= 1)	X	X	0	0	0	0	0	X	0	0	0	0	0
(P= 2)	X	X	X	0	0	0	0	X	0	0	0	0	0
(P= 3)	0	X	0	0	0	0	0	X	0	0	0	0	0
(P= 4)	0	X	0	0	0	0	0	0	0	0	0	0	0
(P= 5)	0	X	X	X	0	X	0	0	0	0	0	0	0
(P= 6)	0	X	0	X	0	X	0	0	0	0	0	0	0
(P= 7)	X	X	0	0	X	X	0	X	0	0	0	0	0

--

Since we are only interested in "significant" EACF values, the above table is summarized. We note the "triangle of insignificant" values appear to originate from the location p=1, q=1. We recognize there are some significant values in this triangle, but we realize that this is always a possibility. Hence we will entertain an ARMA(1,1) model.

The model is specified and held under the name "ARMA11". Here we specify the model of the form $(1-\phi B)Z_t = C + (1-\theta B)a_t$. We use the label "THETA2" for θ here since "THETA1" was used above and we wish 0.1 as initial values for all ARMA parameters.

```
~~UTSMODEL  NAME IS ARMA11.  NO SHOW.                                   @
~~          MODEL IS (1-PHI*B)SERIESA = CNST+(1-THETA2*B)NOISE.
--
```

Model "ARMA11" is estimated using CLS. The residuals will not be retained.

```
~~UESTIM  MODEL ARMA11.

ITERATION TERMINATED DUE TO:
RELATIVE CHANGE IN THE OBJECTIVE FUNCTION LESS THAN 0.1000D-03

TOTAL NUMBER OF ITERATIONS IS . . . . . . . . . .          8
RELATIVE CHANGE IN THE OBJECTIVE FUNCTION . . . . 0.1855D-04
MAXIMUM RELATIVE CHANGE IN THE ESTIMATES  . . . . 0.2123D-01

SUMMARY FOR UNIVARIATE TIME SERIES MODEL --  ARMA11

TIME SPAN FOR MODEL ESTIMATION :    1 TO  197
-----------------------------------------------------------------------------
VARIABLE  TYPE OF  PARAMETER-  ORIGINAL    DIFFERENCING
          VARIABLE   IZATION   OR CENTERED

SERIESA   RANDOM     OUTPUT    ORIGINAL
-----------------------------------------------------------------------------

 PARAMETER     VARIABLE    NUM./  FACTOR  ORDER  CONS-     VALUE       STD      T
   LABEL         NAME      DENOM.                TRAINT               ERROR   VALUE

 1   CNST                  CNST      1      0    NONE     1.6102      .7871    2.05
 2  THETA2     SERIESA      MA       1      1    NONE      .5667      .0876    6.47
 3   PHI       SERIESA      AR       1      1    NONE      .9058      .0461   19.63

TOTAL SUM OF SQUARES . . . . . . . . .   0.312420D+02
TOTAL NUMBER OF OBSERVATIONS . . . .              197
RESIDUAL SUM OF SQUARES. . . . . . . .   0.192690D+02
EFFECTIVE NUMBER OF OBSERVATION. . .              196
RESIDUAL VARIANCE ESTIMATE . . . . . .   0.983111D-01
RESIDUAL STANDARD ERROR. . . . . . . .   0.313546D+00
--
```

The model "ARMA11" is re-estimated using the exact algorithm. Initial estimates for the parameters are the final estimates from above. The residuals will again be held under the label "RESI".

```
~~UESTIM  MODEL ARMA11.  METHOD IS MLE.  HOLD RESIDUALS(RESI).

ITERATION TERMINATED DUE TO:
RELATIVE CHANGE IN THE ESTIMATED VARIANCE LESS THAN 0.1000D-03

TOTAL NUMBER OF ITERATIONS . . . . . . . . . . .          2
RELATIVE CHANGE IN THE OBJECTIVE FUNCTION. . . . 0.1613D-04
MAXIMUM RELATIVE CHANGE IN THE ESTIMATES . . . . 0.4792D-02
RELATIVE CHANGE IN THE VARIANCE. . . . . . . . . 0.2344D-07

SUMMARY FOR UNIVARIATE TIME SERIES MODEL --  ARMA11

TIME SPAN FOR MODEL ESTIMATION :    1 TO  197
```

```
------------------------------------------------------------------------
VARIABLE  TYPE OF   PARAMETER-  ORIGINAL    DIFFERENCING
          VARIABLE  IZATION     OR CENTERED

SERIESA   RANDOM    OUTPUT      ORIGINAL
------------------------------------------------------------------------
```

PARAMETER LABEL		VARIABLE NAME	NUM./ DENOM.	FACTOR	ORDER	CONS- TRAINT	VALUE	STD ERROR	T VALUE
1	CNST		CNST	1	0	NONE	1.6030	.7520	2.13
2	THETA2	SERIESA	MA	1	1	NONE	.5793	.0846	6.84
3	PHI	SERIESA	AR	1	1	NONE	.9063	.0441	20.56

```
TOTAL SUM OF SQUARES . . . . . . . .  0.312420D+02
TOTAL NUMBER OF OBSERVATIONS . . . .           197
RESIDUAL SUM OF SQUARES. . . . . . .  0.191237D+02
EFFECTIVE NUMBER OF OBSERVATION. . .           196
RESIDUAL VARIANCE ESTIMATE . . . . .  0.975701D-01
RESIDUAL STANDARD ERROR. . . . . . .  0.312362D+00
--
```

Note the results here are similar to those obtained using the model "IMA11" but the identification stage was simpler.

Diagnostic checking of the residuals is again performed using only the sample ACF. Although the sample autocorrelation at lag 7 is significant, it may be ignored unless additional evidence indicates otherwise. The ARMA(1,1) model is also acceptable.

```
~~ACF   VARIABLE IS RESI.  MAXLAG IS 12. @
~~      OUTPUT NOPRINT(PLOT).
   .
   .
   .
AUTOCORRELATIONS
```

1- 12	.05	.00	-.08	-.08	-.08	.03	.17	.04	.07	.03	-.07	-.09
ST.E.	.07	.07	.07	.07	.07	.07	.07	.08	.08	.08	.08	.08
Q	.4	.4	1.9	3.2	4.6	4.8	10.8	11.2	12.1	12.2	13.2	14.8

```
--

~~UFORECAST  MODEL ARMA11.
```

Forecasts using the model "ARMA11" (and the final estimates of it) are now requested. The default number of forecasts is 24.

```
----------------------------------
 24 FORECASTS, BEGINNING AT  197
----------------------------------
```

TIME	FORECAST	STD. ERROR	ACTUAL
198	17.3820	0.3124	
199	17.3556	0.3286	
200	17.3316	0.3414	
201	17.3099	0.3516	
.	.	.	
.	.	.	
.	.	.	

Example 2: The series in this example are logarithms of monthly retail apparel sales data from those retail stores the U.S. Department of Census classifies as "other apparel stores" and "shoe stores" during the time period January 1967 through June 1978. We use this example to illustrate multivariate ARMA model building in the SCA system.

Sales data for both the "other apparel stores" and "shoe stores" series are read from a file identified as 10.

```
~~INPUT  VARIABLE ARE OTHERS, SHOES.  FILE IS 10.
```

```
REAL     VARIABLE  OTHERS ,  A  138  BY    1 MATRIX, IS STORED IN THE WORKSPACE
REAL     VARIABLE  SHOES  ,  A  138  BY    1 MATRIX, IS STORED IN THE WORKSPACE
--

  LNOTHER = LN(OTHERS)
--

  LNSHOE  = LN(SHOES)
--
```

Take logarithmic transformation for the series "OTHERS" and store the result in a variable labelled "LNOTHER". Similarly the logarithmic transformation of "SHOE" is stored in "LNSHOE".

Following the procedures developed by Tiao and Box (1981), sample cross correlation matrices and autoregressions should be examined in order to tentatively specify an ARMA model. It is necessary to difference each series to achieve stationarity. Each series is differenced according to $W_t = (1-B)(1-B^{12})Y_t$.

First the sample cross correlation matrices are examined.

```
--MIDEN  VARIABLES ARE LNOTHER, LNSHOE. DFORDERS ARE 1,12. MAXLAG IS 24.
                                                   1       12
DIFFERENCE ORDERS. . . . . . . . . . . . . . . (1-B  ) (1-B  )
TIME PERIOD ANALYZED . . . . . . . . . . . . .  1  TO    138
EFFECTIVE NUMBER OF OBSERVATIONS (NOBE). .               125

SERIES   NAME               MEAN        STD. ERROR

   1     LNOTHER            0.0003      0.0805
   2     LNSHOE             0.0007      0.1234

SUMMARIES OF CROSS CORRELATION MATRICES USING +,-,., WHERE
     + DENOTES A VALUE GREATER THAN 2/SQRT(NOBE)
     - DENOTES A VALUE LESS THAN -2/SQRT(NOBE)
     . DENOTES A NON-SIGNIFICANT VALUE BASED ON THE ABOVE CRITERION

BEHAVIOR OF VALUES IN (I,J)TH POSITION OF CROSS CORRELATION MATRIX OVER
ALL OUTPUTTED LAGS

              1                2
  1    -.........+-     -.........+-
  1    ............     +.........-

  2    -.........+-     -.........+-
  2    +...........     +.........+.
```

For a discussion of significance symbols see Tiao and Box (1981).

```
CROSS CORRELATION MATRICES IN TERMS OF +,-,.

LAGS  1 THROUGH  6
       - -      . .      . .      . .      . .      . .
       - -      . .      . .      . .      . .      . .

LAGS  7 THROUGH 12
       . .      . .      . .      . .      + +      - -
       . .      . .      . .      . .      + +      - -

LAGS 13 THROUGH 18
       . +      . .      . .      . .      . .      . .
       + +      . .      . .      . .      . .      . .

LAGS 19 THROUGH 24
       . .      . .      . .      . .      . .      . -
       . .      . .      . .      . .      . +      . .

--
```

The significance of the 1st and 12th lags suggest a possible model may include an MA(1) and MA(12) term. Perhaps a multiplicative MA model that incorporates both seasonal and nonseasonal behavior could suffice.

Fourteen stepwise autoregressions are computed, but initially only the summary results will be examined.

```
~~MIDEN  VARIABLES ARE LNOTHER, LNSHOE.  DFORDERS ARE 1,12. @
~~       ARFITS ARE 1 TO 14.  NO CCM.  NO DESCRIBE.
                                             1       12
DIFFERENCE ORDERS. . . . . . . . . . . . (1-B  ) (1-B  )
TIME PERIOD ANALYZED . . . . . . . . . .  1  TO   138
EFFECTIVE NUMBER OF OBSERVATIONS . . .            125
      .
      .
      .
```

========== STEPWISE AUTOREGRESSION SUMMARY ==========

AIC refers to the Akaike Information criterion.

LAG	RESIDUAL VARIANCES	EIGENVAL. OF SIGMA	CHI-SQ TEST	AIC	SIGNIFICANCE OF PARTIAL AR COEFF.
1	.348E-02	.139E-02	55.68	-11.049	. -
	.873E-02	.108E-01			. -
2	.289E-02	.137E-02	27.57	-11.252	. -
	.690E-02	.841E-02			. -
3	.283E-02	.137E-02	3.59	-11.229	. .
	.669E-02	.816E-02			. .
4	.280E-02	.134E-02	4.42	-11.215	. .
	.650E-02	.796E-02			. .
5	.266E-02	.129E-02	5.56	-11.213	. +
	.644E-02	.781E-02			. .
6	.264E-02	.128E-02	1.33	-11.168	. .
	.641E-02	.776E-02			. .
7	.262E-02	.126E-02	1.81	-11.129	. .
	.640E-02	.776E-02			. .
8	.253E-02	.123E-02	3.54	-11.109	. .
	.634E-02	.764E-02			. .
9	.246E-02	.115E-02	7.40	-11.132	. .
	.626E-02	.758E-02			. .
10	.204E-02	.107E-02	19.29	-11.290	- .
	.557E-02	.653E-02			- .
11	.192E-02	.107E-02	8.55	-11.329	. .
	.508E-02	.593E-02			. .
12	.186E-02	.844E-03	24.07	-11.553	. .
	.467E-02	.568E-02			. -
13	.172E-02	.833E-03	6.80	-11.577	- .
	.441E-02	.530E-02			- .
14	.172E-02	.818E-03	1.57	-11.538	. .
	.440E-02	.530E-02			. .

NOTE: CHI-SQUARED CRITICAL VALUES WITH 4 DEGREES OF FREEDOM ARE

5% : 9.5   1% : 13.3

It is observed that autoregression lags 1, 2 and 12 are significant. Hence based on the autoregression and CCM information, a model of the form

$$(I-\Phi_1B-\Phi_2B^2)(I-\Phi_{12}B^{12})\nabla_1\nabla_{12}Z_t = a_t$$

or

$$\nabla_1\nabla_{12}Z_t = (I-\Theta_1B-\Theta_2B^2)(I-\Theta_{12}B^{12})a_t$$

is reasonable. The latter will be estimated.

The multiplicative MA model

$$W_t = (I-\Theta_1 B-\Theta_2 B^2)(I-\Theta_{12} B^{12})a_t$$

is specified where

$$W_{1t} = (1-B)(1-B^{12})(\log \text{"other apparel store"})$$

$$W_{2t} = (1-B)(1-B^{12})(\log \text{"shoe store"}).$$

Initial and final estimates of Θ_1 are kept in TH1 and any constraints on the elements of Θ_1 are kept in RTH1. Since no initial estimates for Θ_1 are given, 0.10 is assumed for all matrix elements. Similarly all elements of RTH1 are assumed to be 0, indicating all parameters can be varied. (Note if a value of 1 is specified for an element of RTH1, the corresponding parameter of TH1 will be constrained to its initial value throughout the estimation process.) Θ_2 and Θ_{12} are handled in the same manner.

```
~~MTSMODEL  NAME IS APPAREL.                                                @
~~          SERIES IS LNOTHER((1-B) (1-B**12)), LNSHOE((1-B) (1-B**12)).@
~~          MODEL IS SERIES = (1-TH1*B-TH2*B**2) (1-TH12*B**12)NOISE.  @
~~          CONSTRAINTS ARE TH1(RTH1), TH2(RTH2), TH12(RTH12).

SUMMARY FOR MULTIVARIATE ARMA MODEL -- APPAREL

VARIABLE   DIFFERENCING

 LNOTHER      1 12
 LNSHOE       1 12

 PARAMETER      FACTOR      ORDER    CONSTRAINT

 1       TH1    REG MA        1        RTH1
 2       TH2    REG MA        2        RTH2
 3      TH12    SEA MA       12        RTH12
--
```

```
~~MESTIM  MODEL APPAREL.
```

The model held under the name (label) "APPAREL" is now estimated using CLS.

```
SUMMARY FOR THE MULTIVARIATE ARMA MODEL

SERIES     NAME              MEAN          STD DEV     DIFFERENCE ORDER(S)

  1      LNOTHER            0.0003         0.0805          1  12
  2      LNSHOE             0.0007         0.1234          1  12
                                     12
SPECIFIED MODEL = ( 0, 2) * ( 0, 1)

NUMBER OF OBSERVATIONS =  138  (EFFECTIVE NUMBER = NOBE =  125)

PRELIMINARY MODEL SPECIFICATION WITH INITIAL PARAMETER ESTIMATES

PARAMETER             PARAMETER                  INITIAL
 NUMBER              DESCRIPTION                 ESTIMATE
----------    ----------------------------     -----------
     1          MOVING AVERAGE ( 1, 1, 1)        0.100000
     2          MOVING AVERAGE ( 1, 1, 2)        0.100000
     3          MOVING AVERAGE ( 1, 2, 1)        0.100000
     4          MOVING AVERAGE ( 1, 2, 2)        0.100000
     5          MOVING AVERAGE ( 2, 1, 1)        0.100000
```

MOVING AVERAGE(ℓ,i,j) refers to the (i,j) element of the MA matrix of lag order ℓ.

```
    6          MOVING AVERAGE ( 2, 1, 2)          0.100000
    7          MOVING AVERAGE ( 2, 2, 1)          0.100000
    8          MOVING AVERAGE ( 2, 2, 2)          0.100000
    9     SEAS MOVING AVERAGE (12, 1, 1)          0.100000
   10     SEAS MOVING AVERAGE (12, 1, 2)          0.100000
   11     SEAS MOVING AVERAGE (12, 2, 1)          0.100000
   12     SEAS MOVING AVERAGE (12, 2, 2)          0.100000

-----------------------
ERROR COVARIANCE MATRIX
-----------------------

             1            2
  1      .004688
  2      .005511      .011600

ITERATIONS TERMINATED DUE TO:
  RELATIVE CHANGE IN DETERMINANT OF COVARIANCE MATRIX .LE.  .100D-03
  TOTAL NUMBER OF ITERATIONS IS   10

FINAL MODEL SUMMARY WITH CONDITIONAL LIKELIHOOD PARAMETER ESTIMATES

----- REGULAR THETA MATRICES -----

ESTIMATES OF  REGULAR THETA(  1 ) MATRIX AND ITS SIGNIFICANCE
     .401      .467      + +
     .324      .769      + +

STANDARD ERRORS

     .106      .075
     .137      .099

ESTIMATES OF  REGULAR THETA(  2 ) MATRIX AND ITS SIGNIFICANCE
     .075     -.294      . -
    -.128     -.394      . -

STANDARD ERRORS
     .109      .077
     .134      .098

----- SEASONAL THETA MATRICES OF PERIOD 12 -----

ESTIMATES OF SEASONAL THETA( 12 ) MATRIX AND ITS SIGNIFICANCE
     .413      .155      + +
    -.023      .715      . +

STANDARD ERRORS
     .098      .070
     .120      .083

-----------------------
ERROR COVARIANCE MATRIX
-----------------------

             1            2
  1      .002484
  2      .002018      .004823

        .
        .
        .
--
```

For the significance of a parameter estimate, we use '+' to denote positively significant, '-' negatively significant, and '.' insignificant at the 5% level.

Based on the estimation result, the elements with small values and off-diagonal element of Θ_{12} will be constrained to 0. This is achieved by the following eight analytic statements in which the elements in the matrices holding parameter estimates are set to 0 and the corresponding elements of the constraint matrices are set to 1.

```
~~RTH2(1,1)=1
--

~~TH2(1,1)=0
--

~~RTH2(2,1)=1
--

~~TH2(2,1)=0
--

~~RTH12(1,2)=1
--

~~TH12(1,2)=0
--

~~RTH12(2,1)=1
--

~~TH12(2,1)=0
--

~~MESTIM  MODEL APPAREL.
```

The model with constraints is estimated.

```
SUMMARY FOR THE MULTIVARIATE ARMA MODEL

SERIES     NAME              MEAN          STD DEV     DIFFERENCE ORDER(S)

  1       LNOTHER           0.0003         0.0805          1   12
  2       LNSHOE            0.0007         0.1234          1   12
                                    12
SPECIFIED MODEL = ( 0, 2) * ( 0, 1)

NUMBER OF OBSERVATIONS =  138  (EFFECTIVE NUMBER = NOBE =  125)

PRELIMINARY MODEL SPECIFICATION WITH INITIAL PARAMETER ESTIMATES
      .
      .
      .
ITERATIONS TERMINATED DUE TO:
  RELATIVE CHANGE IN DETERMINANT OF COVARIANCE MATRIX .LE.  .100D-03
  TOTAL NUMBER OF ITERATIONS IS    5

FINAL MODEL SUMMARY WITH CONDITIONAL LIKELIHOOD PARAMETER ESTIMATES

----- REGULAR THETA MATRICES -----

ESTIMATES OF  REGULAR THETA(  1 ) MATRIX AND ITS SIGNIFICANCE
      .471      .430       + +
      .233      .786       + +

STANDARD ERRORS
      .083      .071
      .101      .089
```

```
ESTIMATES OF  REGULAR THETA(  2 ) MATRIX AND ITS SIGNIFICANCE
     .000     -.239        . -
     .000     -.463        . -

STANDARD ERRORS
      --        .064
      --        .076
```

Note that no standard error estimate is given for those parameters fixed at 0. This highlights the constraints imposed.

```
----- SEASONAL THETA MATRICES OF PERIOD 12 -----

ESTIMATES OF SEASONAL THETA( 12 ) MATRIX AND ITS SIGNIFICANCE
     .519      .000       + .
     .000      .648       . +

STANDARD ERRORS
     .067       --
      --       .063

-----------------------
ERROR COVARIANCE MATRIX
-----------------------

              1           2
  1      .002610
  2      .002103     .004868

-----------------------------------
CORRELATION MATRIX OF THE PARAMETERS
-----------------------------------

         1      2      3      4      5      6      7      8
  1    1.00
  2    -.50   1.00
  3     .24    .     1.00
  4     .      .46   -.47   1.00
  5     .     -.52    .     -.50   1.00
  6     .     -.41    .     -.57    .54   1.00
  7     .      .      .      .      .      .     1.00
  8     .      .      .      .      .      .      .28   1.00
--
```

The '.' symbol in the correlation matrix denotes insignificant correlation.

The model is estimated by the maximum likelihood method in order to obtain more accurate results. The residual series for "LNOTHER" and "LNSHOE" are stored under the names "ROTHER" and "RSHOE". They will be used for diagnostic checking later.

```
~~MESTIM  MODEL APPAREL.  METHOD IS MLE. @
~~        HOLD RESIDUALS(ROTHER,RSHOE).

SUMMARY FOR THE MULTIVARIATE ARMA MODEL

SERIES    NAME               MEAN          STD DEV    DIFFERENCE ORDER(S)

  1     LNOTHER           0.0003         0.0805          1  12
  2     LNSHOE            0.0007         0.1234          1  12
                                   12
SPECIFIED MODEL = ( 0, 2) * ( 0, 1)

NUMBER OF OBSERVATIONS =  138   (EFFECTIVE NUMBER = NOBE =  125)

PRELIMINARY MODEL SPECIFICATION WITH INITIAL PARAMETER ESTIMATES
```

```
     .
     .
     .
ITERATIONS TERMINATED DUE TO:
  CHANGE IN   (-2*LOG LIKELIHOOD)/NOBE   .LE.   .100D-03
  TOTAL NUMBER OF ITERATIONS IS    11

FINAL MODEL SUMMARY WITH MAXIMUM LIKELIHOOD PARAMETER ESTIMATES

----- REGULAR THETA MATRICES -----

ESTIMATES OF  REGULAR THETA(  1 ) MATRIX AND ITS SIGNIFICANCE
     .509     .398       + +
     .262     .793       + +

STANDARD ERRORS
     .079     .070
     .101     .088

ESTIMATES OF  REGULAR THETA(  2 ) MATRIX AND ITS SIGNIFICANCE
     .000    -.234       . -
     .000    -.452       . -

STANDARD ERRORS
      --      .063
      --      .077

----- SEASONAL THETA MATRICES OF PERIOD 12 -----

ESTIMATES OF SEASONAL THETA( 12 ) MATRIX AND ITS SIGNIFICANCE
     .872     .000       + .
     .000     .878       . +

STANDARD ERRORS
     .051      --
      --      .052

-----------------------
ERROR COVARIANCE MATRIX
-----------------------

               1           2
  1      .001987
  2      .001556     .003747

LOG LIKELIHOOD AT FINAL ESTIMATES IS          -1240.022427
     .
     .
     .
--
~~MIDEN   VARIABLES ARE ROTHER,RSHOE.  MAXLAG IS 24.

TIME PERIOD ANALYZED . . . . . . . . . . 14  TO   138
EFFECTIVE NUMBER OF OBSERVATIONS . . .            125

SERIES   NAME            MEAN       STD. ERROR

   1    ROTHER         -0.0011      0.0408
   2    RSHOE          -0.0015      0.0575
      .
      .
      .
```

A diagnostic check of the residual series is now performed.

```
CROSS CORRELATION MATRICES IN TERMS OF +,-,.

LAGS  1 THROUGH  6
      . .      . .      . .      . .      . .      . .
      . .      . .      . .      . .      . .      . .

LAGS  7 THROUGH 12
      . .      . .      . .      . .      . +      . .
      . .      . .      . .      . .      . .      . .

LAGS 13 THROUGH 18
      - .      . .      . -      . .      . .      . .
      . .      . .      . .      . .      . .      . .

LAGS 19 THROUGH 24
      . .      . .      . .      . .      . .      . -
      . .      . .      . .      . .      . .      . -
--
```

The sample CCM's for the residual series are computed. Although there are still a few significant elements, the correlations are small. The last model is acceptable.

```
~~MFORECAST  MODEL APPAREL.  NOFS IS 12.
```

Twelve future values are forecasted based on the last model.

```
--------------------------------------
 12 FORECASTS, BEGINNING AT ORIGIN = 138
--------------------------------------
```

SERIES:	LNOTHER		LNSHOE	
T	FORECAST	STD ERR	FORECAST	STD ERR
139	6.474	0.045	6.122	0.061
140	6.621	0.050	6.285	0.062
141	6.577	0.053	6.360	0.071
142	6.614	0.056	6.275	0.080
143	6.698	0.059	6.303	0.087
144	7.207	0.061	6.604	0.094
145	6.379	0.064	6.107	0.100
146	6.290	0.067	5.974	0.106
147	6.577	0.069	6.343	0.112
148	6.581	0.071	6.337	0.117
149	6.580	0.074	6.298	0.122
150	6.578	0.076	6.284	0.127

4. A BRIEF DESCRIPTION OF THE SCA PARAGRAPHS

Below we briefly describe the SCA paragraphs and their functions.

Data input and output

```
INPUT    -- inputs data into workspace
PRINT    -- prints the content of variables
SAVE     -- saves data onto a file
DISPLAY  -- prints text and values according to user specified format
```

Data manipulations and editing

```
GENERATE -- generates a vector or matrix
CHANGE   -- changes values in a vector or matrix
JOIN     -- joins variables together to form a new variable
OMIT     -- omits values in a variable or set of variables
SELECT   -- selects values in a variable or set of variables
```

```
DIFFEREN -- creates a series by differencing a series in the workspace
LAG      -- creates a series by lagging a series in the workspace
```

Plots and Histograms

```
PLOT     -- plots one pair of variables in a frame (i.e. a set of axes)
MPLOT    -- plots multiple pairs of variables in the same frame
TPLOT    -- plots one or several time series in different frames
MTPLOT   -- plots one or several time series in the same frame
HISTOGRAM-- computes and prints the histogram(s) of one or several variables
```

Regression analysis

```
REGRESS  -- performs linear regression analysis
NREGRESS -- performs nonlinear least squares estimation
```

Univariate time series analysis

```
ACF      -- computes the sample autocorrelation function for a time series
PACF     -- computes the sample partial autocorrelation function for a time
            series
EACF     -- computes the sample "extended" autocorrelation function for a time
            series
IACF     -- computes sample inverse autocorrelation function for a time series
CORNER   -- computes the Corner Table for autocorrelation function or transfer
            function weights
CCF      -- computes sample cross correlation function between two time series
UIDEN    -- computes sample ACF, PACF, and EACF for a time series
UTSMODEL -- specifies or modifies a univariate ARIMA or transfer function
            model
UESTIM   -- estimates the parameters in a univariate ARIMA or transfer function
            model
UFORECAST-- forecasts future values of a time series based on a univariate
            ARIMA or transfer function model
UFILTER  -- filters a time series by a univariate ARIMA or transfer function
            model
FFILTER  -- applies bandpass or bandreject filter
PSPECTRA -- estimates spectra or cross-spectra for one or two time series by
            smoothing the periodograms
CSPECTRA -- estimates spectra or cross-spectra for one or two time series
            based on covariance and autocovariances
USPECTRA -- computes spectra or cross-spectra for a univariate ARIMA or transfer
            function model
```

Multivariate time series analysis

```
MIDEN    -- computes cross correlation matrices and performs stepwise
            autoregression fitting for vector series
MTSMODEL -- specifies or modifies a multivariate ARMA model
MESTIM   -- estimates the parameters in a multivariate ARMA model
MFORECAST-- forecasts future values for a vector time series based on a
            multivariate ARMA model
STFMODEL -- specifies a simultaneous transfer function model
SESTIM   -- estimates the parameters in a simultaneous transfer function
            model
SFORECAST-- forecasts future values for a set of time series based on a
            simultaneous transfer function model
MSPECTRA -- computes spectra and cross-spectra for a multivariate ARMA model
```

Distribution and model simulation

SIMULATION-- generates data according to the user specified distribution or univariate time series model

SCA macro procedure

PARAMETER-- specifies the symbolic variables in an SCA procedure
RETURN -- signifies the end of an execution flow in an SCA procedure and subsequent action to be taken
CALL -- invokes an SCA procedure

Miscellaneous

PROFILE -- specifies the environment features for an SCA session
WORKSPACE-- displays the current status of user's workspace, and/or manages the workspace by deleting unused variables or consolidating unused space
REVIEW -- reviews the previous output
HELP -- provides HELP information for a paragraph or specific keyword
RESTART -- restarts an SCA session
TIME -- prints date, clock time, and CPU time
STOP -- terminates an SCA session

5. ACKNOWLEDGEMENTS

Much of the theory and algorithms used in the time series portions of the SCA systems is based on material presented or developed by George Box, George Tiao and Steven Hillmer. The principal architects of multivariate time series software include Michael R. Grupe, William Bell, Ruey S. Tsay, Ih Chang and Juha Ahtola. The work of these people is gratefully acknowledged. We also would like to thank Mervin Muller for his helpful suggestions and comments on this paper. Further information on the SCA system may be obtained from the second author.

REFERENCES

BEGUIN, J.M., GOURIEROUX, C., and MONFORT, A. (1980). Identification of a Mixed Autoregressive-Moving Average Process: The Corner Method. In Time Series (Proceedings of the International Conference held at Nottingham University, March 1979). Ed: O.D. Anderson, North-Holland, Amsterdam & New York, 423-436.

BOX, G.E.P. and JENKINS, G.M. (1976). Time-Series Analysis: Forecasting and Control. Revised Edition. San Francisco: Holden Day.

HANSSENS, D.M. and LIU, L.-M. (1982). Lag Specification in Rational Distributed Lag Structural Models. Working paper 82-06, Department of Quantitative Methods, University of Illinois at Chicago.

HILLMER, S.C. and TIAO, G.C. (1979). Likelihood Function of Stationary Multiple Autoregressive Moving Average Models. Journal of the American Statistical Association 74: 652-660.

LIU, L.-M., and HANSSENS, D.M. (1982). Identification of Multiple-Input Transfer Function Models. Communications in Statistics A 11: 297-314.

LJUNG, G.M. and BOX, G.E.P. (1979). The Likelihood Function of Stationary Autoregressive-Moving Average Models. Biometrika 66: 265-270.

TIAO, G.C. and BOX, G.E.P. (1981). Modeling Multiple Time Series with Applications. Journal of the American Statistical Association 76: 802- 816.

TIAO, G.C., BOX, G.E.P., GRUPE, M.R., HUDAK, G.B., BELL, W.R., and CHANG, I. (1979). The Wisconsin Multiple Time Series Program, A Preliminary Guide. Department of Statistics, University of Wisconsin, Madison.

TIAO, G.C. and TSAY, R.S. (1981). Identification of Nonstationary and Stationary ARMA Models. Proceeding of American Statistical Association -- Business and Economic Statistics Section: 308-312.

WALL, K.D. (1976). FIML Estimation of Rational Distributed Lag Structural Form Models. Annals of Economic and Social Measurement 5: 53-64.

TIME SERIES ANALYSIS: Theory and Practice 4
O.D. Anderson (editor)

RECENT RESULTS IN FORECASTS AND MODELS FOR MULTIPLE TIME SERIES USING THE STATE SPACE FORECASTING METHOD

Alan V. Cameron

State Space Systems Inc.
2091 Business Center Drive, Suite 100
Irvine, Ca. 92715
USA

This paper presents several recent theoretical and practical results obtained in forecasting applications for multiple time series. The applications described are - Interest Rate forecasts and efficient portfolios for Money Market Mutual Funds, and forecasts and optimal control strategies for Inventory and Production control.

The approach used in the paper is based on the State Space Forecasting method for directly identifying models for single or multiple time series from available historical data. Once the models have been developed they are used to identify optimal control and decision strategies to meet particular business objectives.

For the money market mutual fund application, investment decisions based on the interest rate forecasts are obtained through the use of Modern Portfolio Theory. For the inventory and production control application, production control decisions are obtained through an objective function which balances production change costs and inventory costs.

There are two concepts represented in each of these applications. The first concept is that improved forecasts with reduced forecast errors can be obtained by taking advantage of the leading, lagging and feedback relationships that may exist between the multiple times series being analysed. The second concept is that in each of these applications there is an objective criterion which can be used to translate the forecasts into a specific business decision.i.e. there is a direct connection between the forecasts that are produced and the decisions that are taken based on the forecasts.

STATE SPACE FORECASTING AND MODERN PORTFOLIO THEORY

The task of forecasting investment rates of return (yields and interest rates) and at the same time managing an interrelated set of alternative investments is an important function for many businesses, government organizations and individual investors.

If rates of return for investments were known in advance without any uncertainty, it would be easy to choose investments for maximum income. However, in most investment areas a common experience has been that the greater the rate of return the greater tends to be the risk for individual investments. Hence the wise investor will develop a portfolio of investments which balances in a satisfactory way, the high expected returns with the risks associated with the returns.

Because of the rapid changes occurring in money markets, government security markets, foreign currency markets and financial futures markets, the classical methods of technical analysis, forecasting and portfolio management have a difficult time in meeting these demanding investment portfolio requirements.

However a new approach combining two important concepts - multiple time series analysis and modern portfolio theory - is now available and can provide important new tools for the investment analyst and portfolio manager. Multiple time series analysis provides the tools to improve forecasts and reduce risks for individual investments and groups of investments. Modern Portfolio Theory provides the tools so the improved forecasts and cross-correlations between the investments can be used to build efficient portfolios with maximum return and minimum risk.

MULTIPLE TIME SERIES ANALYSIS AND STATE SPACE FORECASTING

Using multiple time series analysis improved forecasts can be prepared for the expected returns from each investment. These improvements are made possible by taking advantage of the relationships between investments and between the investments and other leading indicators of future price and yield changes.

In the past, building these forecasts for interrelated series has been a difficult task because of the complex leads, lags and feedbacks between the series and high correlation between various series (multicollinearity). Specialised regression and econometric models have had to be used to build forecasting models for these applications.

Now however, multiple time series analysis techniques, such as State Space Forecasting, can build forecasting models for these interrelated alternative investments in a relatively straight-forward way. Futhermore, by incorporating leading indicator information, these forecating models appear in particular situations to provide significant economic advantage by reducing the risks and improving the forecasts for the investments.

As an illustration of this situation, analysis of many historical price time series for individual investments (e.g. stock price series) shows that they are close to "random walk" processes. This result implies that based on the single historical series alone, the best forecast for all future values is given by the current value of the series and that future values are equally likely to rise or fall.

Analysis of groups of historical price and interest rate series and related economic series shows however, that there are important relationships between the series. The leading, lagging and feedback relationships which can be identified in these situations by the State Space Forecasting program, are useful in improving forecast accuracy and reducing the risks associated with each investment. These results cannot be obtained using standard multiple regression methods because of the feedback and multicollinearity effects in the historical data.

The multiple time series approach also allows forecasts purchased from professional investment advisory services, or forecasts obtained from other forecasting methods, to be included in an efficient composite forecasting model. These alternative forecasts can be regarded as potential leading indicators for the investments. Any independent information they may contain, which is not already reflected in the historical data, is then included in the resulting composite forecast produced by the State Space Forecasting program.

MODERN PORTFOLIO THEORY

The second major concept in this new approach is the use of Modern Portfolio Theory (and Capital Market Theory) to translate the improved forecasts for the interrelated investments into precise recommendations for the composition of a portfolio.

The portfolio approach is important because even with improved forecasts and reduced risks, the highest yielding investments may still have risks which are unacceptably high on their own. These high yield and high risk investments can however, be an important part of a properly balanced portfolio.

Furthermore, by properly selecting the portfolio, ratio of the resulting rate of return to the risk will be greater than that available for any of the individual investments.

This ability to precisely determine the best portfolio composition for a particular set of investment forecasts and cross-correlations, is the important contribution of Modern Portfolio Theory. A futher advantage of the approach is that it also allows the economic value of alternative forecasts (e.g. those purchased from professional services) or alternative historical leading indicators to be precisely evaluated for each portfolio. The relative contribution of each forecast, new time series or additional investment can be directly measured by its effect on the overall portfolio rate of return and risk.

A PORTFOLIO OF MONEY MARKET MUTUAL FUNDS

In this application we will develop forecasts for two money market fund yields, the Dreyfus Liquid Assets Fund and the Merrill Lynch Ready Assets Fund. These are two of the largest money market mutual funds. At the time of writing the Dreyfus Fund has over $10 billion in assets and the Merrill Lynch Fund has over $22 billion in assets.

Each Monday morning the major Unites States newspapers report the yields achieved by 203 money market funds for the prior 7 days and for the prior 30 days. For example on August 2, 1982 the Los Angeles Times reported that the Dreyfus Liquid Assets Fund had an average yield for the week ending July 28 of 13.0% and the Merrill Lynch Fund had a yield of 12.0%. For the prior 30 days the average yields were 13.8% for Dreyfus and 15.4% for Merrill Lynch. In the same time period the average of the 203 funds reported had a 7 day yield of 12.2% and a 30 day yield of 13.0%.

In contrast to other savings investments in banks and savings and loan associations, where the future yields are quoted (and often guaranteed), these reported money market yields are for past returns which are "not necessarily indicative of future yields". So a common investment decision for both individuals and for institutional investors is to determine which funds will perform best in the near future and also how much money should be placed in each of the available funds so as to reduce risks and improve the rates of return.

As the first step, we will use the State Space Forecasting program to develop an efficient forecasting model for these two funds yields. Then we will apply modern portfolio theory and capital market theory to determine the proper mix to maintain in our portfolio consisting of these two funds an 90 day Treasury Bills. The source of data for the analysis is the Telerate Domestic Data Base on the Rapidata Inc. timesharing network. The data is reported weekly but as our portfolio is assumed to be adjusted only once per month, monthly data will be used in the analysis.

The first step in the analysis is to prepare a forecasting model for the future yields of the money funds we are considering. This forecast is prepared using the State Space Forecasting program as shown below.

```
COMMAND ?INPUT PROGRAM SSMMF             - This program SSMMF
                                           prepares money market
                                           forecasts using State Space.

               STATE SPACE MONEY MARKET FUND FORECAST

DO YOU NEED INSTRUCTIONS (Y/N) ?Y

THIS PROGRAM FORECASTS WEEKLY OR MONTHLY YIELDS OF MONEY MARKET
FUNDS IN THE TELERATE DOMESTIC DATA BASE. ENTER ONE OR MORE OF THE
SERIES NAMES FOR THE 7 OR 30 DAY YIELD SERIES E.G. MDLA07,MDLA30.
THE PROGRAM ALSO PROVIDES THE OPTION OF INCLUDING THE LEADING
INDICATOR - 90 DAY T-BILL MARKET INDICATOR - IN THE FORECASTS
WHICH ARE PRODUCED.

FORECASTS CAN BE PRODUCED FOR A SINGLE YIELD SERIES E.G. MALT07,
OR FOR A GROUP OF UP TO 10 SERIES AT ONE TIME.

IF MONTHLY FORECASTS ARE REQUESTED, THEY ARE BASED ON THE LAST
WEEKLY DATA REPORTED EACH MONTH IN THE DATA BASE.

ENTER THE SERIES TO BE FORECAST ?MDLA30,MMLR30

HAVE THE SERIES ALREADY BEEN SELECTED INTO YOUR WORK AREA (Y/N) ?N

ENTER THE STARTING WEEK DATE FOR THE SELECT ?W1/79

CONVERT TO MONTHLY DATA (Y/N) ?Y

INCLUDE 90 DAY T-BILL MARKET INDICATOR (Y/N) ?Y

                    STATE SPACE FORECAST
```

```
43 OBSERVATIONS,   3 SERIES
RANGE =M1/79-M7/82

DIFFERENCING PERFORMED
SERIES                 REGULAR        SEASONAL
BL3QB                     1               0
MDLA30                    1               0
MMLR30                    1               0
```

```
THE FOLLOWING ARE THE ELEMENTS OF THE STATE VECTOR
BL3QB(T)                                 - a first order State
MDLA30(T)                                Vector model has been
MMLR30(T)                                identified by the program.

RESIDUAL MEAN VECTOR                     - these are the residual
    3 ROWS    1 COLUMNS                  one step ahead forecast
ROW  1 -0.2299E-01                       error mean values.
ROW  2 -0.3514E-01
ROW  3 -0.1240

RESIDUAL COVARIANCE MATRIX               - these are the residual
    3 ROWS    3 COLUMNS                  one step ahead covariances.
ROW  1   1.727       0.3245E-01 -0.7303
ROW  2  0.3245E-01  0.3207       0.1487
ROW  3 -0.7303      0.1487        2.813

GOODNESS OF FIT        ORIGINAL DATA      DIFFERENCED DATA
( 9 D.F.)               129.640             76.716

R SQUARED TEST         ORIGINAL DATA      DIFFERENCED DATA
 BL3QB                    0.64258            0.08296
 MDLA30                   0.95498            0.76293
 MMLR30                   0.73517            0.57294
```

These results indicate that the forecasting model developed by the program can explain 76% of the changes in the Dreyfus fund from one month to the next and 57% of the changes in the Merrill Lynch fund yield. The following graph shows the overall changes that have occurred in yields during this historical period, as well as the forecasts produced by the model.

How do these results compare with forecasts produced by other models or with forecasts based just on the single series for each fund? Repeating the above execise for each of the series by itself gives the following results.

For the single series MDLA30 for the Dreyfus Liquid Assets fund, the State Space Forecasting Program identified a second order model with a residual covariance of .793 and an R-Squared statistic of 88.8% on the original data and 41% on the differenced data. This compares with a covariance of .320 for the multiple time series model and an R-Squared statistic of 95% on the original data and 76% on the differenced data.

STATE SPACE FORECAST FOR SERIES
BL3QB MDLA30 MMLR30

B=BL3QB - 90 day Treasury Bill Interest Rate
M=MDLA30 - Dreyfus Liquid Assets Fund yield
3=MMLR30 - Merrill Lynch Ready Assets Fund yield
4=BL3QBP - Forecast for BL3QBS
5=MDLA30P - Forecast for MDLA30
6=MMLR30P - Forecast for MMLR30

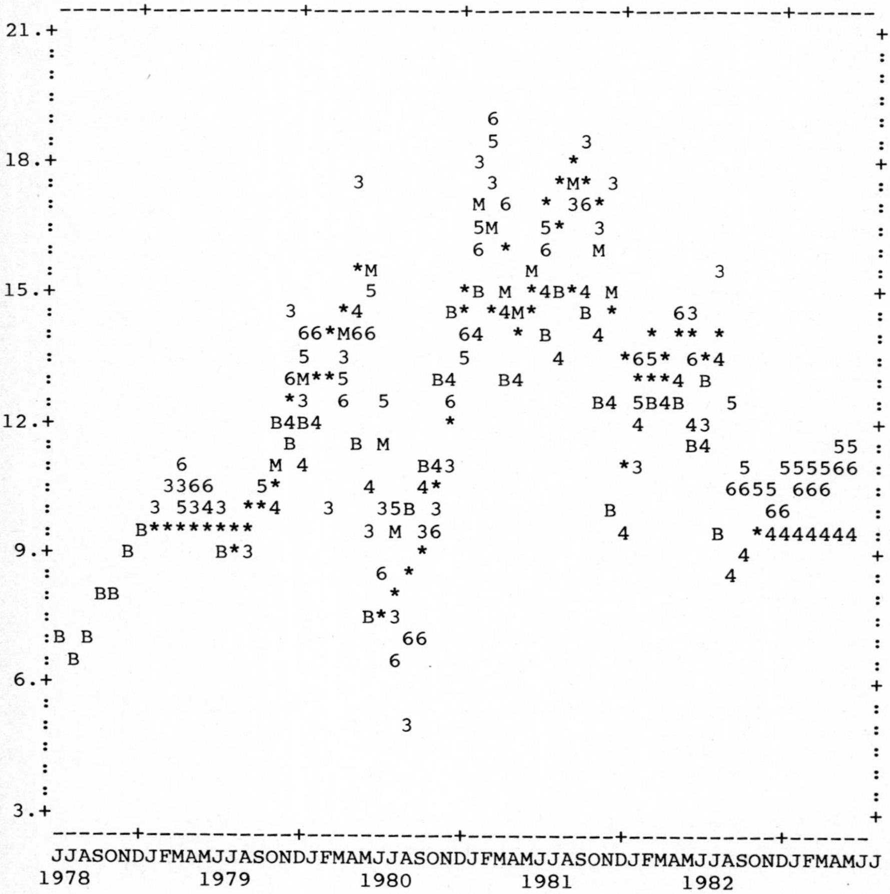

For the single time series MMLR30 for the Merrill Lynch Ready Assets fund, the State Space Forecating program identified a first order model with a residual covariance of 6.35 and an R-Squared statistic of 40% on the original data and only 3% on the differenced data (i.e. close to a random walk model). This compares with a covariance of 2.81 for the multiple time series model and an R-Squared statistic of 73% on the original data and 57% on the differenced data.

These results show that the multiple time series results are significantly better than those obtained from the corresponding single time series models. For both series the residual covariance was dropped to less than half of the corresponding single time series covariances.

Now that we have developed an efficient forecasting model, the next step is to use the results to create an efficient portfolio. The portfolio we will consider will have 3 alternative investments, the two money market funds and also 90 day treasury bills.

If no forecasts were available for the future yields of these investments, a reasonable forecast to use would be the current yield for each investment (i.e. the random walk forecast). Further, a reasonable measure of the risks and the cross-correlations between the investments in the future would be the actual variances and cross-correlations over the historical data. Using these initial assumptions about the future, the following portfolio results are obtained.

```
COMMAND ?PORTM$=BL3QB,MDLA30,MMLR30     - this defines the portfolio
                                          PORTM$ of 3 investments.
COMMAND ?RUN SSMPT PORTM$               - this runs the SSMPT program
                                          on the portfolio PORTM$.
              STATE SPACE PORTFOLIO ANALYSIS

DATE : AUG 15 82 SUN                    TIME : 01:39

         43 MONTHLY    OBSERVATIONS FROM  1/79

PORTFOLIO OF   3 INVESTMENTS
BL3QB     MDLA30     MMLR30

SSMPT CMD ?RUN                          - the risk free rate
                                          determines the
RISK FREE INTEREST RATE      0.00 %       capital market line.
```

SECURITY NAME	EXPECTED RETURN (E)	STANDARD DEVIATION(S)	E/S	PORTFOLIO COMPOSITION
BL3QB	9.60%	1.37	7.00	33%
MDLA30	13.80%	1.16	11.86	64%
MMLR30	15.40%	2.57	6.00	3%
TOTAL	12.46%	0.92	13.58	100%

These results indicate that given only historical data and no forecasts, 33% of the portfolio should be invested in T-Bills, 64% of the portfolio should be invested in Dreyfus Liquid Assets, and 3% of the portfolio should be invested in Merrill Lynch Ready Assets. These results are independent of the risk preferences or utility functions of the individual investor, (because of the Separation Theorem of Capital Market Theory). They are also independent of the size of the portfolio being considered (assuming the size of the portfolio being invested does not change the yields that can be obtained). The results do change however when the risk free interest rate changes. A better measure of the current risk free rate that could be obtained may be 8%. With this assumption the following portfolio is obtained (using only historical data).

```
SSMPT CMD ?CHANGE RFREE 8                - this changes the risk
                                           free interest rate to 8%
SSMPT CMD ?RUN

RISK FREE INTEREST RATE      8.00 %
```

SECURITY NAME	EXPECTED RETURN (E)	STANDARD DEVIATION(S)	E/S	PORTFOLIO COMPOSITION
BL3QB	9.60%	1.37	7.00	15%
MDLA30	13.80%	1.16	11.86	79%
MMLR30	15.40%	2.57	6.00	6%
TOTAL	13.27%	1.03	12.90	100%

```
SSMPT CMD ?PRI COR                       - this command prints out
                                           the correlations between
CURRENT CORRELATION MATRIX                 the investments.
      3 ROWS     3 COLUMNS
  ROW  1    1.000        0.3967E-01 -0.6569E-01
  ROW  2  0.3967E-01   1.000        0.4804
  ROW  3 -0.6569E-01  0.4804          1.000
```

With the change in the risk free interest rate, the portfolio composition has changed. Now there is 79% of the portfolio in Dreyfus Liquid Assets, 15% in T-Bills, and 6% in Merrill Lynch Ready Assets. Notice too the relative changes in the returns and risks between these two portfolios.

As the next step we will now use the forecast results we have obtained earlier using the State Space Forecasting program. These results have provided a new measure of the future yield, a new measure of the risks involved in each individual investment and a new measure of the cross-correlations between the investments in the future. With these new forecast results, the following portfolio is obtained.

```
SSMPT CMD ?RUN

RISK FREE INTEREST RATE      8.00 %
```

SECURITY NAME	EXPECTED RETURN (E)	STANDARD DEVIATION(S)	E/S	PORTFOLIO COMPOSITION
BL3QB	8.74%	1.31	6.65	2%
MDLA30	12.65%	0.57	22.35	96%
MMLR30	10.70%	1.68	6.38	2%
TOTAL	12.53%	0.55	22.75	100%

```
SSMPT CMD ?PRI COR

CURRENT CORRELATION MATRIX
      3 ROWS     3 COLUMNS
  ROW  1    1.000        0.4400E-01  -0.3310
  ROW  2  0.4400E-01      1.000        0.1570
  ROW  3 -0.3310         0.1570         1.000
```

These results show that the best portfolio will be 96% invested in Dreyfus Liquid Assets, 2% in T-Bills and 2% in Merrill Lynch Ready Assets (assuming an 8% risk free yield). If the risk free rate is assumed to be 0%, the following portfolio is obtained.

SSMPT CMD ?CHANGE RFREE 0 — - this changes the risk free rate to 0%

SSMPT CMD ?RUN

RISK FREE INTEREST RATE 0.00 %

SECURITY NAME	EXPECTED RETURN (E)	STANDARD DEVIATION(S)	E/S	PORTFOLIO COMPOSITION
BL3QB	8.74%	1.31	6.65	12%
MDLA30	12.65%	0.57	22.35	81%
MMLR30	10.70%	1.68	6.38	7%
TOTAL	12.04%	0.51	23.63	100%

This portfolio has 12% invested in T-Bills, 81% invested in Dryfus Liquid Assets and 7% invested in Merrill Lynch Ready Assets.

Comparing the portfolios to see the contribution of the forecasts, the return over risk ratio (E/S) has almost doubled through the use of the forecasts (i.e 22.75 for the portfolio using the forecasts and 12.90 for the original portfolio at the 8% risk free rate.) Further the standard deviation in the rate of return has been dropped almost in half by using the forecasts (i.e. 0.55 for the portfolio using the forecasts and 1.03 for the original portfolio, at the 8% risk free rate).

SUMMARY OF PORTFOLIO APPLICATION

The SSMPT program provides powerful new capabilities for the forecasting and analysis of investment time series since it ties together information and concepts from -

State Space Forecasting
- where forecasts for interacting time series are developed using available historical data, leading indicators and alternative forecasts.

Modern Portfolio Theory
- where portfolios of investments are developed to produce the highest level of return for the least risk using forecasts for each individual investment and the correlation matrix of the forecasts.

Evaluation of Alternative Forecasting Methods
- where the value of additional forecasting analysis and professional forecasting services can be assessed by their impact on portfolio performance.

Preparation of Composite Forecasts
- where forecasts prepared by alternative approaches, e.g. econometric forecasts or forecasts from professional services, can be combined to improve both forecast and portfolio performance.

Historical Time Series Data Bases and Analysis Tools
- where accurate and up to date daily, weekly and monthly time series data on financial investments is available for analysis using a comprehensive data management, analysis and display system,(e,g. the Telerate II data base and the PROBE system).

APPLICATIONS OF STATE SPACE FORECASTING IN INVENTORY CONTROL

In this section, we will illustrate a second application of the State Space Forecasting method when there is a specific objective function available to determine the appropriate business decision to be used based on the forecasts. This application is in the area of inventory control & production planning.

In the last decade there has been a major effort by many companies to improve control of their inventories through the use of sophisticated inventory and MRP systems. As inventories have rapidly increased in the current recession however, once again it is clear that more progress needs to be made in overall inventory management.

A glance at the following illustrates the magnitude of the problem in each of the last recessions since 1968. Inventories rise very rapidly during each recession and then they fall sharply as business recovers.

Despite the best efforts of inventory and production planners and system designers, there appears to have been relatively little progress in reducing the size of these inventory cycles. The situation is even more dramatic if inventory carrying costs are also taken into account. The following diagrams show the dollar value of inventories of U.S. manufacturing companies and an estimated dollar value of inventory carrying costs using the prime interest rate as an indicator of the financial carrying cost.

In this paper however we will review some recent developments which give promising new perspectives for inventory control and production planning systems. These new developments can produce improvements in manufacturing efficiency and can reduce inventory carrying costs in many manufacturing companies.

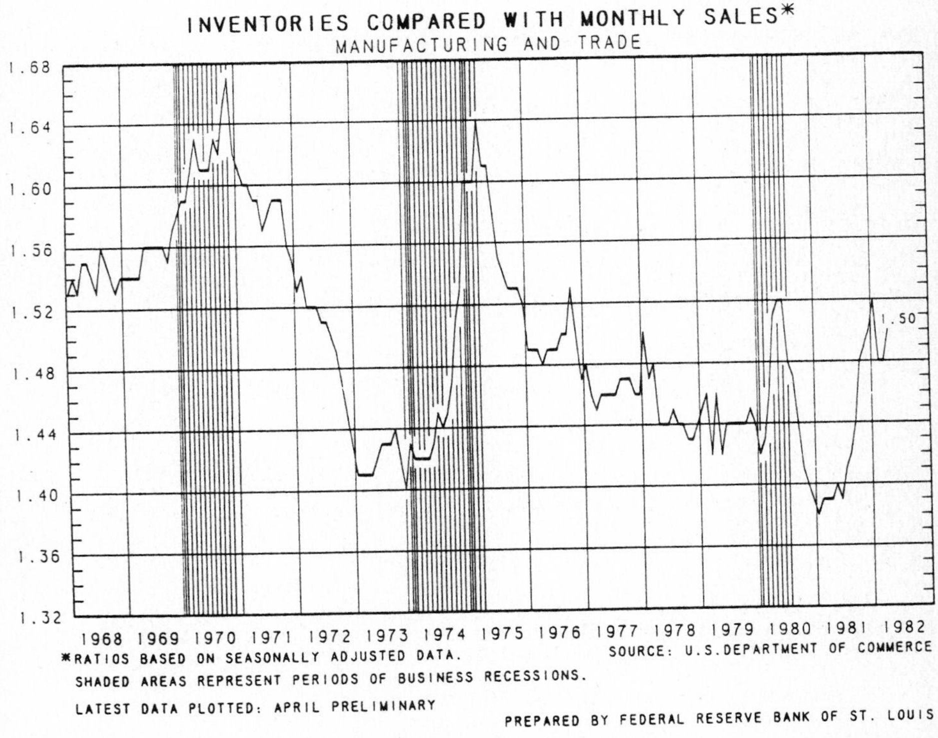

The two major developments we will discuss are -

1) improved shipment forecasting methods which are now available using multiple time series methods and leading indicators.

2) new inventory and production control techniques which can be used to improve profitablity in volatile business environments.

NEW FORECASTING DEVELOPMENTS

Most companies have implemented forecasting programs and procedures to improve their ability to anticipate business sales and shipment changes. Quantitative methods such as exponential smoothing and adaptive smoothing have been widely used for individual product and warehouse forecasts. However as the cost of computing and data bases has declined, more sophisticated multiple time series methods can now be used for improving forecast accuracy. Multiple time series methods are efficient in identifying and using relationships between pairs of data series such as monthly leading economic indicators and product sales, or sales and shipments at the warehouse level, or promotional expenditures and product sales.

The benefit of these new methods is that forecast accuracies can more than double or triple those of the single series methods such as exponential smoothing. Since safety stocks are often the largest component of inventories for each product and since they are directly proportional to forecast errors, these types of forecast accuracy improvements can produce important reductions in total inventory levels and carrying costs.

THE DYNAMICS OF INVENTORY CONTROL SYSTEMS

Most established inventory control systems have several major modules which work in conjunction with each other. The traditional planning modules in inventory systems are -

Forecasting
Inventory Control
Material Requirements Planning

In these types of systems, forecasts are prepared first, inventory targets are constructed based on the forecasts, then material requirements and production plans are determined to meet the target inventories. The forecasts are inputs to the inventory module. The inventory module provides inputs to the MRP system.

A significant problem with these types of systems is their performance in volatile business situations. For efficient operation, the leads, lags and feedbacks which are present in the system must be adjusted to match the dynamics of the marketplace and the physical capabilities of the manufacturing plants and other facilitites. In the actual performance data for U.S.manufacturing companies, this is a major problem area now and it is an important opportunity area for improved profitability. At present for U.S. manufacturing companies, the changes that are taking place in production are almost as great as the actual shipment changes. Many of these changes could be eliminated if proper use is made of the forecasts that are available and better use is made of inventory as a buffer for unexpected shipping activity or production problems. A basic measure of overall efficiency can be defined for any manufacturing & inventory control system to describe this type of dynamic performance-

$$\text{i.e.} \quad \frac{\text{the average production change}}{\text{the average shipment change}} = \text{PSC RATIO}$$

$$= \text{Production/Shipment Change Ratio}$$

If this ratio is close to 1 or higher for your company of your product, (as it is for the average U.S. manufacturing company), then there is room for significant improvement in inventory control. In the next section we will show how the appropriate new control strategies can be derived to gradually improve and maintain the performance of these inventory control systems.

NEW INVENTORY CONTROL STRATEGIES

The problem of choosing the appropriate production and inventory plan in an uncertain and constantly changing business environment requires a balance to be made in several conflicting objectives. On the one hand inventories should be maintained at levels which are as close as possible to the target inventory levels needed to meet the forecast sales, shipments & customer service objectives. On the other hand, production changes which may be required to adjust inventory levels can be very expensive and they can be a continual drag on the profitability of manufacturing operations.

In a profitable inventory & production planning system, a precise balance must be made which reflects these conflicting costs and the real dynamics of the market and the actual leads and lags needed for manufacturing operations to react.

The analysis and planning system which meets these requirements can be surprisingly simple to implement. The analysis uses historical data on sales, shipments, inventory and production (the monthly data already available in most inventory accounting systems is usually sufficient). The implementation typically requires only minor changes in the calculation routines for production requirements in the inventory systems which are currently in use.

In the analysis phase, a statistical model and computer simulations are built to reflect the actual characteristics of the market and manufacturing process. Next an optimal control strategy is developed to minimize the inventory and production change costs. Finally, the new production policy can be gradually implemented by introducing a single new manufacturing cost change parameter into the inventory requirements calculation process.

These new strategies can be applied at the individual product level or at a product group level. The ongoing monthly operation of the models can provide an important tool for management to monitor actual performance of the overall inventory system.

To build a model for a typical U.S. manufacturing company we will make several basic assumptions about the normal planning procedure. The typical manufacturer is assumed to require a fixed one month production plan and no changes in production are allowed till the monthly periods at least two months into the future. Some manufacturers could have longer fixed production plans if long lead times are involved in their manufacturing. Others could have shorter fixed scheduling requirements if their manufacturing operations are very responsive. A one month fixed schedule is assumed to be the average of these cases.

The next requirement in our model is the cost function which reflects a balance between inventory costs & production change costs. The objective we will use in this model can be written as-

$$(Ip(t+2) - Id(t+2))**2 + K * (P(t+2) - P(t+1))**2 \quad (1)$$

where Id(t+2) is the target inventory level at the end of period (t+2), Ip(t+2) is the planned inventory level at the end of period (t+2), P(t+2) is the production in period (t+2) to be decided, i.e. the control variable, and P(t+1) the planned fixed production already decided for period (t+1). t is the current time, t+1 is the end of the next (monthly) planning period, t+2 is the end of the second planning period.

The first component of the objective function measures the difference between the desired & planned inventory levels at the end of the second planning period. The second planning period inventory level is the first one which can be controlled since the production in the first period is already predetermined & fixed. The difference in levels is squared to reflect the fact that the costs of inventory levels which are either above or below the desired target are equally expensive & important.

The second component of the objective function measures the change in production from the predetermined level in period t+1 to the new planned level in period t+2. This cost is again squared to reflect the equal importance of production increases & production decreases from the existing production level.

The parameter K is used to reflect a relative balance between production change costs & the inventory costs. The higher the value of K the greater the importance given to manufacturing change costs.

This objective function is minimised when its first derivative with respect to the control variable P(t+2) is zero, i.e. when -

$$Ip(t+2) - Id(t+2) + K * (P(t+2) - P(t+1)) = 0 \quad (2)$$

where the planned inventory level at the end of period t+2 is -

$$Ip(t+2) = I(t+1) + P(t+2) - S(t+2) \quad (3)$$

I(t+1) is the inventory at time t+1,

S(t+2) is the shipment forecast for period t+2.

i.e. $Ip(t+2) = I(t) + P(t+1) - S(t+1) + P(t+2) - S(t+2)$

Adding & subtracting P(t+1) to the right hand side we obtain -

$$Ip(t+2) = I(t+1) + P(t+1) - S(t+2) - P(t+1) + P(t+2)$$

i.e. $$Ip(t+2) = In(t+2) + (P(t+2) - P(t+1)) \quad (4)$$

where In(t+2) is the projected inventory level at the end of period t+2 assuming no change was made production from period (t+1) to (t+2), i.e. In(t+2) is the inventory level assuming P(t+2) = P(t+1) and

$$In(t+2) = I(t+1) + P(t+1) - S(t+2)$$

Hence from equations 2 and 4, the best value for the objective function is obtained when-

$$K *(P(t+2) - P(t+1)) = Id(t+2) - In(t+2) - (P(t+2) - P(t+1))$$

i.e. $$P(t+2) - P(t+1) = \frac{1}{K+1} * (Id(t+2) - In(t+2)) \quad (5)$$

i.e. the best value of the objective function is obtained when the production change P(T+2) - P(T+1) is equal to-

$$\frac{1}{K+1} * (Id(t+2) - In(t+2))$$

This value for the production change is an important result & we will call it the Economic Change Quantity (ECQ).

To investigate the properties of this ECQ function we will evaluate several special cases with differing values of the parameter K.

When K=0, there is no cost associated with a production change. In this special case, the ECQ is -

$$ECQ0 = (Id(t+2) - In(t+2)).$$

Hence from equation (4), Ip(t+2) = Id(t+2), i.e. the planned inventory level is exactly equal to the desired target level.

When K=1, there is equal weight given to production change costs and inventory differences from the desired target. In this special case, the ECQ is -

$$ECQ1 = 0.5 * (Id(t+2) - In(t+2)) = 0.5 * ECQ0.$$

Hence the production change is half the amount required for the special case when K=0.

Similarly for any value of K, the ECQ is given by -

$$ECQ = \frac{1}{K+1} * ECQ0$$

SUMMARY OF INVENTORY CONTROL APPLICATION

There appear to be major opportunities for improved inventory control and production planning in typical U.S. manufacturing companies. The historical record shows relatively little change in balancing finished goods inventory and production control during the past 15 years despite a significant investment in new computerized systems which has taken place. This paper has shown that new analysis tools are now available which show significant potential for improving the performance of these systems in current business situations. These tools include (a) new forecasting methods which use multiple time series approaches to model the leads and lags in the marketplace for particular products, (b) new models for the dynamics and costs of inventory systems, and (c) new optimal control strategies to control production and inventory in a dynamic and volatile business environment.

REFERENCES

CAMERON, A.V. and MEHRA, R.K., (1982) State Space Forecasting Handbook. State Space Systems Inc. Revised Edition

CAMERON, A.V. (1981) Interest Rate Forecasting and Portfolio Analysis using the State Space Forecasting System, In Applied Time Series Analysis, (Proceedings of the International Conference held at Houston, Texas, August 1981). Eds: O.D. Anderson and M.R.Perryman, North-Holland, Amsterdam and New York, 43-52

TIME SERIES ANALYSIS: Theory and Practice 4
O.D. Anderson (editor)

FITTING JOINED LINE SEGMENTS TO TIME SERIES DATA: URINARY ESTROGENS AS AN EXAMPLE

J. A. Norton

Department of Statistics
California State University, Hayward
California 94542
USA

A study of 37 sequences of one menstrual cycle each for 32 women is performed using a joined line segment model. The amount of urinary estrogens excreted in each 24 hour period is measured from day one of a menstrual cycle to day one of the next cycle. Two consecutive cycles are available for four women and results are compared. Several measurements are missing for some women. Residuals are analyzed for autocorrelation. Results of generalized adaptive filtering analysis are explored for further comparison.

1. INTRODUCTION

Medical data collected for long term study may exhibit periodic behavior. For example, aldosterone levels measured at half hourly intervals from blood samples appear to repeat patterns daily (Katz et al, 1972). Endocrine levels in daily urine samples exhibit cyclic patterns from menstrual cycle to menstrual cycle in women (D'Amour, 1940). If the form of each pattern contains abrupt changes in direction over time, the observations may be modelled as intersecting line segments. Norton (1977) used an intersecting line segment model on excretions in urine to forecast mature ova for future use in artificially inducing ovulation. Smith and Cook (1980) applied an intersecting line segment model when attempting to infer time of rejection for transplanted kidneys.

Hinkley (1971) studied inference and large sample distributions for joined linear models. Hudson (1966) discussed maximum likelihood estimates and Hawkins (1976) described methods which produce approximate maximum likelihood estimates. Norton (1977) and Smith and Cook (1980) developed bayesian methods for incorporating prior information into their forecasts of join locations. Norton (1982) studied bayesian forecasts for joined linear models using Monte Carlo methods.

This paper studies the fits of joined linear segment models to estrogen excretions of 32 women for 37 menstrual cycles. The purpose of this task is to evaluate whether the joined linear segments models fit reasonably well, whether a combination of previous data can aid directly in forecasting for another woman, or whether fits for cycles within the same woman can be used for forecasting a later cycle. These comparisons are made graphically, by computing the mean squared error and by studying the residuals for remaining patterns. As an additional comparative measure, menstrual cycles of women with consecutive cycles are linked and studied using a generalized adaptive filter. The data was collected by Dr Janet McArthur, Massachusetts General Hospital, Boston, Massachusetts from her own studies and from data of other researchers.

2. THE TIME SERIES MODEL

D'Amour (1940) reports that in a normal menstrual cycle urinary estrogen excretion rises from an initial plateau reaching a peak on about the tenth day. This peak is followed by a drop and a secondary rise on about the twenty-first day. Estrogen levels decline to the end of the cycle from the second peak. This description of the estrogen sequence induces an image of five intersecting line segments such as the one pictured in Figure 1.

Figure 1. A Joined Line Segment Model with Five Segments

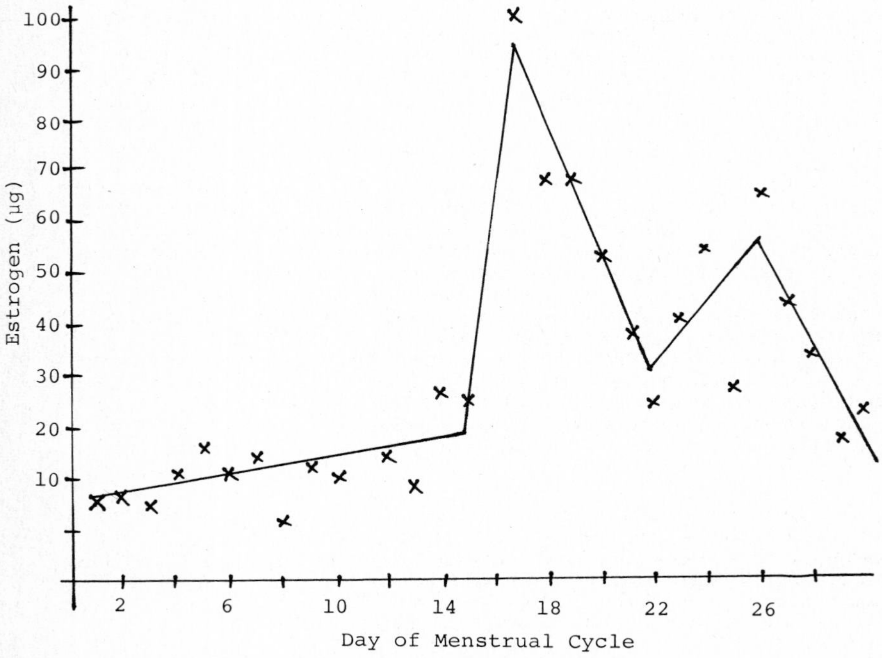

Suppose that the estrogen levels, $\{Y_t\}$, arrive at time $t = 1, 2, \ldots, T$. For a set of unknown integers

$$0 = r_0 < r_1 < r_2 < r_3 < r_4 < r_5 = T ,$$

the sequences $\{Y_t\}$, between integers $\{r_i\}$, $i = 0,1,\ldots,4$, are simple linear functions of the arrival times of the observations. These linear functions are constrained to intersect at four unknown locations τ, so that τ lies between r and $r + 1$. A five line segment model for estrogen is defined by the equations

$$Y_t = a_i + b_i t + e_t \quad (t = r_{i-1} + 1, \ldots, r_i, \quad i = 1, \ldots, 5)$$

with constraints

$$a_i + b_i \tau_i = a_{i+1} + b_{i+1} \tau_i \quad (i = 1, \ldots, 4) .$$

The parameters a and b for each of the five intervals are the customary regression parameters constrained so that the line segments intersect at the four τ. The e_t are independent, identically distributed random variables with mean 0 and variance σ^2 (white noise).

3. APPLICATION TO ESTROGEN

Individually fitting the time series model, using maximum likelihood estimates as presented in Hudson (1966), to the 37 menstrual cycles generates estimates for the five segment model. Stem-and-leaf plots for the first three join locations of the cycles are shown in Figure 2. According to D'Amour (1940), we should expect a peak at day 10. However we observe a bimodal join location from 8 to 9 and from 12 to 13, not for the second join or initial peak, but for the break away from the initial plateau. The peak location seems fairly normally distributed with mean 13.6 and standard deviation 2.9. D'Amour's value of 10 is not within two standard deviations of the sample mean for these 37 sequences. The trough location has mean 16.8 and standard deviation 2.8. The second peak location averages 22.9, with standard deviation 3.2, and is slightly negatively skewed. The length of menstrual cycle ranges from 26 to 30 days.

Figures 3 through 6 give examples of fitted models. These are shown for individual women. Estimates from consecutive cycles agree fairly well although there is considerable variation between women. The sequence pair in Figure 6 has very different estimates of the first join location but agree on the peak location. In booster estrogen regimes, ovulation stimulating hormone is thought best introduced at this peak, so dependable predictions of the second join location are valuable.

Figure 7 displays a stem-and-leaf plot of $\hat{\sigma}$, the standard deviation of the residuals. The three largest values for $\hat{\sigma}$ correspond to sequences which have the highest levels of estrogen, as in Figure 1, with a range in estrogen levels from 8μg to over 100μg.

4. AUTOCORRELATION IN RESIDUALS

Figure 8 shows a correlogram of residuals from joined line segment models formed by linking eight cycles of equal length. Each cycle was scaled to obtain unit standard deviation of the residuals. Repeated calculations on other linked cycles of residuals gave similar results. Except at lag 2, the autocorrelations appear quite small.

Figure 2. Stem-and-Leaf Plots of Join Locations

```
       t1                                 t2

16-17 | 6                        20-21 | 0
14-15 | 44                       18-19 | 88
12-13 | 222222333                16-17 | 666677
10-11 | 0011                     14-15 | 44444555555
 8-9  | 8888999999999            12-13 | 222233333
 6-7  | 666677                   10-11 | 0111
 4-5  | 55                         8-9 | 899
                                   6-7 | 6

                    t3

             22-23 | 233
             20-21 | 011
             18-19 | 8888899
             16-17 | 66677777
             14-15 | 444445555555
             12-13 | 2333
```

5. APPLYING A GENERALIZED ADAPTIVE FILTER

As is clear from the autocorrelation study, the techniques used in this comparative study are outdated and difficult to use. The analysis available requires repetition of cycles for any meaningful results. Women with two consecutive cycles were considered (see Figures 4 to 6). Each sequence cycle 1, cycle 2 was repeated three times for 165 data points. An iteratively fit AR(28) model with all weights present[1] was estimated from the data in Figure 4. Using this model and the data in cycle 1 of Figures 5 and 6 to forecast cycle 2, resulted in a mean squared error of 25.8 and 59.9, respectively. A simple AR(1) forecasting scheme resulted in a mean squared error of 76.4 and 76.1 for the same data.

Fitting an AR(28) model with only the first and twenty-eighth weight present to Figure 4 and then applying that model to the data in Figures 5 and 6, gave a fairly good forecast for the second cycle using the first cycle in the model. The mean squared errors were 46.14 and 54.97. This solution was obtained from the Yule-Walker equation with no iteration.

---

[1]Such a model was necessary, given the available software, which also caused the need for repeating the sequences.

Figure 3. Menstrual Cycles for the Same Woman, Two Years Apart

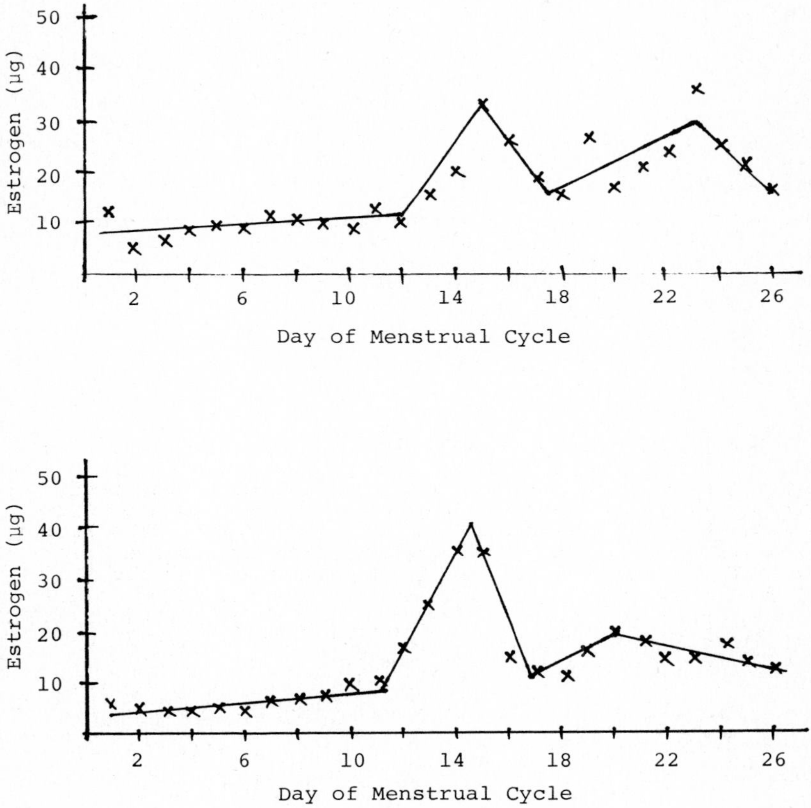

Figure 4. Two Consecutive Cycles for One Woman

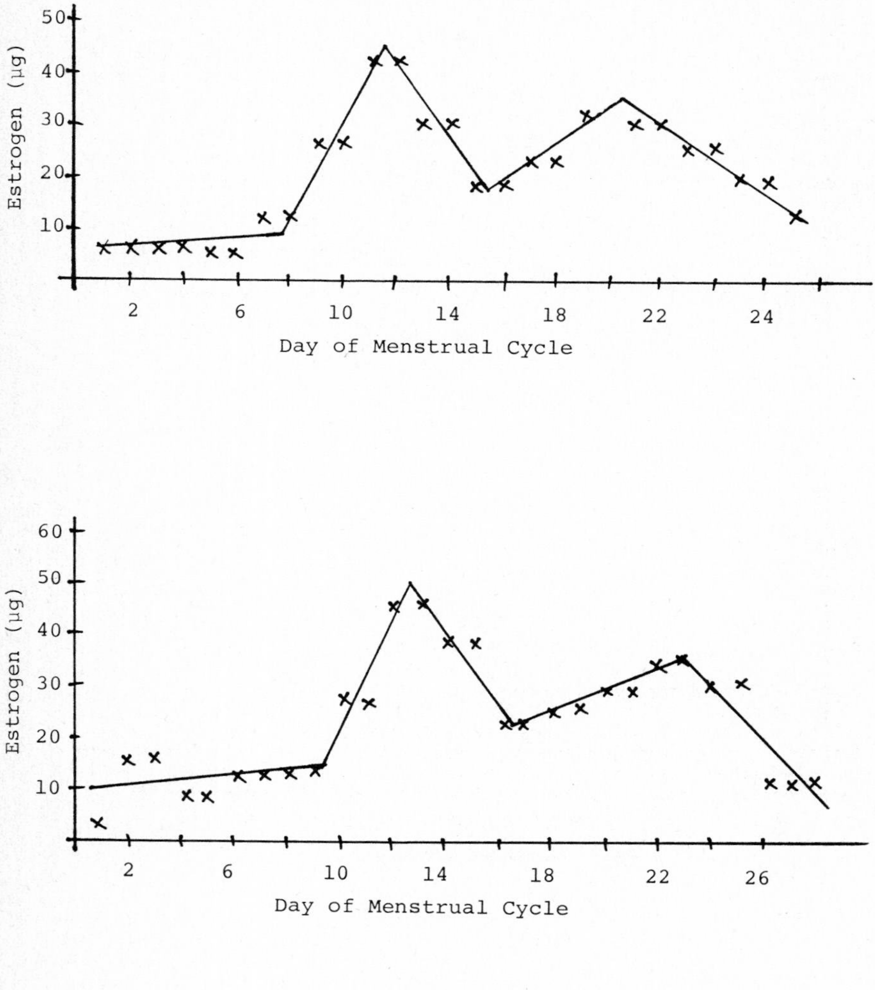

Figure 5. Two Consecutive Cycles for a Second Woman

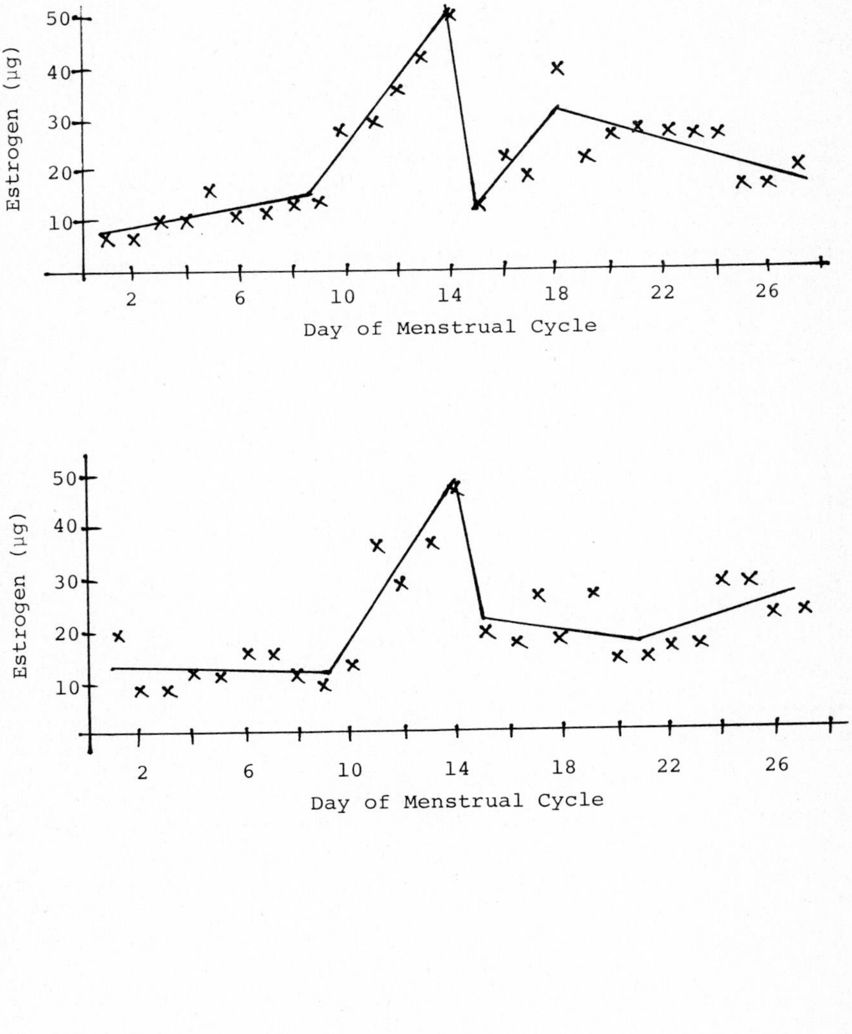

Figure 6. Two Consecutive Cycles for a Third Woman

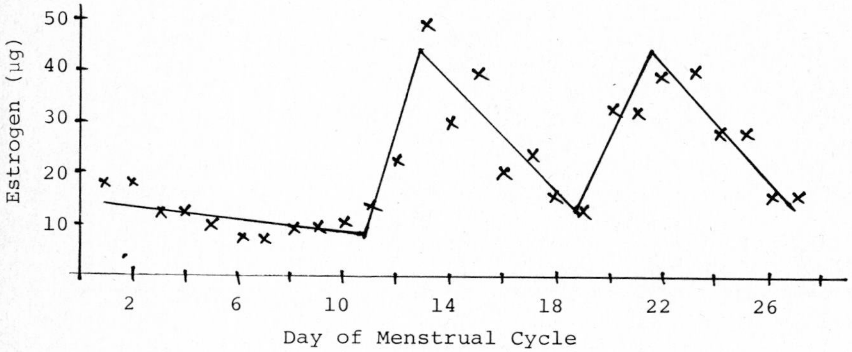

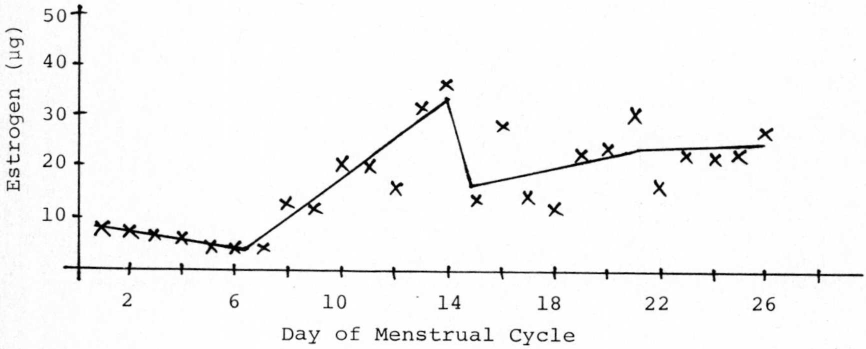

Figure 7. Stem-and-leaf Plot of Standard Deviations for the Residuals

10	8
9	0
8	8
7	
6	0 2 3 9
5	1 1 1 4
4	2 2 3 4 6 8 8
3	1 2 4 5 5 7 8 9
2	0 1 3 6 8 9 9 9
1	5 5
0	9

Figure 8. Correlogram of Residuals

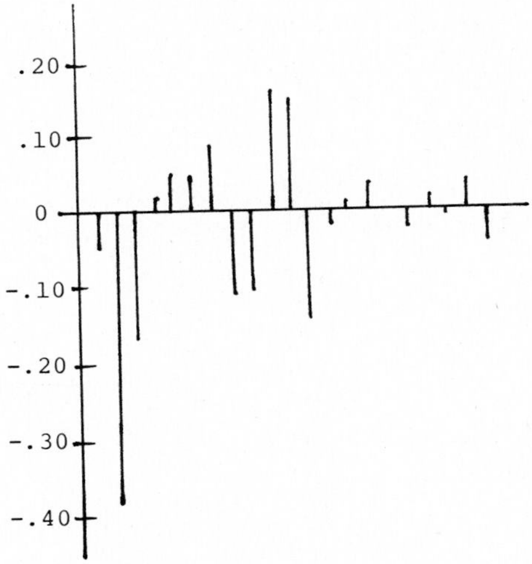

When the joined linear segments time series model based on the data from Figure 4 is applied to the data in Figures 5 and 6, the initial peak is forecast as 12.5 for the four observed peak locations 14, 14, 13, 14 respectively. Forecasting the four cycles results in mean squared errors of 49.6 and 78.3, for Figures 5 and 6 respectively.

One referee suggested several additional approaches to the problem of modelling and forecasting, and one of these is explored. Suppose we begin each cycle on day 1, then look at the differenced series for days 1 to t. Again using the pair of cycles from Figure four to obtain a model, that equation using four iteratively fit weights was used to forecast for the differenced sequences in Figures 5 and 6. These forecasts have mean squared errors of 83.04 and 65.96, respectively.

Table 1 summarizes the approaches taken in this paper by comparing expected mean squares. The forecasts from five joined line segment models appear to be competitive with the simple time series methods explored. Certainly much more study is in order, but briefly, we can conclude that the segmented linear models are suited to estrogen time series data.

Table 1. A Comparison of Methods Considered

Methods	data figure 3	data figure 4	data figure 5	data figure 6
1. Fit from data. 5 line segments	4.2	8.4	13.8	16.0
2. Forecast from first cycle. 5 line segments	34.5	59.5	49.2	99.0
3. Forecast from previous data. 5 line segments			49.6	78.3
4. Forecast equation from previous data. AR(28) all weights			25.8	59.9
5. Forecast equation from previous data. AR(28) two weights			46.1	55.0
6. Forecast equation from previous data. AR(1)			76.4	76.1
7. Fit Cycle 2-Cycle 1	49.0	69.5	65.2	97.6
8. Forecast equation from previous Cycle 2-Cycle 1			41.5	76.7

6. CONCLUSION

When estrogen levels from urine excretions are available for one menstrual cycle of length v, a forecast of the estrogen levels for the following cycles may be made using a joined linear segments model with 5 sections or an AR(v) model. Both models appear to forecast the initial estrogen peak well. An AR(1) model is an adequate forecaster when no long term history is available.

The parameters of the joined linear segments model are easily interpreted in terms of the behavior of estrogen. In addition, forecasts can be set in a Bayesian framework as shown by Norton (1977), Smith and Cook (1980) and Norton (1982), when no history of estrogen levels is available.

REFERENCES

D'AMOUR, F. E. (1940). Further studies on hormone excretion during a menstrual cycle. *American Journal of Obstetrics and Gynecology*, 40, 958-965.

HAWKINS, D. M. (1976). Point estimation of the parameters of piecewise regression models. *Applied Statistics*, 25, 51-57.

HINKLEY, D. V. (1971). Inference in two phase regression. *Journal of American Statistical Association*, 66, 736-743.

HUDSON, D. J. (1966). Fitting segmented curves whose join-points have to be estimated. *Journal of American Statistical Association*, 61, 1097-1129.

KATZ, F. H., ROMFH, P., and SMITH, J. A. (1972). Episodic secretion of aldostrone in supine man: relationship to cortisol. *Journal of Clinical Endocrinology*, 35, 178-181.

NORTON, J. A. (1977). Estimation and sequential prediction for unknown join points in broken line segment models. PhD Thesis, Harvard University, Cambridge, Massachusetts.

NORTON, J. A. (1982). A Monte Carlo study of Bayesian predictions for join locations in segmented linear models. In *Applied Time Series Analysis* (Proceedings of the International Conference held at Houston, Texas, August, 1981). Eds: O. D. Anderson and M. R. Perryman, North-Holland, Amsterdam and New York, 261-272.

SMITH, A. F. M. and COOK, D. G. (1980). Straight lines with a change-point: A Bayesian analysis of some renal transplant data. *Applied Statistics*, 29, 180-189.

TIME SERIES ANALYSIS: Theory and Practice 4
O.D. Anderson (editor)

ON A SIMPLE MODEL FOR POPULATION DYNAMICS IN STOCHASTIC ENVIRONMENTS

Oliver D. Anderson
TSA&F, 9 Ingham Grove, Lenton Gardens, Nottingham NG7 2LQ, England

This paper discusses some work on population dynamics, by Roughgarden (1975), where the methods of univariate time-series analysis were applied to a simple linear approximation of the logistic model, in discrete time. The approximation is reformulated, in order to clarify some of Roughgarden's conclusions, and also to make his ideas more amenable to generalisation. An attempt is made to explain Roughgarden's simulation experience, and to show how the relatively complicated models, which might result from sophisticated statistical analysis, of the type developed by Box and Jenkins (1970), can perhaps be explained in biological terms. Finally, we indicate that even more impressive results can be expected, if the methods of multivariate time-series modelling are employed.

1. INTRODUCTION

Roughgarden (1975) has given an interesting and important application of time-series modelling to the ecological problem of determining population dynamics in stochastic environments. A comparatively simple but powerful analysis was presented, which is applicable to very general environments. A linear approximation to the logistic model in discrete time, for small variations around a large average carrying capacity, was introduced and shown to provide results in close agreement with those obtained from certain simulation experiments. In this paper, we will extend Roughgarden's models by using the time series methodology of Box and Jenkins (1970).

One version of the logistic equation in discrete time, for a fluctuating environment, is

$$N_{t+1} = (r + 1 - rN_t/K_t)N_t \qquad (1)$$

where K_t is the environment's carrying capacity at time t; $\{N_t\}$ is the population size series; and r is a constant, measuring the responsiveness of the population to temporally varying carrying capacities, (or, more strictly, the per capita growth rate), which is restricted to between 0 and 2 - thus excluding the pathological behaviour of the model, as described by, say, May (1976). $\{K_t\}$ is taken to follow a second-order stationary process (satisfying $E[K_t]$ constant and $Cov[K_t,K_{t+h}]$ a function of h only) with a necessarily *fixed* mean $\bar{K}$.

Then, working with deviations from $\overline{K}$, defined by $k_t = K_t - \overline{K}$ and $n_t = N_t - \overline{K}$, the logistic equation (1) becomes

$$n_{t+1} = \left[1 - r\left(\frac{\overline{K}}{\overline{K}+k_t}\right)\right]n_t + r\left(\frac{\overline{K}}{\overline{K}+k_t}\right)k_t + r\left(\frac{k_t-n_t}{\overline{K}+k_t}\right)n_t \tag{2}$$

with $0<r<2$. If it is next assumed that,

$$\overline{K} >> k_t, n_t \tag{3}$$

equation (2) can be approximated by the linear model

$$n_{t+1} = (1-r)n_t + rk_t \qquad 0<r<2. \tag{4}$$

Note that the series $\{k_t\}$, of carrying capacity deviations, does not have to represent a "white-noise" sequence of realisations, from uncorrelated identically distributed zero-mean random variables. Such a restrictive assumption would appear unnecessarily severe in most biological contexts.

Relation (4) shows that, in general, the population deviation depends both on its own dynamics and on those of its environment. As pointed out by Roughgarden, for r approaching zero, the environmental deviation becomes unimportant; whereas, for r near unity, it is all-important. For r close to two, violent oscillations tend to be introduced into the population deviations.

2. REFORMULATING ROUGHGARDEN'S MODEL

If we replace the series $\{n_t\}$ by $\{y_t \equiv n_t/r\}$, and $(1-r)$ by β, relation (4) becomes

$$y_{t+1} = \beta y_t + k_t \qquad -1<\beta<1. \tag{5}$$

Of course, for $\beta>1$, this would represent a potentially "explosive" process; whilst, for $\beta=1$, it would be unstable, in that its local level would wander freely, making extinction eventually inevitable. For $\beta \leq -1$, we would get corresponding oscillatory behaviour. (Evidently, extinction occurs when a y_t value attains $-\overline{K}/r$.)

For $-1<\beta<1$, extinction is still possible, under certain circumstances. But there is a continual pull back to the fixed positive mean of $\{N_t\}$, $\overline{K}$ - or, equivalently, to zero, in (5) - whenever the series wanders from this equilibrium level. However, in many cases, where extinction is theoretically possible, it is unlikely to be observed in practice, due to the relative shortness of the period involved.

Actually, it is evident that the assumption (3) will break down long before there is the obvious threat of extinction, provided by a low N_t, as then the deviations in n_t will be of the same order as the large mean carrying capacity. However, as Roughgarden notes, there is nothing particularly holy or fundamental about the logistic model. So, even when (3) breaks down, our relation (5) may still be found to form an acceptable model for a biological process, since the soundness or otherwise of a *reasonable* model can only be assessed by testing it with data.

We note that (5) is more general than the analogous density-dependent model given by equation (2) of Bulmer (1975)

$$y_{t+1} = \beta y_t + e_t \tag{6}$$

where, this time, $0<\beta<1$ and $\{e_t\}$ *is* a white-noise series.

3. EXTENDING THE ANALYSIS

If we use Roughgarden's model (16) for the series $\{k_t\}$, which assumes that the stochastic variation in the carrying capacity can be represented by a stable first-order autoregressive process, we have

$$k_t = \alpha k_{t-1} + \zeta_t \qquad -1<\alpha<1 \tag{7}$$

where $\{\zeta_t\}$ is a white-noise series. Then (5) becomes, on eliminating $\{k_t\}$,

$$y_{t+1} = (\alpha+\beta)y_t - \alpha\beta y_{t-1} + \zeta_t \tag{8}$$

which gives a second-order autoregressive representation for $\{y_t\}$. (For $\beta=1$, this would reduce to an "integrated" first-order autoregressive process; whilst, for $\beta>1$, it would be an explosive process.) Thus, when equations (5) and (7) are applicable, relation (8) shows that study of the environment is unnecessary for study of the population, as the former can be immediately eliminated.

Of course, if $\{k_t\}$ follows some model other than (7), a different representation to (8) will be deduced for $\{y_t\}$. In general, the best way of deriving relations, such as (8), is to employ the backshift operator B. This has the property that, for any series $\{x_t\}$ and any integers i and j, $B^j x_i \equiv x_{i-j}$.

If $\{k_t\}$ follows a first order moving average model - which could occur if, say, the effect of weather was not immediately fully assimilated by the environment, then we would have

$$k_t = \zeta_t + \gamma\zeta_{t-1} \qquad -1\leq\gamma\leq 1 \tag{9}$$

which can be rewritten as $k_t = (1+\gamma B)\zeta_t$. Then, eliminating $\{k_t\}$ from (5), we get

$$(1-\beta B)y_{t+1} = (1+\gamma B)\zeta_t \tag{10}$$

which is a mixed first-order autoregressive, first-order moving average model. (Note that, in relation (9), a choice with $|\gamma|>1$ can always be rewritten in a form with $|\gamma|<1$, whilst $|\gamma|=1$ causes no problems of instability. See the details in, say, Anderson (1977a).)

What I would suggest is that the more general methods of the Box-Jenkins approach could be applied to identify, estimate and check linear models for population series. When more complicated models arise, they can often be interpreted in a satisfactory manner. For instance, an apparently unpalatable fitted model, such as (10), reduces to the more acceptable explanation of the pair of relations (5) and (9).

However, let us now see what happens when we introduce the inevitable observation error terms into the model (5). We then have $y^*_{t+1} - \varepsilon_{t+1} = \beta(y^*_t - \varepsilon_t) + k_t$, where $\{y^*_t\}$ is the *observed* population series, obtained from contaminating the *true* series by adding a series of observation errors, $\{\varepsilon_t\}$, which we will take to be a white-noise series, for purposes of exposition.

Then, employing the backshift operator, we get, instead of (10),

$$(1-\beta B)y^*_{t+1} = (1+\gamma B)\zeta_t + (1-\beta B)\varepsilon_{t+1} \tag{11}$$

which yields

$$(1-\beta B)y^*_{t+1} = (1+\delta B)\xi_t \tag{12}$$

using a lemma discussed in Anderson (1975a and 1977b). Here, δ has a particular value, satisfying $\min(\gamma,-\beta)<\delta<\max(\gamma,-\beta)$, and $\{\xi_t\}$ is a white-noise series having a particular variance; as can be shown by comparing the variances and first autocovariances of the right hand sides of equations (11) and (12), in a manner described by Anderson (1975b, chapter 15).

However, in general, the introduction of error terms will result in a more complicated model - that is, in one with extra constants, or *parameters*, to be estimated. There are also many other ways in which more involved Box-Jenkins models can arise from basically simple situations. For instance, consider observing a population, which really consists of the aggregate of two sub-populations, each following distinct first-order autoregressive schemes, such as might be given by Bulmer's density-dependent model (6), with two different values for β. Then the aggregate will follow a mixed second-order autoregressive, first-order moving average scheme; as is easily verified. (Such aggregated series can arise quite naturally. As a rather obvious example, suppose the Channel Islands kept a record of their rabbit population. Then this would mainly consist of the sum of two separated sub-populations on, respectively, Guernsey and Jersey. In practice, the aggregation will usually be more subtle.)

The moral is that, should the Box-Jenkins fit not be immediately explicable mechanistically, there is still a good chance that it can be interpreted in acceptable biological terms. Of course, if this cannot be done, then the fit perhaps deserves to be viewed with suspicion. But, even then, it should be remembered that the method is one of the most sophisticated statistical approaches, currently available, for analysing time-series. So, rejecting a satisfactorily identified, estimated and verified fit, on mechanistic grounds, means going against the statistical evidence. And generally, with cooperation between the subject specialist and the statistician, a competent analysis can be satisfactorily interpreted. The importance of this final stage of the Box-Jenkins approach is only just beginning to be fully appreciated. See, for instance, Anderson (1977c).

Before leaving this section, we briefly note that equation (5) can be rewritten as $y_t = (1-\beta B)^{-1}k_t$. This demonstrates that the population deviations can be obtained, from those of the carrying capacities, through a filter $(1-\beta B)^{-1} \equiv \Sigma_{j=0}^{\infty}(\beta B)^j$, in the sense of Hubbell (1973), by which the environmental dynamics are considered to be translated (albeit with distortion) into those of the population.

4. SOME FURTHER CLARIFICATIONS OF ROUGHGARDEN'S PAPER

In the spirit of the accuracy, asked for by Levin (1975), it is perhaps worth pointing out that a sustained oscillatory departure from exponential (or more correctly, geometric) decay, for the observed autocorrelations, does not necessarily imply a second-order autoregressive process, as was suggested by Roughgarden, p.721. Nor do such processes necessarily display such behaviour, even theoretically. See, for instance, Anderson (1975b, pp.48 and 21, respectively).

Again, on p.723, Roughgarden suggests that any given degrees of population variability and predictability can be produced by many combinations of population responsiveness and environmental predictability In fact, in each case, there can be at most only two combinations. This is seen by rewriting relation (8) in backshift operator form

$$(1-\alpha B)(1-\beta B)y_{t+1} = \zeta_t. \tag{13}$$

As pointed out by Kendall (1971), equation (13) could either arise from equations (5) and (7) or from analogous equations, obtained by interchanging the parameters α and β. But there are no other possibilities.

Note that, using Roughgarden's notation, this equivalence is not very obvious. For instance, his examples are, for pairs (λ,r),

$(.5,1.5) \approx (-.5,.5)$, $(.5, 1.0) \approx (0,.5)$ and $(0,1.5) \approx (-.5,1.0)$.

However, in our notation, we get for pairs $(\alpha=\lambda,\ \beta=1-r)$,

$(.5, -.5) \approx (-.5,.5)$, $(.5,0) \approx (0,.5)$ and $(0,-.5) \approx (-.5,0)$

or, in general, $(\alpha,\beta) \approx (\beta,\alpha)$. Of course, when $\alpha=\beta$, the two possible explanations become identical. In Roughgarden's notation we then have $\lambda=1-r$, giving the situation he discusses on p.735.

Again we note that, since (13) represents a second-order autoregressive process, all the results in Roughgarden's three pages of Appendices A and B, and many others in his text, can immediately be written down. All that is needed is to replace y_t by n_t/r and (α,β) by $(\lambda,1-r)$ in well-known and easily referenced results, such as those given by Box and Jenkins (1970). Of course, in (13), the autoregressive part has real zeros, and so it will provide an example of a second-order process, whose theoretical autocorrelations do *not* exhibit a regular (damped) oscillation.

Apart from anything else, observing that the results are all recorded already will save the quite heavy algebra necessary to deduce analogous formulae, when different models are postulated.

5. DISCUSSION OF EXTINCTION

On p.725, Roughgarden's calculations appear a little astray. For $\overline{K} = 1000$, and σ_k either 50 or 150, the carrying capacities lie, with probability .9973 *not* .99, within the intervals of, respectively, 850 to 1,150 and 550 to 1,450 - making a Gaussian assumption.

For the linear model (4), using Roughgarden's result (B8),

$$\sigma_n^2 = \frac{(1-\beta)}{(1+\beta)}\frac{(1+\alpha\beta)}{(1-\alpha\beta)}\sigma_k^2 \quad . \tag{14}$$

Thus, for the most extinction prone situation considered by Roughgarden, with $\alpha=\beta=-.5$, we get $\sigma_n = \sqrt{5}\,\sigma_k$. Then, with $\sigma_k = 150$, $N_t = 0$ is 2.98 standard deviations below $\bar{K}$, and the single-tailed probability for extinction to occur, at any time ti, is about .0014. Thus, for this "worst" case, extinction in a run of length 1000 is distinctly likely.

However, the second most dangerous case, with $\alpha=0$, gives $N_t=0$ at -3.85 standard deviations, with the associated probability of .00006. So, using the model, extinction seems really rather unlikely to occur, in a run of length 1000. For all the other seven combinations of α and β, the possibility of extinction is negligible. With $\sigma_k = 50$, even the worst case would have to drop about nine standard deviations, so extinction need not be considered.

However, as already pointed out in section 2, the linear and logistic models diverge considerably when we approach a situation of low N_t, where one might expect a serious risk of extinction to arise. For $\sigma_k = 50$, the differences between the behaviours of the linear and logistic models may well not be too divergent - compare Roughgarden's simulations, shown in his figures 5, 6, 7 and 8 - but, with $\sigma_k = 150$, it is very likely that low values of N_t will be achieved. For the linear model, these will tend to be pulled back towards the mean $\bar{K}$; but, even here, for the "worst" case, in a run as long as 1000, there will still be a strong risk of extinction. With the logistic model, the pull of $\bar{K}$ becomes much weaker as N_t becomes smaller, and the non-linearity begins to dominate. (See also Boyce and Daley, 1980.)

Now, for values of N_t, where (1) can no longer be closely approximated by (4), the condition necessary for extinction, at time (t+1), is that $r + 1 \leq r\, N_t/K_t$; or, equivalently,

$$K_t \leq \frac{r}{r+1} N_t . \tag{15}$$

This evidently requires a "crash" in the value of K_t to at least as little as $.6N_t$, for the most extinct-prone situation.

Consider a relatively small N_t, say of 500. Then, for extinction *in the model*, it is required that $K_t \leq 300$. This, in turn, necessitates that $k_t \leq -700$; which, for $\sigma_k = 150$, is approaching five standard deviations. I find it difficult to reconcile this sort of calculation with Roughgarden's simulation experience, where extinction occurred for five of the nine situations considered, when σ_k was 150.

However, condition (15) does give the clue as to how extinction can happen, when simulating the logistic model. The only conclusion is that the "crashes" tended to occur from *high* N_t.

For instance, when r=1.5, then β=-.5. So, for $\bar{K}$ = 1000, σ_k = 150 and α=-.5, if a situation was attained with N_t = 1500 and K_t = 900, we would have

$$n_t = 500 + 1.49 \text{ standard deviations} \qquad (16)$$

using (14), and

$$k_t = -100 = -\frac{2}{3} \text{ of a standard deviation.} \qquad (17)$$

A situation "as bad as" that described by (16) and (17) is, indeed, really very likely; but then $K_t = 900 = \frac{1.5}{2.5} \times 1500 = \frac{r}{r+1} N_t$ which, from (15), gives extinction. So, under the most precarious of the conditions considered, the model will lead to a virtually inevitable extinction, during a simulation as long as 1000 generations. The reader can similarly analyse the other situations.

6. IN CONCLUSION

Again, extending Roughgarden's analysis in another direction, we note that relation (5) basically demonstrates that the series, $\{y_t\}$, is dependent both on its own past and, also, on that of an associated "leading indicator" series, $\{k_t\}$. This leading property of $\{k_t\}$ is well illustrated by Roughgarden's figures 5, 6 and 7, which show that virtually all the peaks and troughs of $\{y_t\}$ follow one time interval behind those of $\{k_t\}$, and reflect the corresponding magnitudes.

Thus, in practice, it might well be worth considering whether the population dynamics would be better analysed as a "transfer function" model, incorporating the past dynamics of both the population and that of its environment. When this is in fact appropriate, it will have the marked advantage of resulting in a lower error variance than the "best" corresponding univariate analysis, on the population dynamics series alone. For details of this, see Box and Jenkins (1970).

It is to be expected that many good demonstrations of the potentially even more powerful (though considerably more difficult) multivariate methods, for analysing systems of *several* series, will soon be appearing in the literature. The first satisfactory and openly available case study, known to the author, was again an application to population dynamics. This is the masterly study by Jenkins (1975), modelling a bivariate muskrat-mink ecosystem; and indicates, unequivocally, a prey-predator mechanism, easily appreciated by ecologists.

It is hoped that this paper will be taken as a constructive extension of Roughgarden's excellent discussion, and that biologists will continue to investigate the possibilities of applying the ideas of time-series analysis to appropriate situations.

ACKNOWLEDGEMENTS

This paper was originally written for the American Naturalist in 1975. Although the referees were enthusiastic about its statistical content, the editors requested more biological discussion. Collaboration with a biologist was planned but did not materialise. Later, the contribution was submitted to Theoretical Population Biology, who accepted it subject to extending the arguments to other case-studies. This we did not have time to do. We are, however, most grateful to the four referees and two associate editors of these journals for their careful reviews.

I am also grateful to a further referee for the Boyce and Daley (1980) reference; and, prompted by a comment* from another reader, now include Addendum 2. Addendum 1 is also given, as the American Naturalist had requested that its content should be incorporated into the revised paper. Apart from these points, the paper is unchanged from its initial form, with just the references, that were unpublished at the time, being appropriately updated.

ADDENDUM 1

Extracted from a Letter to the Editors of the American Naturalist (Winter 1975)

ON MATHEMATICAL MODELLING

... very rarely have I seen such a lucid exposition of the basic concepts, for the time and frequency domains, as in Roughgarden's article (1975), where the mathematical implementation is of the highest quality. If only the educators would start using such

----

excellent examples for motivating their students towards a career *using* mathematics, as opposed to one imprisoned within the subject - then the sooner would we have a numerate society.

The idea behind Roughgarden's figures 4 and 8, though implicit in Bartlett (1946), is most interesting and, for all intents and purposes, novel. In these autocorrelation plots, Roughgarden draws in confidence bands. It is common for analysts, in the time-domain, to insert such bounds after some lag, under an assumption of no further autocorrelation. Then they can pick out those later lags at which the observed autocorrelation spikes are significantly different from zero. But the visual impact of inserting bands for the *estimated* model seems to be most useful, and would appear to provide a valuable diagnostic check on the adequacy of the model-building.

I suspect that the reason, for it not having been previously demonstrated, is that it does not appear in the writings of the front-runners in the time-domain, Professors Box and Jenkins, whose approach for univariate analysis culminated in their book (1970) - the major reference source for most subsequent work in this field. Roughgarden's idea is however simple, easy to implement, and will (I believe) prove to be a most effective diagnostic tool. I shall certainly do my best to spread it abroad.

.
.
.

Again, I was much impressed by Levin's letter (1975), which echoes many of my own sentiments on implementing mathematics. It is important to emphasise that the actual black magic box of mathematical techniques and tricks in trade is not the vital part in any *applied* science. After all, the scientist can employ a "technician" for this. What is important is the abstracting of a manipulative but meaningful model, from the real situation; and, after the relatively mechanical mathematical analysis, the translation back of the results and the interpretation of the conclusions, in their actual context.

Often the modelling can only be considered as a very rough approximation to reality. Then it will undergo successive refinements; each one reflecting some extra aspect, previously ignored. Thus, attention will be first focused on the most obvious characteristics of the situation, and the initial effort will go into explaining these. Only when the general form has been successfully outlined, will one start to shade in the finer detail. And, often, one never reaches this later stage - either their is not time, or the difficulties of representation become too great. But the real value of model-building is that it gives one a structured way of approaching problems; of looking at and thinking about them. In Levin's words "mathematical models ... can generate useful insights".

As a statistician, I would also like to emphasise the fact that the effective analysis of data does not depend merely on what their numerical values are. It also depends, crucially, on what they represent. Thus biologists, like others, must be very careful to avoid the cook-book approach; whereby they look for a set of data, which superficially resembles their own, and then proceed to copy the analysis. This is frequently seen in the fitting of least squares regression lines to series of observations, that are not independent but serially correlated. Roughgarden's contribution is refreshing in that it treats a time-series as a time-series, and does so competently.

ADDENDUM 2

To the Editor of Biometrics (Spring 1976)

BACK TO BULMER'S STATISTICAL ANALYSIS OF DENSITY DEPENDENCE - SOME COMMENTS ON A REPLY, WITH FURTHER FOREBODINGS OF INTELLECTUAL CENSORSHIP

I was indeed surprised to read Bulmer's reactions (1976) to my remarks (1976) on his originally very interesting paper (1975). I know that, as a mere statistician, I am invading a biomathematician's patch and so have to face the attendant dangers of territorial defence behaviour. But, even so, surely we all do want some interdisciplineary cross-fertilisation of ideas? I feel that Bulmer's reply (1975) short changes me considerably.

(i) His first paragraph claims to disprove my comment (b). His case rests on his very last sentence "it was perhaps not clear in my paper that σ^2 referred to the variance of the uncontaminated process, ξ_t, and not to the variance of the perturbations, e_t".

But in that paper, immediately after his equation (1), he writes "where the e_t's are independent random variables with zero mean and variance σ^2". The reader can judge whether it was only "perhaps not clear".

However, I also note that, in the second paragraph of my own comment (b), I am guilty of a slight sin of omission. The intention was to write "the constant of proportionality is a more complicated expression", rather than to omit the word *more*. Of course, using Bulmer's original definition (1974) and his analysis (1976), we get $\rho_k = \sigma^2\beta^k/\{\sigma^2 + (1-\beta^2)\sigma_d^2\}$, $k\geq 1$. And, as already pointed out in my letter (1976), this yields $\rho_k \equiv 1$, all k, when $\beta=1$, which is the case of a random walk contaminated by uncrosscorrelated measurement error.

(ii) Bulmer (1976) then red-herrings out of my discussion of how more sophisticated modelling can make not only more statistical sense, but can also often be interpreted in simple mechanistic terms. In his second paragraph, he starts by stating his scepticism of fitting unnecessarily complicated models. There is no suggestion of this from Box-Jenkins (1970) practitioners. The approach always goes for the simplest satisfactory fit to the data. This is the statistician's well-established *principle of parsimony*. No-one describes a pine tree by estimating how many cones it might have. However, it is surely playing ostriches when one ignores substantial statistical evidence, for a more complicated model, by saying that it does not suit the simple theory. Certainly, such evidence must either be reconciled with the theory, or one should consider the possibility of modifying the latter.

In my letter (1976), I indicated how apparently puzzling fits could frequently be interpreted in terms of justified theory. But, if the statistical evidence does not fit the current mechanistic explanation then, at very least, this should begin to sow some seeds of doubt, concerning the theory's veracity, in any open mind.

There can be really little debate that the Box-Jenkins approach does provide a most sophisticated means for statistically analysing ecological time series. Researchers who reject the evidence it unearths, in favour of their own theories, surely do so at considerable risk.

The very real danger of such new evidence being suppressed has recently been made by Van Valen and Pitelka (1974). I would like to remind Bulmer that his ideas are very largely based on selected statistical evidence. It is really most unscientific to reject other such evidence, for fear it might mean modifying those ideas. It is rather like the man who insists it is raining *because* his umbrella is up. Simple theories are always attractive and often form a valuable starting point for further thought - but facts should not be suppressed for the sake of protecting theories.

Bulmer is unfortunate in bringing the Jenkins (1975) mink-muskrat study into the arena. He has picked on very much the wrong man to support his views, I am afraid. Professor Jenkins, more than virtually anyone else, skilfully builds models which *do* make sense. If Bulmer rereads this typically brilliant discussion, he will find that the fourth-order autoregressive parts of the models quoted are very successfully explained, in terms of established ecological thinking. For instance, factorising the autoregressive operators into pairs of quadratic factors (with complex roots) gives periods, for the longer (randomly disturbed) mink and muskrat cycles, as 10.2 and 9.9 years, respectively. I cannot see that the conclusion of Bulmer (1974), which forces all the Canadian wildlife series into the straightjacket mould of basically following a sinusoidal pattern of period 9.63 years, is either simpler or intuitively more appealing.

Again, Bulmer seems to imply that the value of Box-Jenkins methodology is in prediction rather than in mechanistic understanding. But surely, if the statistical models do better than the theoretical ones for forecasting, one then has *prima facie* evidence that the statistical model is doing a better job. Then, somehow, the extra statistical evidence must be incorporated into the mechanistic explanation, or at least some thought given to this possibility.

Finally, a last misconception: Bulmer (1976) says that "if one is interested in forecasting, it does not matter how complicated the model is". Oh, but it does! One must try to avoid extrapolating any purely fortuitious patterns in the data. Thus a statistician, faced with four roughly linear points, might well fit a straight line not passing through any of them. He could of course fit a cubic, with all four points lying on it, and so have a more complicated, although apparently perfect, model. However, such a fit would usually be very silly and would, if extrapolated, give rise to absolutely precise but extremely inaccurate forecasts.

REFERENCES

ANDERSON, O.D. (1975a). On a lemma associated with Box, Jenkins and Granger. *J. Econometrics 3*, 151-156.

ANDERSON, O.D. (1975b). *Time Series Analysis and Forecasting: The Box-Jenkins Approach*. Butterworths, London & Boston. (Reissued with printing corrections 1976, 1977 & 1979.)

ANDERSON, O.D. (1976). On Bulmer's statistical analysis of density dependence. *Biometrics 32*, 485-486.

ANDERSON, O.D. (1977a). The time series concept of invertibility. *Math. Operationsforsch. Statist., Ser. Statistics 8*, 399-406.

ANDERSON, O.D. (1977b). An inequality and a lemma revisted. *J. Econometrics 6*, 135-140.

ANDERSON, O.D. (1977c). The interpretation of Box-Jenkins time series models. *Statistician 26*, 127-145.

BARTLETT, M.S. (1946). On the theoretical specification of sampling properties of autocorrelated time series. *J. Roy. Statist. Soc. B8*, 27-41. Correction (1948) *B10*, 1.

BOX, G.E.P. and JENKINS, G.M. (1970). *Time Series Analysis, Forecasting and Control*. Holden-Day, San Francisco. (Revised Edition, 1976)

BOYCE, M.S. and DALEY, D.J. (1980). Population tracking of fluctuating environments and natural selection for tracking ability. *Amer. Natur. 115*, 480-491.

BULMER, M.G. (1974). A statistical analysis of the ten-year cycle in Canada. *J. Anim. Ecol. 43*, 701-718.

BULMER, M.G. (1975). The statistical analysis of density dependence. *Biometrics 31*, 901-911.

BULMER, M.G. (1976). A reply to some comments by Anderson. *Biometrics 32*, 487-488.

HUBBELL, S.P. (1973). Populations and simple food webs as energy filters: (1) one-species systems. *Amer. Natur. 107*, 94-121.

JENKINS, G.M. (1975). The interaction between the muskrat and mink cycles in North Canada. In *Proceedings of the 8th International Biometric Conference*. Eds: L.C.A. Corsten and T. Postelnicu, Editura Akademici Republicii Socialiste Romania, 55-71.

KENDALL, M.G. (1971). Book Review: Time Series Analysis, Forecasting and Control; by G.E.P. Box and G.M. Jenkins. *J. Roy. Statist. Soc. A134*, 450-453.

LEVIN, S.A. (1975). Letter to Editors: On the care and use of mathematical models. *Amer. Natur. 109*, 785-786.

MAY, R.M. (1976). Simple mathematical models with very complicated dynamics. *Nature 261*, 459-467.

ROUGHGARDEN, J. (1975). A simple model for population dynamics in stochastic environments. *Amer. Natur. 109*, 713-736.

VAN VALEN, L. and PITELKA, F.A. (1974). Intellectual censorship in ecology. *Ecology 55*, 925-926.